# Water Quality Modeling

## Volume IV
## Decision Support Techniques for Lakes and Reservoirs

Editor

### Brian Henderson-Sellers

Associate Professor
School of Information Systems
University of New South Wales
New South Wales, Australia

Series Editor

### Richard H. French

Associate Executive Director and Resource Professor
Water Resources Center
Desert Research Institute
University of Nevada
Las Vegas, Nevada

CRC Press
Boca Raton   Ann Arbor   Boston

**Library of Congress Cataloging-in-Publication Data**
(Revised for volume 4)

Water quality modeling.

Includes bibliographical references.
Contents: v. 1.   1. Transport and surface exchange in rivers / author, Steve C. McCutcheon
v. 4. Decision support techniques for lakes and reservoirs / editor, Brian Henderson-Sellers.
1. Water quality—Mathematical models.   2. Water chemistry—Mathematical models.
I. French, Richard H.   II. McCutcheon, Steve C.
TD370.W3955        1989        628.1′61—dc20                                        89-25333
ISBN 0-8493-6971-1 (v. 1)
ISBN 0-8493-6974-6 (v. 4)

This book represents information obtained from authentic and highly regarded sources. Reprinted material is quoted with permission, and sources are indicated. A wide variety of references are listed. Every reasonable effort has been made to give reliable data and information, but the author and the publisher cannot assume responsibility for the validity of all materials or for the consequences of their use.

Direct all inquiries to CRC Press, Inc., 2000 Corporate Blvd., N.W., Boca Raton, Florida, 33431.

© 1991 by CRC Press, Inc.

International Standard Book Number 0-8493-6971-1 (v. 1)
International Standard Book Number 0-8493-6974-6 (v. 4)

Library of Congress Card Number 89-25333
Printed in the United States

# SERIES PREFACE

Authoring a technical reference book is a labor of love; for unlike novelists, the technical author has no prospect of becoming rich and famous and his work will be subject to the critical review of his peers. As Series Editor, I salute the authors that have made the series *Water Quality Modeling* a reality. I would also thank the families, colleagues, and staff who supported and encouraged the authors through the creative process.

The ambitious goal of this series is to bring together the collective knowledge and experience of engineers, biologists, chemists, and other scientists to present the state of the art of water quality modeling. The cost of not having available the capability of forecasting the effects of our actions on essential and limited resources such as water is very high in both economic and human terms. Numerical modeling is the only cost-effective method of forecasting the future, and the art of constructing a numerical model causes us to view the environment in which we live as a system of interrelated components.

Water quality modeling is not a static subject, and I would hope that the information and knowledge presented in this series will be a part of the foundation on which new developments are made. Water quality modeling is a complex combination of education, experience, and creativity — it is science and art — and our understanding of the system continues to develop.

Dr. Steve McCutcheon and I were very fortunate in attracting a number of outstanding engineers and scientists to this series. The credit for this series belongs to the authors, not the editor. I am privileged to be associated with these authors and the publisher.

**Richard H. French**

# PREFACE

Worldwide, many freshwater lakes and reservoirs are undergoing accelerated eutrophication as a result of mankind's activities: increased use of nitrogen- and phosphorus-based fertilizers, which are leached into water bodies, and increased volumes of treated wastewater. Management of the quality of water stored in impoundments is therefore gaining importance whether the water is to be used for potable supply, irrigation and/or recreation. The potential range of conflicting interests contrives to provide the manager with a loosely structured problem with elements from many disciplines, including engineering, economics, physics, chemistry, biology, dependent upon the quality requirements for each particular use for the stored water.

Operational management of lakes and reservoirs, as well as longer term planning for new water resources, utilizes concepts not only from a range of disciplines but also requires the implementation of computer models to represent the processes involved in water quality changes. For the last couple of decades, much research effort has been devoted to the development of water quality simulation models. As a research tool, it has commonly been the case that managerial applications have been rare and that their use in decision support has been essentially ad hoc. Over the last few years, however, model builders have begun to consolidate the reliability and, perhaps more importantly, the accessibility of their simulation models. This is to some degree based on confidence obtained by extended case studies.

Consequently this book is both retrospective (in terms of summarizing past work) and prospective (in terms of forecasting the greater use of such models as part of much needed environmental decision support systems). In Chapter 1 the concepts of both lake and reservoir simulation modeling are introduced and also the concepts of decision support systems, formalized within the information systems discipline. Decision support systems (DSS) rely on an integration of computer models, data bases and a user interface which can give the non-computer specialist access to the knowledge encapsulated within the data and models. Although the aim of most simulation modelers has always been to provide decision support, it has only been possible to do this following the consolidation of the simulation modeling component of the DSS. Each chapter here, then, presents the simulation modeling component, whilst indicating its potential future use in decision support. The chapters progressively develop the theme of simulation through to decision support. In the earlier chapters, case studies are presented which concentrate more on the physical (dynamical and thermodynamical) parameters; in later chapters the presented models stress the need for a more detailed representation of the biology and chemistry. The case studies of the last chapters place a greater emphasis on the management use of the model. In the final, short chapter some new tools and concepts are presented in order to stimulate discussion and evaluation of original ideas to permit the case studies of the sort presented here to be successfully transferred from the arena of research to that of operational and planning management.

I wish to thank all the authors whose work is presented here. The work involved in preparing the text of these chapters was perhaps the least arduous; their effort having been devoted over the last several years in undertaking the research work and model development in order to bring to fruition these simulation models for incorporation into environmental decision support systems, the use of which we are likely to see increase over the next decade.

Finally, I wish to thank my wife, Ann, for her sound advice at various stages of this project.

**B. Henderson-Sellers**
**Sydney, 1989**

# FOREWORD

The middle age aristocracy's perfumed handkerchiefs protected them from the pervasive smell of decaying rubbish in the streets. They did not, however, prevent the spread of diseases which decimated thousands of people irrespective of "blood colour". Canalization was invented resulting in disruption to the natural cycle of organic matter: fields $\rightarrow$ crops $\rightarrow$ man and his domestic animals $\rightarrow$ back to land. The decaying organic matter was now re-directed to water. The consequence was dramatic — increasing pollution, both direct and indirect, the latter created by the need to substitute for organic matter and nutrients lost from fields. Hence mankind started early the habit to rob Peter to pay Paul by creating new, more serious problems in place of the earlier ones. This is the unfortunate attitude we are continuing up to now due to ignorance, carelessness, or both. Or is this our unavoidable fate?

Scientists, as well as artists, are trying to gain an understanding of nature's complexity and to predict the consequences of mankind's increasing intervention with natural processes as a prelude to prevention. The engineer's goal is to develop procedures for satisfying man's expanding demands without destroying natural systems.

In aquatic sciences historically the primary goal was to supply water in the *quantities* needed by man and his increasing agricultural, industrial and other activities. This was a relatively simple goal in hydrologically more balanced temperate regions. Increased difficulties were encountered in water deficient regions like Australia, leading to the construction of systems of dams and to water transfers from remote regions. Still more complicated became the task of coping with increasing deterioration of water *quality*. The present focus is on problems much more cumbersome: to provide *good* water in *adequate* amounts. The quantitative and qualitative aspects are coupled not only in respect of human needs, but also because they are intrinsically interrelated. For water, its quality and quantity are inextricably linked to man's activities. Water is a transporting and concentrating element and all kinds of landscape disturbances are reflected in natural waterbodies. Historically the scale to be considered from this point of view is increasing, from the drainage basin of the given groundwater aquifer, river, lake or reservoir to remote localities of industrial production (air pollution) up to the global scale of Mother Earth (global warming, ozone hole). Natural physical, chemical and biological processes of pollution transformation and assimilation are determining the degree to which man's activities are reflected in water quality. The subtle character of these processes is in contrast with the brute force approach dominating present-day engineering: create new industry, dig out new holes, go to remote areas for new resources, put in more chemicals etc. New activities, however, create new problems, more conflicts between users of the same resource, increased degradation of natural systems. Water being used for drinking, washing, irrigation, power generation, cooling, processing, food production, fishing, swimming, boating etc. is a particularly conflicting resource. Therefore, the solution of water quality problems needs to encompass both natural and technological processes.

Last but not least, the problem is not solved simply by finding scientifically and technically feasible solutions. It is necessary to convince people by means of the existing infrastructure (with its history, attitudes, likeness or disgust for certain approaches, antagonism between different decisive bodies etc.) to accept the appropriate long-term solutions. In this respect the dimension of the problem is further increased. What we have to deal with are fairly complex questions. We cannot expect that for such complicated tasks simple approaches can be adopted.

The essence of the problem is that of information coupling amongst various scientific disciplines and of information transfer between scientists via practicing engineers to decision makers.

Three basic items can be distinguished.

a) how to organize scientific knowledge towards interdisciplinary coupling of hydrophysics, aquatic chemistry, hydrobiology and other branches towards a higher level of abstraction within the system "natural water";
b) how to transfer complex interdisciplinary scientific knowledge to engineers dealing in addition with technologies of distribution and use of natural water for humans; and
c) how to transfer scientific and engineering knowledge to management decisions including social and political problems.

The present book is directed to all these topics. The editor's major goal is to provide acceptable techniques for applying scientific knowledge dealing with lake and reservoir water quality to management decisions. The methodologies used throughout the book for the three above items are as follows:

a) **Systems approach** for solution of interdisciplinary problems. This encompasses not only straightforward relations of physics → chemistry → biology but also the important feedbacks biology → chemistry → physics. In this way secondary and indirect effects are also taken into consideration.
b) **Simulation modeling** for simplifying the interdisciplinary knowledge gained about the natural system into a shape tractable within the framework of technology.
c) **Decision support systems** for implementing the knowledge within the political framework.

The first two steps are presently being developed rapidly and the book focusses on directing them towards management decisions.

One drawback of present simulation modeling of ecological and environmental systems is their direction towards models for particular systems (individual lakes or reservoirs with their specific characteristics and problems). It is the feeling expressed by the editor and shared by myself that for both scientific purposes and management applications we need models of a much broader scope. Scientifically this is dictated by the variable nature of aquatic systems, which not only do not easily adapt to changing environments, but also undergo rapid natural evolution. A specific model taking into account only what is present in a given situation is easily constructed and rationalized to observations. It may, however, become useless for any future state. Natural aging, succession of species or man-made changes alter aquatic systems profoundly. In respect of management, a specific model may prove useless for the simple reason that managerial decisions change the nature of the system to such an extent that in fact another system needing a different model is created. During planning we want to examine different alternatives, for which no prototypes exist on which the model can rely.

Another drawback of present system simulations is the inadequacy of methods for verification and validation. Moreover, there are no ways developed to suggest where in the system the disagreement between model and reality originates and for what reason. The methods of parameter fitting have the disadvantage that they relate only to the prescribed model structure — in fact good approximation can be obtained with a completely irrelevant model. We rely on the intuitive use of theoretical knowledge to search for model structure improvement. This is why simulation is considered both science and art.

The topic treated until now with lowest intensity is the one of information transfer to manager. In respect of management both scientists and engineers have focused their efforts primarily on development of quantitative, objective models. The expectation was that this is the best basis for reasonable decisions. For management, however, subjective evaluations based on people's feelings and attitudes have also to be considered. An interesting experience presented in this volume suggests that the last may gain as much weight as objective reasoning. A principal question seems to arise here: should we stop doing science and technology, and instead change people's attitudes towards scientifically-based decisions? In any case, we cannot shut our eyes and leave others to cope with the situation.

This book is a first step in this process.

Milan Straškraba
České Budějovice
May 3, 1990

# THE EDITOR

**Brian Henderson-Sellers, Ph.D.,** is Associate Professor in the School of Information Systems at the University of New South Wales in Australia.

Dr. Henderson-Sellers graduated in 1972 from Imperial College of Science and Technology (London University) with a B.Sc.(Hons) and A.R.C.S. in Mathematics. He obtained an M.Sc. in meteorology from Reading University in 1973 and a Ph.D. in engineering from Leicester University in 1976. From 1976 to 1983 he was a tenured lecturer in the Department of Civil Engineering at the University of Salford in England and in 1983 moved to the Department of Mathematics and Computer Science in the same university. In 1988 he emigrated to Australia and took up his present post at the University of New South Wales.

Dr. Henderson-Sellers is a Fellow of the Institute of Mathematics and Its Applications; the Institute of Engineers, Australia; and the Royal Meteorological Society and is a Member of the Institution of Water and Environmental Management, the American Society of Civil Engineers, the Australian Computer Society, the American Geophysical Union, the International Society for Ecological Modeling, the Societas Internationalis Limnologiae, the Simulation Society of Australia, the Australian Water and Wastewater Association. He has been on the organizing committee for several international conferences, especially in the area of environmental modeling and is currently on the editorial board of *Ecological Modeling* and *Environmental Software*. He is also series editor for Wiley for *Principles and Techniques in the Environmental Sciences*. He received the McNaughton Award in 1981 and a Sir Peter Kent Conservation prize for his book (with H.R. Markland) *Decaying Lakes. The Origins and Control of Cultural Eutrophication* in 1989.

Dr. Henderson-Sellers is the author of over 150 publications, including 6 books, two of which have been translated into Russian, a decision support software package and has presented over 50 conference papers (several by invitation). His current major interests are object-oriented systems development, with especial application to the water industry; and simulation and decision support tools for reservoir management.

# CONTRIBUTORS

**John J. Cardoni, M.S.**
Associate
Metcalf and Eddy
Meriden, Connecticut

**R. I. Davies, B.Sc.**
Department of Geography
University of Liverpool
Liverpool, England

**S. Dhamotharan, Ph.D.**
Vice President and Managing Principal
Woodward Clyde Consultants
Baton Rouge, Louisiana

**Alec Y. Fu, M.S.**
Department of General Services
Bureau of Waterwaste Treatment
New York, New York

**Dirk C. Grobler, Ph.D.**
Manager, Corporate Environmental Program
CSIR
Pretoria, South Africa

**Donald R. F. Harleman, Sc.D.**
Ford Professor of Engineering
Department of Civil Engineering
Ralph M. Parsons Lab
Massachusetts Institute of Technology
Cambridge, Massachusetts

**Brian Henderson-Sellers, Ph.D.**
Associate Professor
School of Information Systems
University of New South Wales
Kensington, New South Wales
Australia

**Sven E. Jørgensen**
Professor
Department of Environmental Chemistry
DFH
Copenhagen, Denmark

**David C. L. Lam**
Department of Environment
NWRI/CCIW
Burlington, Ontario, Canada

**Richard A. Luettich, Jr., Sc.D.**
Associate Professor
Institute of Marine Sciences
University of North Carolina
 at Chapel Hill
Morehead City, North Carolina

**J. N. Rossouw, M.Sc.**
Division of Water Technology
CSIR
Pretoria, South Africa

**William M. Schertzer**
Department of Environment
NWRI/CCIW
Burlington, Ontario, Canada

**Frank R. Schiebe, Ph.D.**
Director
Water Quality and Watershed
 Research Laboratory
U.S. Department of Agriculture
Agricultural Research Service
Durant, Oklahoma

**Peter Shanahan, Ph.D., P.E.**
HydroAnalysis, Inc.
Acton, Massachusetts

**Heinz G. Stefan, Dr.Ing.**
Professor and Associate Director
St. Anthony Falls Hydraulic Laboratory
Department of Civil and
 Mineral Engineering
University of Minnesota
Minneapolis, Minnesota

**Alexey A. Voinov, Ph.D.**
Senior Research Worker
Laboratory of Mathematical Ecology
Institute of Atmospheric Physics
U.S.S.R. Academy of Sciences
Moscow, U.S.S.R.

## Dedication

*for*
*Eve, Alice, Laurence, Nicholas, Stephen, and Philip*

# TABLE OF CONTENTS

Chapter 1

# WATER QUALITY SIMULATION MODELS FOR DECISION SUPPORT

## B. Henderson-Sellers

## TABLE OF CONTENTS

# I. INTRODUCTION

Good engineering management of lake and reservoir water quality may be facilitated by the increasing availability of mathematical models and software tools.[1] These can be used as part of the design procedure, for example, to assist in making decisions relating to the implementation of destratification devices and the location of inlets and outlets,[2-4] as well as in day-to-day predictions of the quality characteristics of the water, so that appropriate treatment can be assessed or so that in-lake management can be undertaken in order to minimize treatment costs. For river regulation reservoirs, models will similarly predict outlet water characteristics (e.g., temperature, dissolved oxygen) so that the impact on downstream fisheries, for example, may be assessed.[5]

Environmental decision support systems (EDSSs) are beginning to emerge, utilizing concepts from the discipline of information systems.[6] This formalism, relatively new to water quality management, requires the synergism of numerical models (usually simulation models) with large databases, front-ended by a man-machine dialogue component (Figure 1).

To date, user experiences are few but rapidly increasing (as evident in the remaining chapters of this book). Decision support systems (DSSs) with user-friendly interfaces certainly provide problem-solving tools to the manager that were not previously available. However, the provision of such tools must, to some degree, limit the flexibility and growth/evolution of such information systems,[7] although this problem is likely to disappear with the advent of more flexible and extendible software engineering methods (see Chapter 9).

This book is set in the context of a set of case studies of models that are well established and have been or are likely to be used in a decision support context for water quality management. In this introductory chapter, the concepts of DSSs are introduced and a review of the "model" component of a DSS is presented, stressing the breadth of available mathematical and computer model types (with examples) for water quality analysis and simulation. Further chapters then expand on the details of particular models and demonstrate their utility by describing specific case studies in which these models have performed as a component of a DSS for water quality management.

# II. DECISION SUPPORT SYSTEMS METHODOLOGY

Mathematical/computer modeling of ecosystems has come of age. Software simulation packages are well enough developed to be utilized in management decision making in an increasing number of environmental areas. In order for modeling results to be understandable by and useful to environmental managers, they must be packaged in a more "palatable" fashion. The research program, which tends to be high on concept and low on "user-friendliness", must take notice of the methodologies of the relatively new discipline of information systems.

It seems likely that the number of applications identifiable as DSSs will increase in environmental management fields and the impact of information technology in general will become more evident. Once environmental managers can realize the full support of a software DSS, their decisions are likely to become more effective, especially since the large database, which is a vital component of a DSS, will give them access, and hence surrogate experience, to a much larger range of case studies than those within their direct experience. At the same time, problem identification is likely to become easier and the basis of a decision to become more objective.[6]

A frequent description of a DSS is an interactive computer system that assists decision makers to solve *unstructured* (or loosely structured) problems[8] by utilizing both data and models. To this Guariso and Werthner[6] also add a rule-based "knowledge" component in their development of a prototype EDSS, called EDSS-1. However, the more traditional DSS[9] differentiates strongly between a DSS and an ES (expert system) on the grounds that a DSS is

Decision Support System

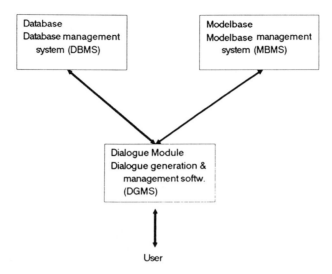

FIGURE 1. Architecture of DSS according to Sprague and Carlson[9] (From Guariso, G. and Werthner, H., *Environmental Decision Support Systems*, Ellis Horwood, Chichester, England. 1989. With permission.)

to provide *support for the expert*, while the expert system is intended much more to *replace the expert* — or at least the need for the local involvement of an expert.[8] In either guise, an EDSS combines the state-of-the-art software technologies in several fields: management DSSs, database systems, simulation modeling, and, possibly, artificial intelligence (as exemplified by ES). Each of these components is outlined briefly in this section.

There are several conceptual approaches to DSSs. Sprague and Carlson[9] describe a DSS in terms of three modules only (Figure 1): a database management system (DBMS), a modelbase management system (MBMS), and a dialogue generation and management software (DGMS) module. In this formalization of a DSS, the three individual components are clearly identified, whilst specific procedural tasks and knowledge representation remain hidden within the three modules. In some respects this seems to have arisen as a consequence of the database-oriented origins of such a conceptualization. Furthermore, the MBMS is modeled on the DBMS, serves a similar purpose, and gives a DSS its special characteristic of a synergistic (or integrated) software system across these three different modules. The MBMS is thus able to cross-reference models within the modelbase, possibly creating new models by prototyping. Similarly, the DGMS must be flexible enough to support a wide variety of user-preferred dialogue styles, contain several options in presentation format for the results, and perform error checking on all inputs.

A second approach is presented by Bonczek et al.,[10] who take a more functional approach. They identify process-oriented submodules (rather than the type of software implementation, as in Figure 1): a language system, providing the interface to the user in a similar way to the DGMS of Figure 1, a problem processing subsystem, and a knowledge subsystem. In this approach, neither modeling nor data retrieval are included explicitly, and both database and modeling components are subsumed in the knowledge subsystem, without differentiation.

An alternative approach is taken by Guariso and Werthner,[6] who attempt to synthesize the ideas of Sprague and Carlson[9] and Bonczek et al.[10] in the context of a proposed EDSS. In their system architecture (Figure 2), a knowledge base is added to the Sprague and Carlson[9] architecture, or, conversely, the components of the Bonczek et al.[10] model are considered to be

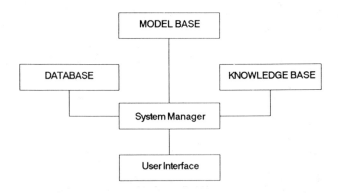

FIGURE 2. Architecture of an environmental decision support system proposed by Guariso, G. and Werthner, H. in *Environmental Decision Support Systems,* Ellis Horwood, Chichester, England, 1989. (With permission).

subdivided. Since it clearly contains a procedural modelbase, it is seen as different from an ES. Each of these manifestations could be considered to define a conceptualization of a DSS, whilst it should also be noted that specific implementations of any one of these conceptual systems models could add further differentiation between individual DSSs.

As noted above, the modelbase is an important component of a DSS. Consequently, it is perhaps appropriate to give a brief overview of the modeling process here (see also following section). The modeling process can best be represented as a simple flow chart (Figure 3). The process can be divided into four sections, some of which overlap: model formulation, solution, interpretation (of the results), verification and validation. In the first stage, the problem is identified, analyzed, and formulated. This is equivalent to the software-engineering life-cycle stages of requirements analysis, systems analysis and design, including an analysis of data modeling and functional decomposition (see, e.g., ref. 11; although cf. Chapter 9). In terms of simulation modeling, data collection must often be initiated, and conversely the impossibility of obtaining certain data may contribute to determining the form of the model itself.

Following the systems analysis and design stages (in simulation modeling, often undertaken by the scientist who will also be responsible for the implementation — and often the maintenance — of the code itself), any remaining parameter values must be estimated, the simulations undertaken, and the results interpreted and understood. Results analysis is often the most interesting and yet complex portion of the whole modeling cycle and includes the most important components of model verification and validation — techniques that are less well developed in a prescriptive sense than earlier stages of the modeling cycle. These concepts are discussed in detail below.

In terms of their incorporation into an EDSS, other characteristics of simulation models must be noted. Possibly the major feature of environmental problems (and hence of the simulation models) is their dynamic nature, both in terms of the time-series solution required (usually) from the set of simultaneous (partial) differential equations and also in terms of changing boundary conditions (for example, baseline pollutant concentrations). In addition to a temporal dimension, simulation models also have a number of spatial dimensions (although many of the model case studies in this book are restricted in their spatial analysis). It is important, however, to make sure that the time and space scales of the model are commensurate not only with the time and space scales of the phenomena being simulated, but also with the scales of the data available for model verification and validation, as noted also by Jørgensen[12] (p85). Periodicities and trends (both statistical and process oriented) must be identified and stochastic influences incorporated at any appropriate level in the model. Large lake models can have immense data requirements, so that the database component of the EDSS contributes a valuable role as an information

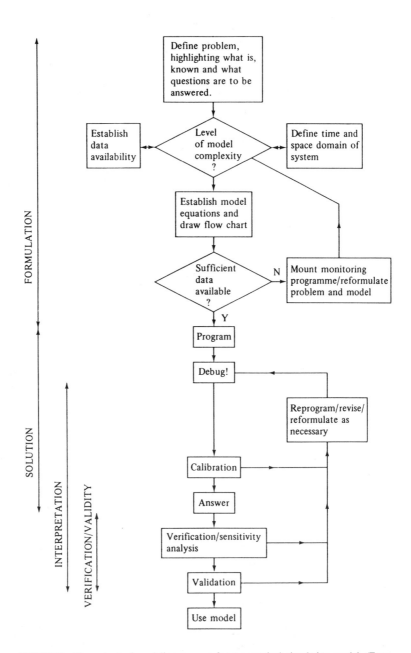

FIGURE 3. Flow chart of modeling process for a numerical simulation model. (From Henderson-Sellers, B., *Mathematical Modelling Courses,* Berry, J. S., Burghes, D. N., Huntley, I. D., James, D. J. G., and Moscardini, A. O., Eds., Ellis Horwood, Chichester, England, 1987. With permission.)

depository and management system, not only for the EDSS as a whole, but also for individual simulation models within the modelbase.

DBMSs have a solid foundation in software engineering. Although several schema for data modeling have been used, such as network and hierarchical data models, the most favored data model today is the relational model.[13,14] Future DBMSs are likely to be based on either this or, in perhaps 5 to 10 years, on object-oriented database models (see Chapter 9). In a database, individual bits of information are stored in identifiable locations and never duplicated. This ensures data integrity. Access by users is via the DBMS, which provides a transparent interface. In other words, if it is necessary to change the form of data storage, only the part of the DBMS

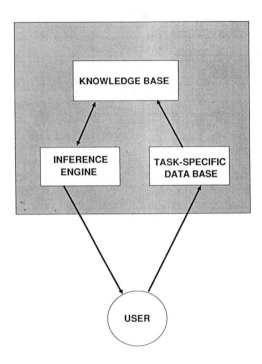

FIGURE 4.  The components of an expert system. (From Ford, L. N., *Information and Management*, 8, 21-26, 1985. With permission from Elsevier Science Publishers, Physical Sciences and Engineering Division.)

accessing the data needs to be changed to be compatible, whilst applications programs remain unaffected. Consequently different applications programs access the same data elements, even if the format required by the several applications programs is very different.

In the context of an EDSS, particularly as applied to managing lakes and reservoirs as discussed here, the database would provide a coherent information base. The several models in the modelbase could then be "driven" with these data,[7] so that simulation results are directly compatible and comparable. Secondly, the immediate availability of a large database that is transparently accessible to the user (here the simulation model) readily permits the implementation of a thorough validation exercise (see also Chapter 5). In a third context, model selection could be undertaken for management application by accessing the database/modelbase through the DGMS in order to ascertain model features appropriate to the specific current case study. These features would include model name, purpose, limitations, availability, hardware requirements, extent of validation, extent of available documentation, source, costs, etc.[15]

In the ES component of the EDSS (if included), knowledge is represented by rules rather than procedures (as is much more frequently the case in simulation models). Logical inferences are drawn from this set of rules, supported by an initial data set. The system attempts to mimic the expert's reasoning skill in problem solving *in a very specific domain*. Data and rules are required and analyzed by an "inference engine" (Figure 4). It is important that the user interface is in a natural language. Also, most ESs will not only deliver an answer, but also a line of reasoning (see also discussion in Chapter 9).

Finally the DGMS that provides the user interface is a crucial component of a DSS, not insofar as it ensures the DSS will be efficient, but that it will be an *effective* tool for the manager. Only with an appropriate and readily understandable interface[16] will the decision support tool be used by managers who are initially unfamiliar with the concepts of a software DSS. A good DGMS should offer the user the option of dialogue style, for example, using menus, forms to fill in, command languages, direct manipulation (mouse, touch screen, etc.) or natural language interface, and, in the future, voice. Interface design is thus not simply a technical question but could involve, for example, psychologists, ergonomists, and cognitive scientists, as well as information systems professionals and computer scientists. (Examples of user interfaces can be found in, for instance, Guariso and Werthner,[6] and discussion of the graphic art and typographic concepts in, for instance, Simmonds and Reynolds.[17])

## III. MODELING CONCEPTS

In attempting to simulate lake and reservoir water quality, it is imperative that a multidisciplinary, integrated systems approach be adopted. Not only is it vital to represent the chemistry[18] and biology, but it is also important that these complex processes be amalgamated with a representation of reservoir physics in such a way that feedbacks between biota and physical characteristics are possible. This amalgamated approach may, in respect of reservoirs, be termed *engineering limnology.*[19] Furthermore, any such integrated study program must include not only numerical modeling, but also field observational programs and laboratory studies. All three aspects must play an equal and complementary part.

Modeling is the representation of the characteristics of a real system by mathematical functions, often realized on a computer, in a way that approximates reality.[20-22] This approximation may be because of constraints in our understanding of the physical interrelationships between constituent parts of the system (e.g., zooplankton predation rate) or be a limitation in the way that the real, continuous, natural system must be represented by a finite difference or finite element grid of a relatively large length scale, compared to the real system (for example, representing the continuous nature of the circulation in a reservoir by cells each of several meters or even tens of meters in length).

Modeling studies may serve not only as an aid to understanding the nature and behavior of lakes and reservoirs, but also as planning and management tools. As such, scientific detail may be sacrificed in the short term to gain the advantage of an engineering tool to solve real problems. The most appropriate model is determined by many factors, from the evaluation of the current state-of-the-art models (often within a predetermined computational budget) to an appreciation of the implicit temporal and spatial scales that any model builder will have necessarily encapsulated within the numerical algorithm, and that, unfortunately, are not always "visible" to the reservoir manager (as the model user). Indeed, it is not always the case that the most complex model will provide the greatest advances in understanding. For example, Sera and Straškraba[23] present an empirical model of temperature stratification for the Klicava Reservoir, Czechoslovakia in which the difference between the surface and deep-water temperature permits the authors to gain insight into the effect of different retention times on the strength and stability of thermal stratification.

Mathematical models can thus provide a useful tool to supplement insights gained from both field and laboratory observations. In water quality modeling at the present time, there is a wide range of available models. Categorization is difficult since there are many attributes that could be used (e.g., internal structure, mathematical character, mode of application).[22] One method of classification of such a set of continuous models is by the number of spatial dimensions in which a number of physical, chemical, and biological variables can be evaluated.

The commonly used and relatively simplistic trophic state assessments are in fact a simple example of a *discrete* model. Assessments of the trophic state as oligotrophic, mesotrophic, eutrophic, etc. have demonstrated that it is not possible to assign a single discrete category to most lakes.[24,25] In the assessment of trophic indices of lakes and reservoirs, the use of regression and correlation techniques is widespread. In the Organization for Economic Cooperation and Development (OECD) program from 1973 to 1976, quantitative analysis of the very large and geographically widespread database almost exclusively concentrated on searching for correlations. Since in many instances causality was indeed present, the results of the program[26] provide useful guidelines, although it is doubtful whether these guidelines can be rigorously applied outside the domain of the sites used in the survey.

In any such simple model, where there is, in effect, a discontinuity in the total range of attainable values, the "grey" range associated with the borderline is not recognized within the classification scheme. Furthermore, with respect to trophic state classification, these problems were recognized in the OECD[26] report, in which categories with 100% certainty of assignment were superceded by a two-part trophic state assessment of probabilities plus the trophic state category. However, despite this, as well as other attempts to smooth over these problems in terms of a compound trophic state index,[27,28] it is still clear that there is a limited usefulness to such simple approaches.

At the same time, it is becoming increasingly feasible to provide reservoir managers with software packages of *continuous* simulation models that encapsulate higher temporal and spatial resolution, yet that require little additional time or user education in comprehension and application in decision making.

Simulation models are those mathematical/computer representations of reality that mimic the real world and *how it changes with time*. In other words, they include a time dependency. As such, simulation models may be the solution of partial differential equations (continuous simulation models) or be the solution of a set of discrete simultaneous equations whose behavior is governed by some type of stochastic behavior. (For a wider ranging discussion, see Watson and Blackstone.[29])

Computer simulation of an aquatic ecosystem is becoming an increasingly important tool for the water engineer and water quality manager.[30] In potentially water-stressed, semi-arid countries like Australia and southern Africa, small changes in urban drainage patterns, agricultural land use, and even marginal changes in meteorological regimes as a result of climatic change and climatic variability may be sufficient to change the ecological (and therefore trophic) status of the water, rendering it potentially less potable and of less use for irrigation. Simulation offers the possibility of forecasting such adverse changes and of providing management information that will permit strategic planning for the water resource. Consequently, most of the chapters in this book deal with ecosystem simulation models, being the best represented in the literature and the most widely used in both scientific as well as engineering and planning case studies. These models are presented both as simulation models per se and also as important components of DSSs.

## A. A MODEL HIERARCHY

A wide range of complexity exists in model formulation for continuous models. As noted above, one simplistic categorization is in terms of the dimension of the model. Many of the more biologically and chemically orientated eutrophication models use the concept of the continuously stirred reactor (CSR). This model, which is essentially a zero-dimensional model, assumes that all state variables are homogeneously distributed throughout the water body. Consequently, there is no possibility of understanding phenomena that are heterogeneous with depth or across the lake or that occur on seasonal or diel time scales, yet the models are very useful for a longer time scale, for example, year-to-year changes of phosphorus.[31]

One-dimensional models in which variations in the vertical direction are represented (or sometimes, in one-dimensional riverine reservoir models, the longitudinal variability of some depth- and width-averaged parameter of interest) are most represented in the literature. They have been used most extensively in both physical (e.g., stratification) as well as chemical (e.g., dissolved oxygen) models, although in ecosystem and eutrophication models (which attempt to describe the biota in more detail), vertical variability is seldom well resolved — at best only two or three layers in the vertical may be utilized.

Two-dimensional models are used to describe lateral and longitudinal variations in variable values but are averaged over the depth. These may be useful for run-of-the-river reservoirs[32] that have (1) a shallow depth and hence are unlikely to stratify and (2) a strong throughflow, which tends to ensure that the water is well mixed in the vertical direction.

Three-dimensional models have been developed more recently, largely in hydrodynamical investigations of reservoir currents.[33-35] In most cases, these models are applied to homogeneous lakes or assume the vertical temperature to be structured into a small number of finite levels, i.e., not including a prognostic stratification model. It is therefore important to note that although the state-of-the-art in physical/thermodynamical models[4] centers on one-dimensional models and in physical/ dynamical models centers on three-dimensional models, there is an urgent need to fuse these apparently disparate approaches into fully three-dimensional representations of thermodynamics plus dynamics plus biochemical processes.

One further difference between the different modeling approaches presented can be identified in terms of the division of the variables in the model between *state variables* (which are the prognostic dependent variables within the model) and the driving (or forcing) variables, which are prespecified, either from data or from some sort of prior knowledge. For example, if it is intended ultimately to model biological interactions, some representation of the epilimnetic depth is needed. Should this be calculated (with either a dynamic thermocline or a mixed-layer model) or is it possible to prespecify this depth, for example, as a linear function of time of year over the summer stratification period? The former approach appears to be more adaptable to unusual circumstances, yet for a typical year's weather and for a reservoir with a long time series of observations and managerial expertise, the latter approach may be as (or even more) successfully applied *to the specific study area*. However, it is most unlikely that such a model could be transferred to other locales,[36] a requirement discussed in more detail in Chapter 5.

## B. MODEL ASSESSMENT

Model construction requires not only mathematical insight and accurate implementation as computer code, but also the acquisition and use of experimental data, initially for calibration of the model, and finally for verification of the model before reliable prognostic simulations can be undertaken for management planning.

Both empirical theories and conceptual theories (although the former predominantly) can give rise to the need to include "constants of proportionality", the values of which require experimental evaluation, either in the field or the laboratory. In some cases, the value is really a constant, independent of any specific case study, i.e., all the values deduced from a wide range of experiments give the same numerical (or functional) value. However, there are many coefficients for which there is no unique value and that must be reevaluated for each specific case study. Frequently the value is chosen so as to minimize errors between the simulation and a restricted data set. This minimization is often referred to as *model tuning or calibration*. Once the calibration has been undertaken, the model should be tested against a second data set, still restricted in size (for example, a second year's data for the same water body). If these parameter values hold, then the model user can have some faith in the model results.

In all of the models discussed here, rather informal methods of "parameter tuning" are normally utilized in calibration. More formal statistical methods of time-series analysis are not normally considered, both because of the complexity of the models and the difficulty of obtaining appropriately large data sets, which would allow for statistical identification and estimation. While this is justifiable in the circumstances, it should be realized that the lack of a formal statistical approach and the use of the more arbitrary parameter tuning can limit the credibility of the model.[37] Even when combined with sensitivity analyses and constraints on parameter values, such deterministic tuning is controlled, to some extent, by the modeler's prior prejudice and can be abused. In particular, the development of a model that matches reasonably a rather meager data set is no guarantee that the model is "valid" in wider terms, although it can still possess considerable practical utility.

In some instances, once calibrated, say for a specific lake, those parameter values can be retained whenever that lake is simulated, either in hindcast or forecast mode. However, the range of validity for such parameters is not clear, e.g., if the phosphorus level of the inflow suddenly (say, because of a new source on the catchment) increases by an order or magnitude, will the model still be valid or will some of the parameters need retuning? In many instances, the same lake model applied to another situation in another part of the country, perhaps where the climate or the water quality has a different bias, will necessitate a full retuning of the whole model.

Ideally, calibration of a model should be restricted to parameters that are likely to have a value of "global" validity, and *process* models should be used preferentially to empirical regression models. However, with such a complex system as an aquatic ecosystem, it is perhaps inevitable that conceptual understanding in one subdiscipline will lag behind that in another, such that the level of empiricism in various parts of the large ecosystem model will be different (see Section V.C).

## IV. STRATIFICATION SIMULATION MODELING

Thermal stratification is a seasonal and diurnal phenomenon common to a large range of water bodies of significant depth (viz., not ponds, puddles, marshes etc.), although for turbid water bodies (e.g., many billabongs, oxbow lakes) stratification becomes possible in relatively shallow aquatic systems.[38] The temperature patterns and the way they change over both seasonal and diel (i.e., over a 24-hour period) cycles are determined by a wide range of influences. Under different circumstances, these forcing processes may be an imbalance in the surface energy budget, penetration of short-wave radiation, convection, turbulent mixing, advection, currents, and physicochemical characteristics of the water (e.g., turbidity, salinity).[38,39] (See further discussion and description in Chapter 2 of the stratification cycle). Knowledge of the temperature characteristics of water bodies may be of importance directly (e.g., controlling the temperature, and hence the acceptability, of potable water in reservoirs), in the influence on chemical transports (e.g., in eutrophication studies[40]) and as a component part of a larger scale model (e.g., ocean stratification submodels as a component of global climate modeling studies[41]). Mathematical models are available on different time and space scales as an aid to understanding the complex interactions occurring between these various processes.

Since, in general, isotherms in a lake form horizontal planes parallel to the surface across the majority of its cross-sectional area, it is reasonable to describe the thermal stratification by means of a one-dimensional equation that models processes occurring along the vertical direction. Methods of solution for the one-dimensional equations used to describe the vertical profiles of temperature and velocity in a water body may be expressed in vector form[42] or component form.[43] Neglecting the horizontal advection terms, the vector form is

$$\frac{\partial \mathbf{v}}{\partial t} + \mathbf{f} \wedge \mathbf{v} = -\frac{\partial}{\partial z}\left[\overline{v'w'}\right] - \frac{\mathbf{F}}{\rho_0} \tag{1}$$

$$\frac{\partial T}{\partial t} + \frac{\partial}{\partial z}\left[\overline{w'T'}\right] = \frac{Q(z,t)}{\rho_w c_p} \tag{2}$$

where $\mathbf{v}$ is the vector velocity with components $(u, v, w)$, $\mathbf{f}$ the Coriolis vector, $z$ the depth, $\mathbf{F}$ a damping term, $\rho_0$ a reference density, $T$ the water temperature, $Q(z,t)$ a heat flux source term, $\rho_w$ the water density, $c_p$ the specific heat, and primes indicate perturbation quantities.

Figure 5 illustrates the major pathways of energy transformations within a water body. However, it does not illustrate the comparative magnitudes of these various processes, which will differ as a function of time and space (vertically downwards in the one-dimensional

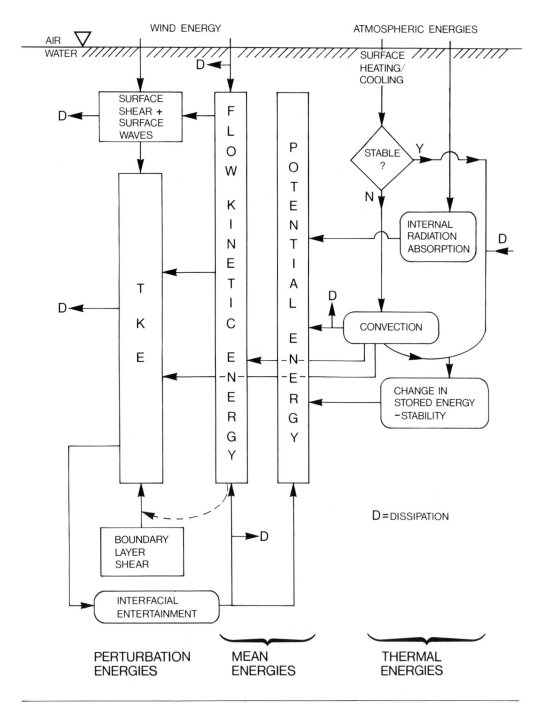

ENERGY BALANCE OF THE UPPER MIXED LAYER OF A LAKE

FIGURE 5. Energy balance of the upper mixed layer of a lake. (From Henderson-Sellers, B., *Environ. Software*, 2, 78–84, 1987. With permission.)

structure imposed here for modeling simplification). It should be noted initially that the forcing of largest magnitude on the long time scale in this system is the energetics associated with the surface energy budget and short-wave penetrative component, a point noted by Kraus and

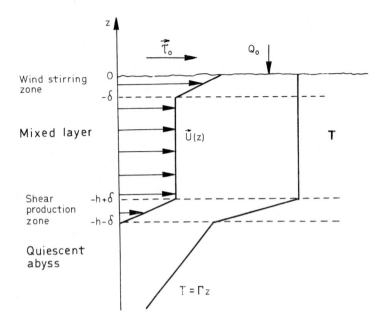

FIGURE 6. Definition sketch showing the vertical structure of the mixed-layer model. (From Niiler, P. P., *J. Mar. Res.*, 33, 405-422, 1975. With permission.)

Rooth,[44] and it is often only the smaller time and space scales that are influenced by the transformations shown on the left-hand side of Figure 5. Nonetheless, these energy transformations (and their source terms at the surface and at the thermocline interface) can be of paramount importance in the deepening of the thermocline and cannot be neglected if the phenomena are to be included in a two- or three-dimensional model or if such effects (e.g., large-scale lateral advection of water masses and hence of energy) are to be incorporated in the one-dimensional model by some type of appropriate parametrization.

Specification of the turbulent kinetic energy (TKE) budget is a common method of attaining closure of Equations 1 and 2 in order to provide a parametrization of the perturbation quantity,[43] or, more usually, in terms of the interfacial entrainment velocity, $w_e$. There are three major identifiable sources for TKE: at the surface as a result of wind-wave induced current shear; at the base of the upper mixed layer (Figure 6), i.e., the interface located at the thermocline, and from convection. As can be seen from Figure 5, there is also a sink to potential energy and via dissipation. A further sink is energy lost by internal gravity waves through underlying stable waters.

The relative importance of these different source terms has been analyzed by many authors; for example, Niiler[42] first reconciled the oceanic modeling approaches of Kraus and Turner[45] and Pollard et al.[46] as asymptotic limits of one generalized theory, an interpretation echoed by de Szoeke and Rhines[47] and Price et al.[48] Further discussion[49] has elucidated that not only are these different approaches limiting forms for the generalized TKE equation, but the processes are likely to dominate on different time scales.

The methods of solution that are discussed briefly below refer to the different methods of closure of these equations (essentially via the correlation terms $\overline{w'v'}$, $\overline{w'T'}$), the neglect of differing energy transfer processes, as well as differing parametrizations of the terms retained.

The mode of closure selected can be used to divide the models into those using an eddy diffusion closure scheme[43,50] and those in which closure is expressed in terms of the rate of descent of the thermocline, $w_e$, given simply as the rate of change of mixed layer depth, $h$. However, a more useful division is between *differential* models that solve the primitive equations in a multilevel finite difference scheme and *bulk* or *integral* models that make an *a priori* assumption about the existence of a homogeneous mixed layer (ML). Consequently differential models are able to *predict* the existence and depth of a mixed layer, and, in contrast, the assumed existence of this homogeneous ML by the bulk model permits the equations to be simplified by integration over the depth, $h$, of the ML. The characteristics of the ML are thus represented by a single equation, which decreases the computational time required for solution at the expense of resolution within the ML; for example, they may not be as able to represent biological parameters, such as diel vertical migration of blue-green algae.[51] These model types can be exemplified by (1) the ML model DYRESM[52] and (2) the eddy diffusion model EDD1 (Eddy Diffusion Dimension 1).[53] These two types of models have been developed in parallel for some years, and both have been shown to be more than adequate for simulating a wide range of lakes and reservoirs. Clearly, a direct comparison of these models, already undertaken analytically,[54] needs to be repeated numerically using common data sets (cf. comparable oceanic testing of Martin[55]).

## A. MIXED LAYER MODELS

Observations of the phenomenon of thermal stratification in oceans as well as lakes and reservoirs suggest that, once the stratification is established, the upper layer of water is quasi-homogeneous in nature. Consequently, it may be reasonable to take as a simplifying assumption in the mathematical model that there exists a homogeneous ML (also called the upper quasi-homogeneous layer [UQL][56]), whose temperature and thickness may vary throughout the seasons; although Kundu[57] cites several sets of observational data that do not support the detail of such a simplification.

Introducing this assumption permits a single differential equation to be used for the whole of this ML. Furthermore, it is a common assumption[42] that the waters underlying the interface are a "quiescent abyss", i.e., no energy "leaks" into these waters from above (nor also by direct radiative penetration) (Figure 6). It is also worth noting here that such restrictions have never been necessitated in the alternative (eddy diffusion) approach to stratification modeling.

It should also be noted that some authors simplify the vertical temperature not into two regions (upper ML and deep water), but into three layers (as exemplified by the case study of Chapter 2). In this approach, observations are reinterpreted as suggesting the existence of an upper ML, a quasi-isothermal deep layer, and an intermediate zone where there exists a steep temperature gradient. In this classical limnological approach, the upper and lower layers are identified as the epilimnion and hypolimnion, respectively. The intermediate layer is known variously as the metalimnion or mesolimnion or, more loosely, as the thermocline region, although strictly the thermocline depth is defined as that surface over which the second derivative of temperature (with respect to depth) vanishes (Figure 7).

The Niiler and Kraus[58] model (dubbed the "energy conservation" model [ECM] by Price et al.[48]) appears to be the most comprehensive in its inclusion of the various active processes (neglecting large-scale advection and inhomogeneities within the ML). Although originally devised for oceanic applications, the modelled processes are, in general, also valid for lakes and reservoirs. The ECM is described by the equation

$$\frac{w_e}{2}\left(q^2 + c_i^2 - \bar{\mathbf{v}}^2 + c_e w_*^2\right) = m_1 w_*^3 + \frac{1}{2}hB_0 + \left(\frac{h}{2} - \frac{1}{\eta}\right)J_0 - \frac{1}{3}C|\bar{\mathbf{v}}|^3 - \int \varepsilon dz \qquad (3)$$

$$\quad\ \text{A}\quad\ \text{B}\quad\ \text{C}\quad\ \text{D}\qquad\ \text{E}\qquad\ \text{F}\qquad\quad\ \text{G}\qquad\quad\ \text{H}\ \cdots\ \text{I}$$

The interpretation of the terms is as follows:

A = rate of energy needed to agitate the entrained water

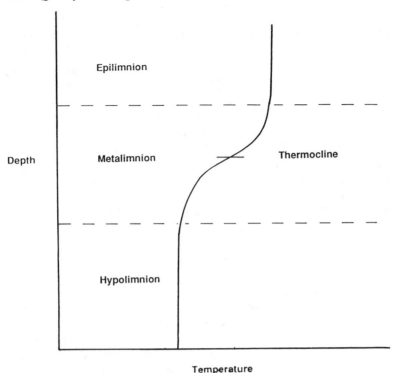

FIGURE 7.  Typical summer temperature profile in a stratified lake. The thermocline is located at the depth  at which the second derivative of temperature with respect to depth equals zero.

B = rate of work to lift entrained water and mix it
C = reduction of mean kinetic energy (KE) by entrainment
D = rate of change of turbulent kinetic energy (TKE)
E = rate of work by wind
F = rate of potential energy (PE) change by surface fluxes
G = rate of potential energy (PE) change by solar radiation
H = work from internal waves
I = dissipation

Niiler and Kraus[58] consider this equation for the *steady state*, thus assuming that there is no change in overall TKE, $\bar{q}$, which might occur at the expense of the main KE or PE fields.

As noted by Niiler and Kraus,[58] whilst the specification of the mixing length scale is the most arbitrary part of a turbulent diffusion closure model,[43] the most arbitrary part of an integral ML model is in the specification of the dissipation term. Niiler and Kraus[58] compare several formulations and advocate that a fraction, $m$, of the wind energy, a fraction, $s$, of the mean KE reduction as a result of entrainment, and a fraction, $n$, of the convective flux all contribute to the dissipation (cf. Figure 5). Hence

$$w_e\left(c_i^2 - s\bar{v}^2 + c_e w_*^2\right) = 2mw_*^3 + 0.5h\left[(1+n)B_0 - (1-n)|B_0|\right] + \left(h - \frac{2}{\eta}\right)J_0 \qquad (4)$$

where, it should be stressed, the coefficients $s$, $m$, $n$, and $c_e$ must be empirically derived for each specific study and are hence "tuning coefficients".

Modeling approaches to thermal stratification that utilize the integral energy (ML) concept differ in their inclusion, neglect, or difference in parametrization in each of these different

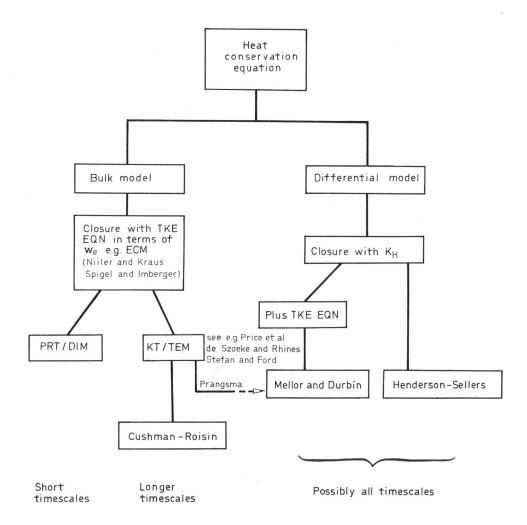

**FIGURE 8.** Flow chart describing various forms of mixed-layer and stratification models, differentiating between ECM, DIM, TEM, and differential models. (From Henderson-Sellers, B., *Environ. Software*, 2, 78-84, 1987. With permission.)

sources of TKE and in how the net TKE is partitioned between enhancing waves, currents, and the descent of the thermocline (Figures 5 and 8). Three basic modeling approaches have been identified by Price et al.[48] and others as

1.    Niiler and Kraus[58] (ECM), which is also used in the original Niiler[42] format
2.    Pollard, Rhines, and Thompson[46] (PRT model or dynamic instability model [DIM])
3.    Kraus and Turner[45] (KT model or turbulent erosion model [TEM])

These different approximations are in fact valid over different time scales.[47,49] Although the DIM and TEM are subsumed in the ECM, since many authors use the original KT model, the PRT model, or the Niiler model formulations, these original identities are often retained for discussion purposes.

In the lacustrine environment, similar approaches have been used in ML modeling, although often with grosser approximations and frequently starting from first principles, rather than by developing appropriate models from the oceanic ML models described above. Essentially such lake models are TEMs. Since it is intended to utilize such models to simulate seasonal time scales and longer, this is appropriate. For example, Stefan and Ford[59] derived their lake model by considering that input kinetic energy (from the wind) will change the potential energy of the water column or will be dissipated. A similar approach is discussed by Sherman et al.[60] and Imberger et al.,[61] but using four empirical constants. Spigel and Imberger[52] synopsize this in the form of an ECM, expressed in finite difference form (named DYRESM). This number of empirical constants is typical of such TKE-based ML models. The central equation to DYRESM is

$$\frac{1}{2}\left(C_T q^2 + \frac{\Delta\rho}{\rho_0} gh\right)\Delta h = \frac{C_K}{2} q^3 \Delta t + \frac{C_S}{2} U^2 \Delta h - \Lambda_L \Delta t \tag{5}$$

where $q^2$ is the turbulent kinetic energy, $\Delta\rho$ is the density difference between the upper and lower layers, $g$ is the acceleration due to gravity, $h$ is the ML thickness, $\Delta h$ is the change in ML thickness, $\Delta t$ is the time step, and $U$ is the wind speed.

In Equation 5, the values of the four empirical coefficients, $C_T$, $C_K$, $C_S$, and $\Lambda_L$, must be found experimentally. However, unlike the tuning or calibration coefficients of models like CE-QUAL-R1,[62] which require separate reevaluation, it is suggested that when the model DYRESM is transferred to another experimental site, the coefficient values can be established once and will remain valid for all cases. Thus, despite being empirical in nature, they are of a fundamentally different ("global") character than the typical tuning coefficients found in many other models.

## B. DIFFERENTIAL MODELS

Eddy diffusion models of the thermocline start with the same set of equations (Equations 1 and 2) but parametrize the Reynolds' stress terms using an eddy diffusion coefficient, $K$. This closure can be undertaken at any one of a hierarchical set of "levels".[63] At the lowest level, a relationship is invoked between the stress terms and the eddy diffusion coefficients for heat and momentum of the form

$$\text{Heat:} \quad -\overline{w'T'} = K_H \frac{\partial T}{\partial z} \tag{6}$$

$$\text{Momentum:} \quad -\overline{w'v'} = K_M \frac{\partial v}{\partial z} \tag{7}$$

and it is this formulation that was originally applied extensively. Differences between models relate to the mode of parametrization of the value of the eddy diffusion coefficients, $K_H$ and $K_M$, as functions of both the ambient meteorological and oceanographical conditions and as a function of the stability. Two basic alternative formulations are current: that of the Mellor and Yamada[63] level-2 (MYL2) scheme in which

$$K_M = l q S_M; \qquad K_H = l q S_H \tag{8}$$

and the "classical" representation of an eddy diffusion coefficient as the product of a neutral value (denoted by subscript 0) and a function of the stability (often expressed in terms of the Richardson number $R_i$):[50,63-66]

$$K_M = K_{M_o} g(R_i); \qquad K_H = K_{H_o} f(R_i) \tag{9}$$

In these equations, $q^2/2$ is the TKE; $l$ a turbulent length scale; and $S_M$, $S_H$, $f(R_i)$, and $g(R_i)$ are functions of stability. It should be noted that both approaches require a stability relationship, but in the MYL2 approach additional equations for the TKE, $q$, and length scale, $l$, are needed in comparison with a relationship for $K_{M_o}$ and $K_{H_o}$ in the classical approach. In the MYL2 model, this requirement produces a link with the integral ML models, since the TKE equation used in Mellor and Yamada[69] and Mellor and Durbin[43] is a form of the TKE equation for the steady state. In essence, then, the Mellor and Durbin[43] model neglects the vertical diffusion of TKE and balances the local rate of TKE production by the local rate of dissipation.[57]

In these models, the dynamics and thermodynamics are effectively decoupled by this means so that Equation 2 is solved in the form

$$A(z)\frac{\partial T}{\partial t} = \frac{\partial}{\partial z}\left(A(z)\left(\alpha_H + K_H\right)\frac{\partial T}{\partial z}\right) + \frac{\frac{\partial}{\partial z}\left(A(z)\phi(z)\right)}{\rho_w c_p} \tag{10}$$

where the molecular diffusivity of heat, $\alpha_H$, is included so that for low wind speed or large depth situations, there is still a minimal diffusion occurring. The cross-sectional area term, $A(z)$, must be included for lake basins, whilst in the ocean a water column is usually considered, whereby $A(z)$ = constant, thus permitting simplification of Equation 10. Using representations such as Equation 9, there is no need to specify the TKE, as in the Mellor and Durbin[43] diffusion model, since these forms of closure for $K_H$ are sufficient. Since $K_H$ can be related theoretically and conceptually to meteorological and hydrological variables, the value is calculated internally to the stratification model. Hence there is no opportunity for the model user to use it as a calibration parameter.

## C. SURFACE ENERGY BUDGET

Almost all energy exchanges at the surface occur within the top few millimeters[67] and can thus be considered as occurring at the air-water interface, the exception being that a portion of the short-wave radiation penetrates this surface layer. (For parametrization of this penetrating component see Equation 19 of Chapter 2 and Equation 16 of Chapter 6.) It is important to determine fairly accurately the total heat budget of the lake, as well as the feedbacks associated with the surface energy fluxes, which can be of the order of ten to several hundred watts per square meter (on average). The net available energy, $\phi_N$, can be given as the sum of the nonreflected incoming short-wave radiation $\phi_0$, the nonreflected incoming atmospheric long-wave radiation $\phi_{ri}$, the outgoing long-wave radiation $\phi_{ro}$, the evaporative energy flux $\phi_e$, and the convective (or sensible) heat loss $\phi_c$. (In this calculation the heat flux associated with precipitation is negligible, because there is no phase change). Thus

$$\phi_N = \phi_0 + \phi_{ri} - \phi_{ro} - \phi_e - \phi_c \tag{11}$$

The portion of the short-wave radiation that penetrates (about 60%) can have a marked effect, not only on the subsurface heat budget, but also upon the potential biotic niches. The differentiation between wavelength regions of this fraction of solar radiation has been examined by several authors.[68-70]

Each of the flux terms in Equation 11 can be related to the meteorological (and astronomical) forcing parameters. Parametrization in terms of cloud cover, air temperature, etc. have been sought over the last few centuries.[71,72] A recent review of the wide selection of currently utilized parametrizations is given in Henderson-Sellers.[73]

## D. AVAILABILITY OF STRATIFICATION MODEL PACKAGES

Software packages for thermal stratification have not been made generally available but can often be obtained by direct contact with the authors. (For details of sources, see the cited papers themselves.[43,50,59,61,74]) In some cases, the original stratification model codes have been later incorporated into larger models. For example, the model CE-THERM[62] is part of CE-QUAL-R1 (a lake water quality model of the ML type); the Minnesota Lake Temperature Model (MLTM) of Stefan and Ford[59] is now part of the University of Minnesota model RESQUAL II.

Other models known by acronyms are DYRESM[52] and EDD1.[53] Indeed, one of the purposes of this book is to bring to the attention of the environmental manager the various modeling packages currently available and useful in decision support.

# V. MODELING THE EUTROPHICATION PROCESS

Modeling has now been accepted as an essential part of any lake restoration program. However, as no two water bodies are alike, models often have to be adjusted to suit the local circumstances or may require substitution by a more appropriate model schematization.

## A. INPUT-OUTPUT (PHOSPHORUS) MODELS

Phosphorus models may be divided broadly into two classes: those that consider the lake as a black box and deal only with inputs, outputs, and the total phosphorus mass in the lake; and those in which differential equations represent rates of change of phosphorus at different spatial locations and of different fractional forms. Each class may contain steady-state models, seasonal models, or annual models; models that include exchanges with the sediments, well-mixed models, and multibox models. The utility of each model depends upon the question to be answered, together with a statement of uncertainty with which the model predictions are made. For example, for a lake that freezes in winter, the annual variability in phosphorus loading becomes more marked, since phosphorus may be retained in the ice cover until it is released "quasi-instantaneously" at ice melt in the spring.

Input-output models are often based on the original work of Vollenweider.[75,76] This model was the first to relate all the applicable parameters concerned in eutrophication, by assuming all nutrients to be well mixed. As with many more sophisticated models, its basis is the assumption of continuity (or mass balance), applied here to the net nutrient balance, calculated both from nutrient inputs and outputs from the system associated with stream inflows and outflows, and from nutrients released (e.g., from sediments and decaying algae) into the water column. Based on data from only 20 lakes, he found a correlation between annual areal nutrient loading and the mean lake depth for a given trophic status. In this model it is assumed that: (1) the input phosphorus is immediately mixed throughout the lake (obviously inapplicable during the summer stratification, although on the annual time scale this limitation produces no immediate problems); (2) the outflow concentration is equal to the concentration prevailing in the lake; (3) the sedimentation rate of phosphorus is proportional to its concentration; and (4) seasonal fluctuations in loading may be neglected. The model is thus most applicable for calculations of the steady-state or year-to-year changes.

An alternative approach is to parametrize the mass balance equation in terms of an apparent settling velocity, $v_s$. In this approach, the deposition rate becomes a function of the bottom surface area. This model was used by Dillon and Rigler[77] for oligotrophic and mesotrophic lakes in Southern Ontario; they found high correlations ($r \sim 0.85$) between the predicted and actual phosphorus values.

Such models imply that the phosphorus concentration can be decreased by increasing the annual sedimentation rate (for example, using a sedimentation rate proportional to the square of the concentration) or the flushing rate. However, the advantages of a higher flushing rate are partially negated by the accompanying decrease in the phosphorus retention coefficient. Input-output models such as these are discussed in detail by, e.g., Reckhow and Chapra.[78] Vollenweider points out that the model neglects the possibility that there may be an internal phosphorus loading (from the sediments) and suggests that a modified model should be used in which the loading is plotted, not against mean depth, but against $\bar{z}/t_w$, where $t_w$ is the mean residence time of the water defined by $t_w$ = lake volume/yearly water discharge.

Although the solution sought is that of a criterion of trophic status for a lake of given annual phosphorus loading, mean depth, and hydraulic residence time, it should be noted that the trophic states are in effect a continuum from oligotrophy to eutrophy, and, hence, the idea[26] of associating a probability with each of the trophic states is to be recommended. Defining a "critical" level, which may be needed as part of any water quality legislation, thus cannot be done easily. The concept of critical loading is thoroughly discussed by Vollenweider,[79] and a new model is developed to give $L_{crit}$ in terms of the hydraulic load, $q_n (= \bar{z}/t_w)$:

$$L_{crit} = 10q_n\left(1 + \frac{\sqrt{\bar{z}}}{q_n}\right) \qquad (12)$$

Vollenweider's models of the input-output type appear reasonably well validated in terms of temperate zone lakes. However, to gain a wider use and acceptance, it is necessary to validate the models for lakes in other climatic regions, although preliminary results from Australia and South Africa have not been overconvincing.[80,81]

## B. MULTILAYER MODELS

To study seasonal variations it is necessary to formulate a model using at least two compartments – corresponding to the epilimnetic and hypolimnetic water masses. Two differential equations are formulated to describe the phosphorus fraction transformation rates. The model has two modes of operation – summer stagnation (two layers) and winter circulation (one layer). As with any temperature model that is similarly formulated, this is somewhat unrealistic, especially during the spring warming period. A similar, yet alternative, model stipulates a stratification period of exactly 6 months, when the hypolimnion is assumed to remain oxic (a supposition not supported by many observations) and inputs of phosphorus from the sediments are neglected. Useful guidelines are given for utilization of such a model for management purposes. This model has been applied by the model developers to lakes with a wide range of detention times, who claim "excellent" results for their calculations of average phosphorus concentrations (see Chapter 5 for a fuller discussion on model validation). However, application of the model by others to Lake Shagawa[82] underestimated the epilimnetic phosphorus concentrations.

It should be noted that a model containing only two layers (often of prespecified depths), representing epilimnion and hypolimnion, respectively, must, *de facto*, only be applicable to a stratified situation, viz., during summer. The period of the year when the lake is not stratified (approximately September/October until March/April in the northern hemisphere) cannot be realistically simulated in this way. The justification for the use is that it is during the stratification period that the problems of algal blooms occur. However, for a more reliable model, which can predict phosphorus levels both throughout the year and from year to year, it is necessary to utilize multilayer models, which incorporate the temporal development of thermal stratification. Consequently, one of the major lines of development has been that of multilayer stratification models by the addition of chemical and biological species. Since these were originally developed by agencies with an engineering bias, they typically "lump together" the various identifiable biotic species into a smaller number of variables. For example, equations for dissolved oxygen (DO) and biological oxygen demand (BOD) were added to the one-dimensional Massachusetts Institute of Technology (MIT) thermal model.[83] Assuming the surface to be saturated, they proposed two alternative models: (1) the entire euphotic zone is maintained at saturation by photosynthetic production and surface reaeration; (2) the saturation exists to a smaller depth. Field tests in Fontana Reservoir were used to test the model's sensitivity to assumed values of inflowing BOD and the BOD decay rate constant. (A further discussion of a more recent version of this model and its application is to be found in Chapter 3.)

The Minnesota Lake Temperature Model[59] has been recently adapted to describe phosphorus cycling and phytoplankton blooms following a destratification event in Lake Calhoun[40] (see also case study of Chapter 6). Homogeneity is maintained within the surface layer, sediment releases are specified as input and not calculated within the model, and the destratification is modelled on a simple vertical advection/entrainment of mass, without consideration of feedbacks to turbulent diffusion alterations. The authors demonstrate that a large algal bloom follows the release of hypolimnetic nutrients upon the initiation of destratification, which would otherwise not have occurred (implying that the phytoplankton system was phosphorus limited). Reorientation of the model towards the calculation of turbidity in shallow lakes,[84,85] in which the model was renamed RESQUAL II, permitted simulations to be undertaken for Lake Chicot in Arkansas. The model does, however, possess a relatively large number of site-specific coefficients, which must be calibrated separately for each new case study. This model, RESQUAL II, is described in full detail in Chapter 6. A newer, research version, renamed MINLAKE (Minnesota Lake Water Quality Management Model),[86] not only includes more processes, but is intended to have the generic properties advocated here in Chapter 5.

Work at the Waterways Experiment Station of the U.S. Army Corps of Engineers has also concentrated on the development of dynamical models. Using two different assumptions regarding the parametrization of vertical mixing processes, they have developed two models for water quality simulation: WESTEX and CE-QUAL-R1, the latter being the ecosystem extension of the thermal model CE-THERM. The model CE-QUAL-R1[62] contains 34 variables, including nitrogen and phosphorus cycles, carbon cycling, and several trophic levels, in addition to the physical factors of CE-THERM, but, as noted above, requires "tuning" to a specific lake before use. It is a one-dimensional model intended to be appropriate to the deepest part of the lake only. Although originally developed from a strictly thermal model, it now has many of the attributes of the more biologically orientated models, developed, e.g., at Rensselaer Polytechnic Institute in New York State. Although the DO profiles are in reasonable agreement with observations, the other chemical parameters are not. This could perhaps reflect the problems of (1) incompatibility between the time scales necessitated by the model and the sampling interval and (2) the assumptions and simplifications made in developing the model, especially that of one dimensionality.

## C. ECOSYSTEM AND EUTROPHICATION MODELS

At about the time that stratification-based models were being developed, groups with more biologically oriented backgrounds were attacking the problem from their viewpoint. As their emphasis has been on the accurate representation of the biochemical characteristics of an ecosystem (usually at the expense of the dynamics and thermodynamics), such models, often with large numbers of "tuning coefficients", will be referred to collectively as ecosystem and eutrophication models (although some researchers differentiate between ecosystem models as placing more emphasis on detailed aspects of ecosystem dynamics than eutrophication models).

The calculations in ecosystem and eutrophication models are based on equations for mass balances. Kinetic formulations are needed for growth rates of phytoplankton, etc., but it is not necessary to calculate growth for each individual species in the lake; indeed, other sources and sinks may well be combined. Furthermore, many of the models do not try to predict the development of the thermal stratification but assume the epilimnion to be well mixed, restricting their analysis to times of the year when the lake can be assumed to be well represented by two static layers or by a single homogeneous layer (the "well-mixed reactor").

A large simulation model has been developed at Manhattan College in New York and utilized for Lakes Huron[87] and Erie.[88] This model (which could almost be called an ecosystem model) utilizes 15 variables. Versions of this model possessed initially only three layers (prespecified, not predicted), then seven layers, and hence straddle the divide between extended multilayer models and comprehensive ecosystem models. The model contains a wide range of "tunable" parameters and has been shown to be useful in simulations of Lake Erie and Lake Ontario.[89] The calibrated results for chlorophyll-*a*, phosphorus, and orthophosphorus show only fair agreement across the lake basin (although no statistical measures of goodness of fit are used; see Chapter 5). However, they do show the silica depletion in May, the nitrogen depletion in the western basin in July, and the phosphorus depletion in the central and eastern basins in July, all of which can be connected with the termination of the two annual observed blooms. Furthermore, this complex model is expensive in terms of computer time.[90]

The ecosystem model of Scavia et al.[91] has been applied successfully to several of the Great Lakes.[92] In a further study,[93] the model was used to examine the phytoplankton dynamics in Lake Ontario. The two-layer model includes parametrization of available P, $NH_3$, $NO_x$, dissolved N, soluble reactive Si, particulate Si, five groups of phytoplankton, six groups of zooplankton, the inorganic C system, sedimentary C, detrital C, and benthic macroinvertebrates. This model has also been compared by Scavia and Chapra[94] with the loading models of Dillon[95] and others. The results are comparable under most circumstances. It is stressed that the choice of model should be determined by the temporal and spatial resolution required, balanced against the additional computational time that would be required to attain such resolution.

The ecosystem approach adopted at the Rensselaer Polytechnic Institute led to the development of the models CLEAN, CLEANER (developed for Lake George, NY, USA but also applied to several European lakes)[96] and then, after a modification to a multisegment model, MS.CLEANER[97] (including versions for use on mini- and microcomputers[98]). These models attempt to include a large number of variables, especially biological, which necessitated sacrificing some representativeness of the dynamic formulations. More recent versions incorporate more dynamical concepts and should therefore be more widely applicable without extensive recalibration. CLEANER utilizes the Steele[99] light limitation function, nutrient limitation, and full temperature dependence on all parameters. In addition, MS.CLEANER includes intracellular storage of nutrients and Michaelis-Menten kinetics, which is also formulated as a function of light and temperature. It has 40 state variables and provisions for horizontal and vertical transport. An application of MS.CLEANER to Övre Heimdalsvatn in Norway has also been undertaken, with results for phytoplankton and zooplankton in relatively good agreement with observations.

The fine details of these large eutrophication models are beyond the scope of this introductory chapter, and the interested reader is encouraged to consult the original published papers and reports for further details. In addition, recent reviews appear in books on water quality modeling.[12,19,78,100-104] In this book, the application of an ecosystem model in a management context to the upper Nile lake system is given in Chapter 7. Other case studies described in this book utilize parametrizations of biochemical processes to a lesser extent.

# REFERENCES

1. **Henderson-Sellers, B. and Gallagher, D. R.,** Modelling tools for water management, *Mathematics and Computers in Simulation,* 32, 143–148, 1990.
2. **Burns, F. L. and Powling, I. J., Eds.,** *Destratification of Lakes and Reservoirs to Improve Water Quality,* Australian Government Publishing Service, Canberra, 1981.
3. **Smalls, I. C. and Petrie, L. G.,** Low cost destratification in small upland reservoirs, in *Procs. Tenth Federal Convention,* Australian Water and Wastewater Association, 1983, 20.1–20.14.
4. **Henderson-Sellers, B. and Markland, H. R.,** *Decaying Lakes. The Origins and Control of Cultural Eutrophication,* John Wiley, Chichester, England, 1987.
5. **Graham, D. S. and Willer, D. C.,** Design of retrofit hydro plant for water quality, in *Applying Research to Hydraulic Practice,* Smith, P. E., Ed., ASCE, New York, 1982, 65–75.

6. **Guariso, G. and Werthner, H.,** *Environmental Decision Support Systems*, Ellis Horwood, Chichester, England, 1989.

7. **Fedra, K.,** A modular interactive simulation system for eutrophication and regional development, *Water Resour. Res.*, 21, 143–152, 1985.

8. **Ford, L. N.,** Decision support systems and expert systems: a comparison, *Information and Management*, 8, 21–26, 1985.

9. **Sprague, R. H., Jr. and Carlson, E. D.,** *Building Effective Decision Support Systems*, Prentice-Hall, Englewood Cliffs, NJ, 1982.

10. **Bonczek, R. H., Holsapple, C. W., and Whinston, A. B.,** *Foundations of Decision Support Systems*, Academic Press, Orlando, FL, 1981.

11. **Sommerville, I.,** *Software Engineering*, 3rd ed., Addison-Wesley, Wokingham, England, 1989.

12. **Jørgensen, S. E.,** *Lake Management*, Pergamon, Oxford, 1980.

13. **Martin, J.,** *Principles of Data-Base Management*, Prentice-Hall, NJ, 1976.

14. **Date, C. J.,** *An Introduction to Database Systems*, Addison-Wesley, Reading, MA, 1983.

15. **Guariso, G. and Werthner, H.,** A computerised inventory for water resources models, *Environ. Software*, 1, 40–46, 1986.

16. **Fedra, K. and Loucks, D. P.,** Interactive computer technology for planning and policy modelling, *Water Resour. Res.*, 21, 114-122, 1985.

17. **Simmonds, D. and Reynolds, L.,** *Computer Presentation of Data in Science*, Kluwer, Dordrecht, Holland, 1989.

18. **Stumm, W.,** *Chemical Processes in Lakes*, John Wiley, Chichester, England, 1985.

19. **Henderson-Sellers, B.,** *Engineering Limnology*, Pitman, London, 1984.

20. **James, D. J. G. and McDonald, J. J.,** *Case Studies in Mathematical Modelling*, Stanley Thornes, Cheltenham, England, 1981.

21. **Smith, I. R. and Henderson-Sellers, B.,** An introduction to mathematical modelling for ecologists and environmental scientists, *Env. Educ. Inf.*, 1, 143–151, 1981.

22. **Neelamkavil, F.,** *Computer Simulation and Modelling*, John Wiley, Chichester, England, 1987.

23. **Straškraba, M. and Sera, Z.,** An empirical model of temperature stratification for Klicava Reservoir, Czechoslovakia, in *Proc. Int. Conf. Reservoir Limnology and Water Quality*, Schweizerbart'sche Verlagsbuchhandlung, Stuttgart, Germany, 1989, 69.

24. **Hannan, H. H., Barrows, D., and Whitenberg, D. C.,** The trophic status of a deep-storage reservoir in central Texas, in *Proc. Symp. Surface Water Impoundments*, Vol. I, Stefan, H. G., Ed., ASCE, New York, 1981, 425–434.

25. **Lind, O. T. and Terrell, T. T.,** Trophic classification: some special problems of reservoirs, in *Proc. Int. Conf. Reservoir Limnology and Water Quality*, Schweizerbart'sche Verlagsbuchhandlung, Stuttgart, Germany, 1989, 647.

26. **OECD,** *Eutrophication of Waters. Monitoring, Assessment and Control*, OECD, Paris, 1982.

27. **Carlson, R. E.,** A trophic state index for lakes, *Limnol. Oceanogr.*, 22, 361–369, 1977.

28. **Porcella, D. B., Peterson, S. A. and Larsen, D. P.,** Index to evaluate lake restoration, *Proc. ASCE, J. Env. Eng. Div.*, 106(EE6), 1151–1169, 1980.

29. **Watson, H. J. and Blackstone, J. H., Jr.,** *Computer Simulation*, 2nd ed., John Wiley, New York, 1989.

30. **Henderson-Sellers, B., Young, P. C., and Ribeiro da Costa, J.,** Water quality models: rivers and reservoirs, in *Proc. Int. Symp. Water Quality Modeling of Agricultural Non-Point Sources, Utah, 1988*, U.S. Department of Agriculture, ARS-81, 381–419, 1990.

31. **Chapra, S. C.,** Total phosphorus model for the Great Lakes, *Proc. ASCE, J. Env. Eng. Div.*, 103(EE2), 147–161, 1977.

32. **Edinger, J. E. and Buchak, E. M.,** Developments in LARM2: a longitudinal-vertical, time-varying hydrodynamic reservoir model, Technical Report E-83-1, U.S. Army Engineer Waterways Experiment Station, Vicksburg, MS, 1983.

33. **Paul, J. F. and Lick, W. J.,** A numerical model for three-dimensional, variable-density hydrodynamic flows, U.S. Environmental Protection Agency Report, Washington, DC, 1981.

34. **Strub, P. T. and Powell, T. M.,** Wind-driven transport in stratified closed basins; direct versus residual circulations, *J. Geophys. Res.*, 91, 8497–8508, 1986.

35. **Lee, K., Bonazountas, M., and Kallidromitou, D.,** LIMNO: a limnology water quality model, in *Computer Techniques in Environmental Studies*, Zannetti, P., Ed., Computational Mechanics Publications, Southampton, England, 131–139, 1988.

36. **Henderson-Sellers, B. and Reckhow, K. H.,** Application of a lake thermal stratification model to various climatic regimes, in *Proc. Int. Conf. Reservoir Limnology and Water Quality*, Schweizerbart'sche Verlagsbuchhandlung, Stuttgart, Germany, 1989, 71–78.

37. **Beck, M. B. and van Straten, G., Eds.,** *Uncertainty and Forecasting of Water Quality*, Springer-Verlag, Berlin, 1983.

38. **Kirk, J. T. O.,** Solar heating of water bodies as influenced by their inherent optical properties, *J. Geophys. Res.*, 93, 10, 897–10, 908, 1988.

39. **Kondratyev, K. Ya., Adamenko, V. N., Vlasov, V. P., Pozdniakov, D. V., Rumiantsev, V. B., Tikhomirov, A. N., Filatov, N. N., and Henderson-Sellers, B.,** Using large lakes as analogues for oceanographic studies, *Focus on Modeling Marine Systems*, 2, 299–304, 1990.

40. **Gulliver, J. S. and Stefan, H. G.,** Lake phytoplankton model with destratification, *Proc. ASCE, J. Env. Eng. Div.*, 108(EE5), 864–882, 1982.

41. **Semtner, A. J., Jr.,** Development of efficient, dynamical ocean-atmosphere models for climatic studies, *J. Clim. Appl. Meteor.*, 23, 353–374, 1984.

42. **Niiler, P. P.,** Deepening of the wind-mixed layer, *J. Mar. Res.*, 33, 405–422, 1975.

43. **Mellor, G. L. and Durbin, P. A.,** The structure and dynamics of the ocean surface mixed layer, *J. Phys. Oceanogr.*, 5, 718–728, 1975.

44. **Kraus, E. B. and Rooth, C.,** Temperature and steady state vertical heat flux in the ocean surface layers, *Tellus*, 13, 231–238, 1961.

45. **Kraus, E. B. and Turner, J. S.,** A one-dimensional model of the seasonal thermocline II. The general theory and its consequences, *Tellus*, 19, 98–105, 1967.

46. **Pollard, R. T., Rhines, P. B., and Thompson, R. O. R. Y.,** The deepening of the wind-mixed layer, *Geophys. Fluid Dynam.*, 3, 381–404, 1973.

47. **de Szoeke, R. A. and Rhines, P. B.,** Asymptotic regimes in mixed-layer deepening, *J. Mar. Res.*, 34, 111–116, 1976.

48. **Price, J. F., Mooers, C. N. K., and van Leer, J. C.,** Observation and simulation of storm-induced mixed-layer deepening, *J. Phys. Oceanogr.*, 8, 582–599, 1978.

49. **Niiler, P. P.,** One-dimensional models of the seasonal thermocline, in *The Sea, Vol 6: Marine Modeling*, McCave, I. N., O'Brien, J. J., and Steele, J. H., Eds., John Wiley, New York, 1977, 97–115.

50. **Henderson-Sellers, B.,** New formulation of eddy diffusion thermocline models, *Appl. Math. Model.*, 9, 441–446, 1985.

51. **Reynolds, C. S. and Walsby, A. E.,** Water blooms, *Biol. Rev.*, 50, 437–481, 1975.

52. **Spigel, R. H. and Imberger, J.,** The classification of mixed-layer dynamics in lakes of small to medium size, *J. Phys. Oceanogr.*, 10, 1104–1121, 1980.

53. **Henderson-Sellers, B.,** Sensitivity of thermal stratification models to changing boundary conditions, *Appl. Math. Model.*, 12, 31–43, 1988.

54. **Henderson-Sellers, B. and Davies, A. M.,** Thermal stratification modelling for oceans and lakes, *Ann. Rev. Numerical Fluid Dynamics and Heat Transfer*, 2, 86–156, 1989.

55. **Martin, P. J.,** Simulation of the mixed layer at OWS November and Papa with several models, *J. Geophys. Res.*, 90, 903–916, 1985.

56. **Shelkovnikov, N. K.,** A study of entrainment at the interface in a two-layer liquid, *Soviet Met. Hydrol.*, 1983(1), 40–45, 1983.

57. **Kundu, P. K.,** A numerical investigation of mixed-layer dynamics, *J. Phys. Oceanogr.*, 10, 220–236, 1980.

58. **Niiler, P. P. and Kraus, E. B.,** One-dimensional models of the upper ocean, in *Modelling and Prediction of the Upper Layers of the Ocean*, Kraus, E. B., Ed., Pergamon Press, New York, 1977, chap. 10.

59. **Stefan, H. and Ford, D. E.,** Temperature dynamics in dimictic lakes, *Proc. ASCE., J. Hyd. Div.*, 101(HY1), 97–114, 1975.

60. **Sherman, F. S., Imberger, J., and Corcos, M. G.,** Turbulence and mixing in stably stratified water, *Ann. Rev. Fluid Mech.*, 10, 267–288, 1978.

61. **Imberger, J., Patterson, J., Hebbert, B., and Loh, I.,** Dynamics of reservoir of medium size, *Procs. ASCE, J. Hyd. Div.*, 104(HY5), 725–743, 1978.

62. **Environmental Laboratory,** CE-QUAL-R1: A numerical one-dimensional model of reservoir water quality, User's Manual, Instruction Report E-82-1, U.S. Army Corps of Engineers, Waterways Experiment Station, CE, Vicksburg, MS, 1982.

63. **Mellor, G. L. and Yamada, T.,** A hierarchy of turbulent closure models for planetary boundary layers, *J. Atmos. Sci.*, 31, 1791–1806, 1974.

64. **Rossby, C. C. and Montgomery, B. R.,** The layer of frictional influence in wind and ocean currents, *Papers in Physical Oceanography*, 3, 101, 1935.

65. **Munk, W. H. and Anderson, E. R.,** Notes on a theory of the thermocline, *J. Mar. Res.*, 7, 276–295, 1948.

66. **Hearn, C. J.,** On the Munk-Anderson equations and the formation of the thermocline, *Appl. Math. Model.*, 12, 450–456, 1988.

67. **McAlister, E. D. and Macleish, W.,** Heat transfer in the top millimeter of the ocean, *J. Geophys. Res.*, 74, 3408–3414, 1969.

68. **Effler, S. W., Wodka, M. C., and Field, S. D.,** Scattering and absorption of light in Onondaga Lake, *J. Environ. Eng.*, 110, 1134–1145, 1984.

69.  **Field, S. D. and Effler, S. W.,** Light-productivity model for Onondaga Lake, N.Y., *Proc. ASCE, J. Environ. Eng. Div.*, 109, 830–844, 1983.

70.  **Kirk, J. T. O.,** *Light and Photosynthesis in Aquatic Ecosystems*, Cambridge University Press, Cambridge, 1983.

71.  **Dalton, J.,** Experimental essays on the constitution of mixed gases; on the force of steam or vapour from water and other liquids in different temperatures, both in a Torricellian vacuum and in air; on evaporation and on the expansion of gases by heat, *Mem. Proc. Manchester Lit. Phil. Soc.*, 5, 535–602, 1802.

72.  **Ångström, A.,** Solar and terrestrial radiation, *Q. J. Roy. Meteorol. Soc.*, 50, 121–125, 1924.

73.  **Henderson-Sellers, B.,** Calculating the surface energy balance for lake and reservoir modelling: a review, *Rev. Geophys.*, 24, 625–649, 1986.

74.  **Harleman, D. R. F.,** Hydrothermal analysis of lakes and reservoirs, *Procs. ASCE, J. Hyd. Div.*, 108(HY3), 302–325, 1982.

75.  **Vollenweider, R. A.,** *Scientific Fundamentals of the Eutrophication of Lakes and Flowing Waters, with Particular Reference to Nitrogen and Phosphorus as Factors in Eutrophication*, Organization for Economic Cooperation and Development, DAS/CSI/68.27, OECD, Paris, 1968.

76.  **Vollenweider, R. A.,** Möglichkeiten und Grenzen elementarer Modelle der Stoffbilanz von Seen, *Arch. Hydrobiol.*, 66, 1–36, 1969.

77.  **Dillon, P. J. and Rigler, F. H.,** The phosphorus-chlorophyll relationship in lakes, *Limnol. Oceanogr.*, 19, 767–773, 1974.

78.  **Reckhow, K. H. and Chapra, S. C.,** *Engineering Approaches for Lake Management*, 2 vols, Ann Arbor Sci., Ann Arbor, MI, 1983.

79.  **Vollenweider, R. A.,** Advances in defining critical loading levels for phosphorus in lake eutrophication, *Mem. Ist. Ital. Idrobiol. Dott. Marco de-Marchi*, 33, 53–83, 1976.

80.  **Smalls, I.,** personal communication, 1988.

81.  **Robarts, R.,** personal communication, 1987.

82.  **Snodgrass, W. J. and O'Melia, C. R.,** Predictive model for phosphorus in lakes, *Environ. Sci. Technol.*, 9, 937-944, 1975.

83.  **Markofsky, M. and Harleman, D. R. F.,** Prediction of water quality in stratified reservoirs, *Proc. ASCE, J. Hyd. Div.*, 99, 729–745, 1973.

84.  **Stefan, H. G., Dhamotharan, S., and Schiebe, F. R.,** Temperature/sediment model for a shallow lake, *Proc. ASCE., J. Environ. Eng. Div.*, 108(EE4), 750–765, 1982.

85.  **Stefan, H. G., Cardoni, J. J., Schiebe, F. R., and Cooper, C. M.,** Model of light penetration in a turbid lake, *Water Resour. Res.*, 19, 109–120, 1983.

86.  **Riley, M. J. and Stefan, H. G.,** MINLAKE: a dynamic lake water quality simulation model, *Ecol. Model.*, 43, 155–182, 1988.

87.  **Di Toro, D. M. and Matystik, W. F., Jr.,** *Mathematical Models of Water Quality in Large Lakes, Part 1, Lake Huron and Saginaw Bay*, Manhattan College, New York, internal report, 1977.

88.  **Di Toro, D. M. and Connolly, J. P.,** *Mathematical Models of Water Quality in Large Lakes, Part 2, Lake Erie*, Manhattan College, New York, internal report, 1977.

89.  **Thomann, R. V., Winfield, R. P., Di Toro, D. M., and O'Connor, D. J.,** *Mathematical Modeling of Phytoplankton in Lake Ontario, Part 2, Simulations Using Lake 1 Model*, U.S. Environmental Protection Agency report EPA-600/3-76-065, Duluth, MN, 1976.

90.  **Schnoor, J. L. and O'Connor, D. J.,** A steady state eutrophication model for lakes, *Water Research*, 14, 1651–1665, 1980.

91.  **Scavia, D., Eadie, B. J. and Robertson, A.,** *An Ecological Model for Lake Ontario — Model Formulation and Preliminary Evaluation*, Environ. Res. Lab. NOAA Tech. Rep. ERL 371-GLERL12, 1976.

92.  **Scavia, D., Eadie, B. J., and Robertson, A.,** An ecological model for the Great Lakes, in *Environmental Modeling and Simulation*, Ott, W. T., Ed., U.S. Environmental Protection Agency report ERL 600/9-76-016, Washington DC, 1976, 629–633.

93.  **Scavia, D.,** An ecological model of Lake Ontario, *Ecol. Model.*, 8, 49–78, 1979.

94.  **Scavia, D. and Chapra, S. C.,** Comparison of an ecological model of Lake Ontario and phosphorus loading models, *J. Fish. Res. Bd. Canada*, 34, 286–290, 1977.

95.  **Dillon, P. J.,** The phosphorus budget of Cameron Lake, Ontario: the importance of flushing rate relative to the degree of eutrophy of a lake, *Limnol. Oceanogr.*, 20, 28–39, 1975.

96.  **Park, R. A., Scavia, D., and Clesceri, N. L.,** CLEANER, the Lake George model, in *Ecological Modeling in a Management Context*, Russell, C. S., Ed., Resources for the Future, Washington D.C., 1975, 49–81.

97.  **Park, R. A., Collins, C. D., Connolly, C. I., Albanese, J. R., and Macleod, B. B.,** *Documentation of the Aquatic Ecosystem Model MS.CLEANER*, A final report for Grant No. R80504701, U.S. Environmental Protection Agency, Athens GA, 1981.

98.  **Albanese, J. R., Collins, C. D., Connolly, C. I., Macleod, B. B., and Park, R. A.,** The potential role of ecosystem models in reservoir management, in *Proc. Symp. Surface Water Impoundments*, Vol. I, Stefan, H. G., Ed., ASCE, New York, 1981, 576–584.

99.  **Steele, J. H.,** Notes on some theoretical problems in production ecology: primary production in aquatic environments, Goldman, C. R., Ed., *Mem. Ist. Idrobiol*, Vol. 18, Suppl., Univ. California Press, Berkeley, CA, 1965, 383–398.

100.  **Krenkel, P. A. and Novotny, V.,** *Water Quality Management*, Academic Press, New York, 1980.

101.  **Orlob, G. T., Ed.,** *Mathematical Modeling of Water Quality: Streams, Lakes and Reservoirs*, John Wiley, Chichester, England, 1983.

102.  **Straškraba, M. and Gnauck, A. H.,** *Freshwater Ecosystems, Modelling and Simulation*, Elsevier, Amsterdam, 1985.

103.  **Gray, W. G., Ed.,** *Physics-Based Modeling of Lakes, Reservoirs, and Impoundments*, ASCE, New York, 1986.

104.  **Jørgensen, S. E. and Gromiec, M. J., Eds.,** *Mathematical Submodels in Water Quality Systems*, Elsevier, Amsterdam, 1989.

105.  **Henderson-Sellers, B.,** Modelling for non-mathematicians, in *Mathematical Modelling Courses*, Berry, J. S., Burghes, D. N., Huntley, I. D., James, D. J. G., and Moscardini, A. O., Eds., Ellis Horwood, Chichester, England, 1987.

106.  **Henderson-Sellers, B.,** One-dimensional modelling of thermal stratification in oceans and lakes, *Environ. Software*, 2, 78–84, 1987.

Chapter 2

# MODELING LAKE ERIE WATER QUALITY — A CASE STUDY

## W. M. Schertzer and D. C. L. Lam

## TABLE OF CONTENTS

# I. INTRODUCTION

Coincident with increasing population and industrial pressures within the Great Lakes basin, there emerged an accelerated eutrophication of the lakes and embayments.[29] Symptomatic of the decline in water quality has been the occurrence of high annual primary production, algal blooms, and increased growth of *Cladaphora* along the shorelines. The eutrophication problem has been related to increased production due to high loadings of the nutrient total phosphorus from such sources as detergents and fertilizers.[72,75] Although decreases in water quality were observed in embayments and near-shore zones of many of the Great Lakes, eutrophication was more acute in the case of Lake Erie, its physical dimensions limiting its capacity to cope with the loading stresses. On the basis of trophic status, defined with reference to phosphorus loading and hydraulic retention time,[77] Lake Erie is considered mesotrophic to eutrophic, its physiographic regions experiencing conditions ranging to highly eutrophic (polluted).

Of specific concern in Lake Erie has been the occurrence of widespread hypolimnetic anoxia within the central basin. Under conditions of severe anoxia over large areal extent and long duration, chemically reducing conditions can occur that lead to a release of phosphorus and other deleterious materials from the lake sediments back to the water column. Such internal loadings serve to aggravate the oxygen depletion problem.

Chapra[13] developed a total phosphorus model from which he determined a historical loading of the nutrient to the Great Lakes and specifically to Lake Erie and its basins from 1800 to 1970. Recent measurements by the National Water Research Institute (NWRI) in Canada over the period 1966 to 1982[39] documented the total phosphorus loading, which peaked in the early 1970s decreasing thereafter in response to the implementation of a phosphorus removal program in an international cooperative effort between Canada and the United States.[30] Even with substantially reduced loadings, anoxia is still observed to occur intermittently in the Lake Erie central basin hypolimnion.

The oxygen depletion problem in Lake Erie has been examined through a number of approaches, generally based on oxygen mass balances in which the sediment oxygen demand (SOD) and the water oxygen demand (WOD) are compared to oxygen sources due to physical processes.[8,9,12,18,22] Burns[8] applied a mesolimnion exchange model to the central basin of Lake Erie to calculate the sources and sinks of oxygen between two successive cruises. The modeling procedure was hampered by insufficient temporal resolution, but indicated the necessity of considering the seasonal changes in the thermocline depth in determining the oxygen depletion. On the basis of correlations between corrected oxygen depletion rates and hypolimnion thickness, Charlton[12] concluded that basin morphometry had a pronounced effect on oxygen depletion in the central basin. In another approach, DiToro and Connolly[21] applied a complex 15-variable ecological model to Lake Erie to derive a relation between the oxygen depletion rate and the total phosphorus loading. Upon model verification, it was found that assumptions of constant hypolimnion depths during the summer and fall months had to be relaxed in order to provide good correspondence between computed and observed oxygen depletion values, expecially during anomalous years.

Previous investigations alluded to the influence of physical factors such as basin morphology and thermal structure on the development and progression of anoxia in the central basin of Lake Erie. Indeed, the oxygen depletion problem in this lake was summarized by Barica[4] in a remark that anoxia may occur from time to time as a result of an unpredictable interaction of physical and biochemical processes.

This chapter provides a summary of the water quality modeling efforts on Lake Erie conducted by the NWRI through the application of a detailed modeling framework to analyze the interactions of the physical and biochemical processes affecting oxygen depletion, especially in the central basin. Specifically, a framework is devised to examine the water quality responses of Lake Erie to the influence of loading and weather variability. The model framework combines a nine-box mass balance model[39] with an existing two-variable water quality model[63] and a simple SOD model.[67] The dynamic effects of thermal layer thickness, turbulent diffusion, and vertical entrainment on oxygen concentration are calculated by incorporating a one-dimensional thermocline model[40] to simulate daily vertical temperature profiles. The model framework is applied over the period 1967 to 1982 on Lake Erie, and comparisons are made between computed and observed physical and chemical variables, such as temperature, nutrient, and dissolved oxygen concentrations.

Long-term records in Canada and the United States, and well-documented measurements on Lake Erie from 1967 to 1982, have offered a unique opportunity to develop, verify, and validate a model framework applied to water quality problems. While the individual aspects of this modeling study have been reported separately in Schertzer,[58-59] Fraser,[26] Lam and Schertzer,[40] and Lam et al.,[41-42] the present chapter is an attempt to combine these individual results and to present them as a case study. Thus, the chapter is structured so that observations related to the modeling study are described first. Then the most important physical model (i.e., the thermocline model) is discussed. The thermocline model results have a major impact on the choice of model resolution and the framework appropriate for the analysis of the nutrient and oxygen interactions. Only after the appropriate physical basis is chosen can the biological and chemical models be developed. Comparison of model results to observations is strongly emphasized in stages from calibration and in verification to a post-audit analysis. The postaudit of the model is possible because of the long-term data record and the intervention imposed by the phosphorus control measures. After the accuracy of the models are established, scenario predictions under various management strategies are presented, with a detailed account on the uncertainties. These uncertainties are discussed in the final section of this chapter in the context of probabilities and frequency distributions in relation to the anoxic occurrences. Thus, this chapter offers an evolutionary account of the Lake Erie Water Quality Modeling Study.

## II. BASIN CHARACTERISTICS

Lake Erie forms the southernmost portion of the Laurentian Great Lakes system (Figure 1). It is centered on 42°15′N latitude and 81°15′W longitude. The basin encompasses an area of 103,000 km$^2$ and lies across the international boundary of Canada and the United States. Approximately one-quarter of the basin area, 25,320 km$^2$, represents the lake surface area itself. Compared with the other Great Lakes, Lake Erie contains the smallest volume of water and is the shallowest.[53]

Physical characteristics of Lake Erie are given in Tables 1A and 1B. Figure 1 illustrates the bathymetry. Most of the lake is shallow, with a progressive increase in depth from a maximum of 10 m in the west end to a maximum of 64 m in the east end of the lake. On the basis of bathymetry, the lake can be subdivided into three distinct physiographic regions: the western basin, the central basin, and the eastern basin. The shallow western basin is separated from the relatively flat-bottomed central basin by a rocky island chain extending from Point Pelee, Ontario to Marblehead, OH. A low, wide, submerged sand and gravel ridge, called the

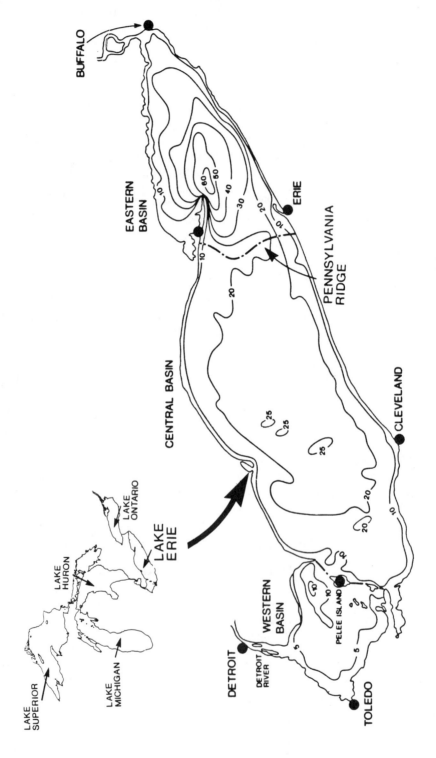

FIGURE 1.  Lake Erie bathymetry and basin boundaries.

**TABLE 1A**
**Physical Characteristics of Lake Erie at Low Water Datum (LWD)**

| | |
|---|---|
| Low water datum (LWD) above sea level | 173.1 m |
| Length | 388 km |
| Width | 92 km |
| Shoreline length | 1,377 km |
| Total surface area | 25,320 km$^2$ |
| Volume at LWD | 473 km$^3$ |
| Mean geometric depth | 18.7 m |
| Depth of half volume | 9.97 m |
| Maximum depth | 64 m |

**TABLE 1B**
**Physical Characteristics of Lake Erie Basins**

| | West | Central | East |
|---|---|---|---|
| Surface area | 2,912 km$^2$ | 16,746 km$^2$ | 5,672 km$^2$ |
| Volume | 20 km$^3$ | 300 km$^3$ | 150 km$^3$ |
| Maximum depth | 10 m | 25 m | 64 m |

Pennsylvania Ridge, extends from Long Point, Ontario to Erie, PA, separating the central and eastern basins. The Pennsylvania Ridge has a maximum depth of 21 m, whereas the central basin is only 25 m deep. Under conditions of a deep thermocline, below the maximum depth of the Pennsylvania Ridge, the exchange of hypolimnion water in the central basin with the eastern basin is curtailed.

## III. OBSERVATIONS

The water quality model incorporates physical and chemical data. Fortunately, in the case of Lake Erie, long-term measurements for most variables are available from federal and provincial/state institutions, both in Canada and the United States.[39] Physical data include air and water temperatures, wind speed, humidity, heat fluxes, light extinction, flows, and water level changes, while the chemical parameters considered are total phosphorus (TP), soluble reactive phosphorus (SRP), and dissolved oxygen (DO). Application of a dynamic water quality model to Lake Erie requires critical consideration as to the combination of model resolution with constraints imposed by the data. Thus, instead of describing the whole database, we have selected the following observational aspects that strongly exert influences over the design and development of the models.

### A. ONE-DIMENSIONAL MODEL RESOLUTION

Models of lake water quality are subject to the proper complexity, preferably intermediate complexity within the limitations of the data and computing facilities. The approach adopted in this case study has been to apply sound limnological analysis to simulate the system beyond pure statistical analysis.[76] In Lam et al.,[39] models of fine spatial resolution were considered. However, more complex models are limited by the available data and knowledge, and by the practicality of using them for long-term prediction.[61-62] The alternative chosen in this investigation was to constrain the spatial complexity as low as possible for some variables (i.e., a six-compartment or nine-compartment three-variable model) and to depend on the results of models of higher complexity (i.e., one-dimensional thermocline model) for those variables requiring higher spatial resolution. Thus, instead of simulating all variables with the same grid resolution, some (e.g., oxygen and phosphorus) can be simulated with a coarse grid and others, such as lake temperature, with a finer grid.[33]

## B. DATA REPRESENTATION

The scale of water quality model resolution requires consideration of the most appropriate procedure to derive representative lake data for a one-dimensional model. Details of the data analyses are given in Lam et al.[39] and are briefly discussed here for the major parameters. Physical and chemical data analysis basically involved measurements from both lake and land-based stations.

During the study period, a total of 99 lakewide surveys of physical and chemical parameters were conducted by NWRI. Measurement surveys occurred primarily between the months of April to November, concentrating on the period of lake thermal stratification. The frequency of surveys varied from a minimum of two to a maximum of eight per year, with station densities on the order of 1:400 km[2].

Profiles of physical and chemical data measured over a grid of stations were processed by digitization. For each cruise, the data from each station were computer interpolated using an inverse distance squared procedure applied over a grid dimension of 2 km by 1 m depth. Chemical data were further analyzed by incorporating information on the thermal structure. The data reduction procedure allows data representation in one, two, or three dimensions.

An example of the data resolution and representation for one-dimensional water quality analysis can be demonstrated by considering the processing of water temperature data, which are reduced to a basin average value.[59] Figure 2a illustrates the original time-series of profile data collected at a fixed station with a depth of 20 m in the central basin of Lake Erie. At this temporal and spatial scale, diurnal cycles and thermocline calculations at a point in the lake are clearly discernible. Alternatively, data from several stations in a lake survey can be processed to derive a two-dimensional representation (Figure 2b) along a cross-section of the lake. The two-dimensional scale results in a loss of the finer temporal information at each location but reveals details of the spatial distribution of the data. Figure 2c illustrates the same temperature data plotted in a one-dimensional representation for each of the lake basins. The nature of the basin average profile is such that some spatial and temporal detail is lost compared to the more detailed data representation.

The advantage of the one-dimensional basin average profile approach is that it allows for model development at a scale of complexity that is relatively simple and offers the essential physical parameters, such as daily hypolimnetic depths and vertical diffusivities, to link up with the biological and chemical submodels.[38] In comparison with more detailed and complex two- and three-dimensional models, the one-dimensional approach has advantages of simplicity, efficiency, and adequacy for addressing the basinwide oxygen depletion problem.

In terms of thermal structure, the one-dimensional data representation and models greatly simplify the basinwide description for the three thermal regimes, i.e., a homogeneous upper layer (epilimnion), a homogeneous lower layer (hypolimnion), and the middle layer (mesolimnion or metalimnion). Chemical data, such as TP and SRP and DO concentrations, can also be readily reduced to the one-dimensional basinwide average.

## C. CHEMICAL LOADINGS

Chemical data include phosphorus and dissolved oxygen. Two forms of phosphorus are considered: TP (soluble and particulate forms of inorganic and organic phosphorus) and SRP (soluble inorganic form). Details of the measurements and processing of these data are provided in Lam et al.,[39] Fraser and Willson,[25] and Fraser.[26] Basically over the period 1967 to 1982, phosphorus concentrations and flows were collated from several agencies in Canada and the United States. Rigorous loadings computations were conducted considering tributary, municipal, industrial, atmospheric, and connecting channels sources. Whereas reliance was placed on published loads for atmospheric contributions, i.e., total yearly loads[24] were assumed constant and spatially homogeneous over the study period, loadings computations for the other parameters involved application of sophisticated estimator techniques[5,23] to derive meaningful values from discontinuous data.[26]

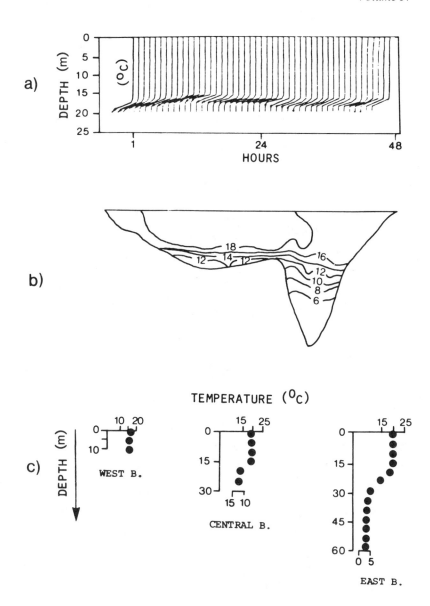

FIGURE 2. (a) Hourly averaged profiles; (b) two-dimensional *(x–z)* contours generated by interpolation of ship cruise temperature data, i.e., 13 to 19 September, 1978; (c) one-dimensional volume-weighted averaged temperature profiles (obtained from two-dimensional contours for 13 to 19 September, 1978). From Lam, D. C. L.and Schertzer, W. M., *J. Great Lakes Res.*, 13(4), 757-769, 1987. With permission.)

## 1. Total Phosphorus

As indicated previously, historical phosphorus loading to Lake Erie has been estimated by Chapra[13] by incorporating the influences of land runoff, atmosphere, upstream sources, human waste, and detergents. Chapra estimates that from 1800 to 1850 the annual loading of phosphorus to Lake Erie ranged from 3 to 4 t/year. The period 1850 to approximately 1945 saw a rapid linear increase in loadings to 11 t/year, attributable to increased agricultural activities in the basin. Increased sewering, population growth, and the introduction of phosphate detergents made a strong impact on loadings to Lake Erie in the period after 1945. Chapra[13] estimated loadings in the late 1960s in the range of 20 to 21 t/year.

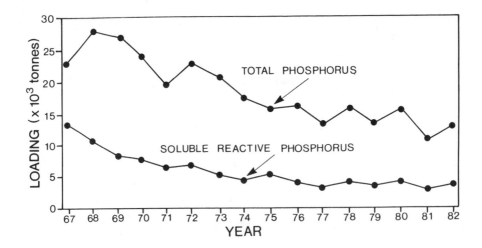

FIGURE 3. Annual total phosphorus and soluble reactive phosphorus loading to Lake Erie. (Based on Fraser, A. S., *J. Great Lakes Res.*, 13(4), 659-666, 1987 and Lam, D. C. L. et al., Simulation of Lake Erie Water Quality Responses to Loading and Weather Variations, NWRI, 1983. With permission.)

Figure 3 illustrates the annual total phosphorus loading to Lake Erie based on measurements and computations from major sources in the period 1967 to 1982.[26,39] The annual loading in 1967 amounted to 23.4 t, increasing to a peak of 27.9 t in 1968. The period 1969 to 1982 saw a significant quasi-linear decrease in loadings to approximately 12.4 t in 1982. Decreases in the TP loads have resulted from the introduction of tertiary water treatment at municipal water plants around the basin for phosphorus removal and from limitations of detergent phosphorus. The Canada/U.S. Water Quality Agreement[30] calls for reductions of phosphorus loading to a target of 11.0 t/year.

The dominant loading component for TP up until 1977 was the Detroit River. Decreases in phosphorus loadings have been attributed largely to substantial reductions from this source.[39] Reductions of the order of 50% occurred in the Detroit River, as well as from municipal and industrial sources, while tributary loading was more variable. Loading along the U.S. shoreline is greater than in Canada due to the larger population and industrial capability. As expected, the nutrient loading has a seasonal component, reflecting the higher runoff in the spring freshet and in the fall season resulting from higher rainfall amounts. The western basin receives more than half of the lakewide phosphorus load.

### 2. Soluble Reactive Phosphorus

SRP, which may be considered as the soluble inorganic form of phosphorus that is most readily available for biological use, shows loading patterns similar to that of TP. The magnitude of the annual loading, however, is substantially lower compared to TP. SRP constitutes approximately 34% of the lakewide TP load (Figure 3).

The bioavailability associated with this form is an important factor in managing the Lake Erie eutrophication problem. A reduction in the SRP load is more efficient in decreasing in-lake phosphorus concentrations than an equivalent reduction in particulate load,[66] as shown by Simons and Lam.[65] SRP loading (shown in Figure 3) has decreased also. This is due largely to reductions in the Detroit River source, and seasonal trends are similar to those observed with TP.[26]

A significant internal source of SRP may come from regeneration at the sediment/water interface. In the loadings computations, this source is unaccounted for. It is assumed here that

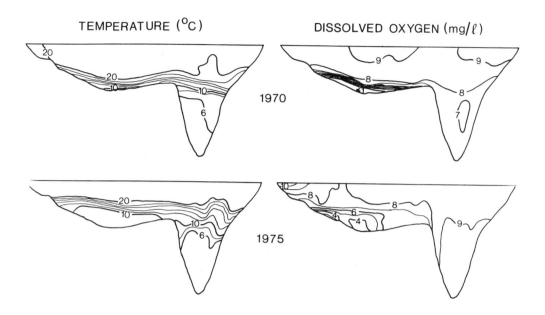

TEMPERATURE (°C)     DISSOLVED OXYGEN (mg/ℓ)

1970

1975

FIGURE 4. Two-dimensional representation of observed data for temperature and dissolved oxygen for 1970 and 1975.

subsequent uptake and sedimentation, followed by the onset of the fall mixing period, may offset much of the impact of such internal loading.

## D. DISSOLVED OXYGEN DYNAMICS

Actual values of the observed TP, SRP, and DO for each of the lakewide cruises in each of the Lake Erie basins and layers is illustrated in the results section in comparison with computations. Although the analysis is applied to one-dimensional considerations, the two-dimensional representations of temperature and chemical parameters can demonstrate the dynamics of the seasonal DO concentration observed in Lake Erie with respect to the influence of the thermal structure. Figure 4 shows two-dimensional representations of observed temperature and DO along a longitudinal cross-section of Lake Erie.

Concentrations of oxygen in aquatic systems are governed by many mechanisms. In epilimnetic waters, the concentration is mainly controlled by ambient temperature and pressure. Under normal conditions, there is sufficient oxygen present to establish 100% saturation at all times. Typically, during the winter-spring isothermal period, oxygen concentrations lie in the range 12 to 13 mg $O_2$/l. As the lake warms, the oxygen concentration in the epilimnion decreases but still maintains 100% saturation. Under fully stratified conditions and when the temperature in the epilimnion reaches 20°C, oxygen saturation is achieved with concentrations of 8 to 9 mg $O_2$/l (Figure 4).

Chemical and microbiological oxidation processes occur continuously in both the water column and the surficial sediments of the lake. These processes comprise the WOD and SOD.[8-9] Hypolimnetic oxygen depletion in the central basin has been the center of international concern for many years.[10,29,72] Anoxia occurs when aerobic decay processes acting on organic material consume the available oxygen in solution. Although anoxia technically is the absence of any oxygen, the International Joint Commission has accepted an analytical value of less than 0.5 mg $O_2$/l to be representative of the condition.[20] In our modeling studies using a three-layer framework, we have chosen a limit of 1.5 mg/l.[39]

Since the thermocline prevents significant diffusion of oxygen from the epilimnion to the hypolimnion, the depth of the thermocline on the central basin is of prime importance in the

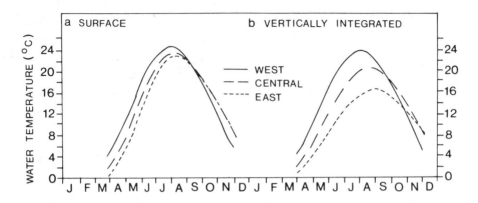

FIGURE 5.   Long-term mean seasonal cycle of surface temperature and vertically integrated temperature for Lake Erie basins (1967 to 1982). (From Schertzer, W. M. et al., *J. Great Lakes Res.*, 13(4), 468-486, 1987. With permission.)

evaluation of the relative severity of the anoxic conditions. Figure 4 shows different levels of oxygen depletion in the central basin hypolimnion for two contrasting years. The strong stratification of 1970 permitted a particularly severe episode of anoxia to occur that, as previously noted, released substantial amounts of nutrients into the hypolimnion. Although some low oxygen concentrations were computed for the western end of the central basin in 1975, only a very small area was involved. The dominant feature of this latter year is the thickness and position of the thermocline relative to the basin morphometry.

## E. PHYSICAL DATABASE

Observations from land-based stations at the lake periphery were collated and formed into daily means by appropriate averaging or interpolation,[39] and are used in heat balance calculations.[58] Details of the hydrometeorological characteristics of Lake Erie are presented in Schertzer[58] and Schertzer et al.[59] Climatological characteristics of the basin are described in Phillips and McCulloch.[49] A brief description of some of the more important parameters is provided below.

### 1. Water Temperature and Stratification Cycle

The seasonal temperature cycle begins with springtime heating of the surface layers by solar radiation, resulting in thermal expansion and a decrease in water density. Wind-generated turbulent mixing forces the warmer surface waters downward, in an attempt to overcome the buoyancy resistance of the colder and denser underlying waters. This process continues until the initial isothermal profile is modified to a stable stratification structure consisting of three distinct vertical layers, i.e., the epilimnion, mesolimnion, and hypolimnion. These equilibrium layers are so thermally stable that vertical transport and turbulent diffusion are all but suppressed across their interfaces, particularly across the thermocline, where the vertical temperature and density gradients are the largest. Schertzer et al.[59] show the development of the thermal layers and formation of intense gradients across the thermocline region in response to meteorological conditions. In the late fall, increased wind-generated turbulence, decreased thermal heat input, and excessive convective heat losses at the air-water interface return the lake to isothermal conditions. The "fall overturn" marks the end of the stratification period and restores high oxygen levels to the entire water column. The precise timing of the fall overturn impacts on the duration of time that oxygen-depletion processes can drive the hypolimnion to an anoxic condition. In wintertime, Lake Erie is often partially ice covered, reaching minimum temperatures.[59] The cycle repeats itself in the springtime.

Generally, the bathymetric characteristics of Lake Erie result in substantially different heating and cooling rates within the three basins. Figure 5 illustrates the progression of spring

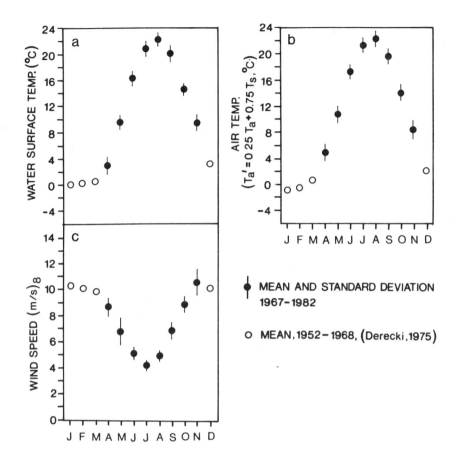

FIGURE 6. Long-term lakewide monthly means of water surface temperature and over-lake representative air temperature and wind speed for Lake Erie. (From Schertzer, W. M., *J. Great Lakes Res.*, 13(4), 454-467, 1987. With permission.

heating in Lake Erie from April to November. The lakewide mean surface water temperature occurs in the period late July to mid-August, but as shown in Figure 5a, the western basin heats to a maximum earlier than the central or eastern basins and also cools more rapidly in the fall. The lag in temperature characteristics is clearly shown in the basin vertically integrated temperature (Figure 5b). Schertzer et al.[59]demonstrate the shoreline lateral heating and general west-to-east heating progression in springtime and cooling patterns in the Lake Erie basins utilizing satellite digital data.[1] Water temperature (Figure 6a) influences the overlying meteorological conditions, such as air temperature (Figure 6b) and humidities. Within the modeling framework discussed below, the heating patterns are generalized using a one-dimensional thermocline model.[40]

## 2. Wind Speed

Wind speed (Figure 6c) varies seasonally. Maximum wind speed occurs in the fall and winter months, and minimum wind speeds are observed in June, July, and August. Wind speeds are collated from meteorological stations at the lake periphery and are referenced to a common height using the logarithmic wind profile relationships. In addition, over-lake values are derived by scaling using monthly lake/land ratios.[48] Other variables, such as air temperature and humidity, are also scaled using similar lake/land adjustments.[43,52] Examination of lake/land transfer functions was conducted by Resio and Vincent,[51] Phillips and Irbe,[48] Schwab and Morton,[60] and Sawchuk and Schertzer.[55]

## 3. Lake Inflow and Outflow

Detroit River inflow and Niagara River outflow[11] are an order of magnitude greater than other hydrological components of the Lake Erie system and have a pronounced effect on the water quality modeling problem in each basin. Figure 7a summarizes the monthly mean and deviations of the inflow and outflow volumes. Average discharges over the stratification period are $5.07 \times 10^{14}$ cm$^3$/d and $5.52 \times 10^{14}$ cm$^3$/d for the Detroit and Niagara rivers, respectively. For purposes of mass balance models discussed in this chapter, the average hydraulic flow is computed and represents the average of the inflow and outflow volumes. Polynomial fit through long-term point measurements of river temperature is used to estimate river heat inputs and outputs.

## 4. Water Level

Water level fluctuations (Figure 7b) are generalized by interpolation through the monthly mean derived from daily water level measurements at 12 shoreline stations.[39,79] Lake Erie experiences seasonal water level fluctuations, with higher levels in early summer and lower levels in winter, usually under significant ice cover.[59] Over the study period, lowest water levels occurred in 1967. High-water years with average summer values greater than 174.5 m above sea level occurred in 1973 to 1976 inclusive. Large annual water-level variations occurred in 1969, 1972, 1973, 1974, 1976, and 1977, the difference between maximum and minimum levels averaging approximately 0.5 m.

## 5. Light and Primary Production

Radiation as a function of depth determined using the mean vertical extinction coefficient[39,57] must be related to the optimum light intensity for phytoplankton growth. A convenient formula for the effects of light on photosynthesis[50,73] was suggested by Steele,[69] namely, $r = R \exp(1-R)$, where $r$ is the relative photosynthesis and $R$ is the solar radiation in the photosynthetic wavelength band, 400 to 700 nm, expressed as a fraction of the saturation light intensity. In a homogeneous layer of water, the light intensity decreases with depth as $R = R_t \exp(-\varepsilon Z)$ where $R_t$ is the incident radiation at the top of the layer and $\varepsilon$ is the extinction coefficient (Figure 7c). In shallow Lake Erie, high snow-melt runoff (erosion) and high wind speeds during the spring and fall contribute to higher turbidity and consequently higher light extinction coefficients. Values during the summer show the influence of lower turbidity and an increased biological influence on light extinction.

Thus, Steele's function can be integrated immediately to give the depth-averaged light factor:

$$\rho = \frac{1}{\varepsilon\delta}\left(e^{1-R_b} - e^{1-R_t}\right) \tag{1}$$

where $R_b$ is the radiation passing through the bottom of the layer and $\delta$ is the layer thickness.

The light factor, $\rho$, is directly incorporated into the photosynthesis formula (see Table 5) and represents a daily average value computed at hourly intervals from the Lake Erie database. The saturation light intensity in the photosynthetically active (400 to 700 nm) band was taken to be 4.75 cal/cm$^2$/m by recourse to the light effect curves presented by Stadelmann et al.[68] Approximately 48% of the incoming solar radiation is photosynthetically active.[56]

## IV. MODEL FRAMEWORK

As indicated previously, the main objective of the Lake Erie case study was to examine the water quality responses of Lake Erie to the influence of loading and weather variability. On the

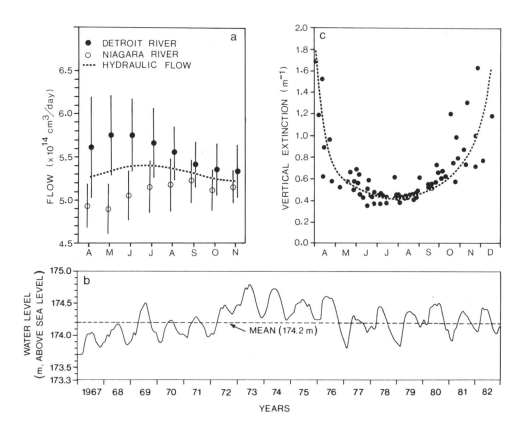

FIGURE 7.   Seasonal variation of selected physical components for Lake Erie for (a) Detroit River and Niagara River flows, (b) water level, and (c) mean vertical extinction coefficient.

basis of previous studies on the lake, it was hypothesized that the oxygen depletion problem, in the central basin in particular, was influenced in part by the meteorological conditions at the air-water interface, which affects the development of thermal stratification. Therefore, within the water quality framework discussed below, the dynamical nature of the thermocline is parameterized and forms an integral part of the model framework. The primary model elements include the surface forcing in terms of heat flux and wind speed; the thermal structure determined through the one-dimensional thermocline model, which incorporates the surface meteorological conditions; and the nine-box water quality model.

This stage of model development is a crucial step. In order to maintain efficiency in the computations, there must be a balance between having all of the necessary key processes included in the model and achieving the simplest possible model resolution to describe the system. In earlier attempts,[35,36,64] detailed hydrodynamical and biological/chemical mechanisms were incorporated to simulate episodic events. While the detailed approach was useful for event analyses, the description of the long-term changes, particularly those associated with water quality variables, did not warrant such complex model structures. However, not all processes can be simplified. In the case of Lake Erie, as is demonstrated below, the water quality model framework requires a sufficiently detailed description of the thermocline, albeit a one-dimensional model. Conversely, we can afford to reduce the water quality model into a nine-box structure with only three key variables. The following presents a description and rationale for our Lake Erie modeling framework.

## A. THERMOCLINE MODEL

There are a number of one-dimensional thermocline models[6,32,46,63,70,78]; each has its own physical basis and area of application. Few are designed for the areally averaged, whole-lake profile. Even fewer of them, if any, have been verified with long-term data, i.e., data spanning 10 or more years.

We have developed a one-dimensional thermocline model using the representation (Figure 2c), the details of which have been discussed in Lam,[34] Lam et al.,[39] and Henderson-Sellers and Lam.[27] Briefly, the heat balance equation for water column or a lake basin is given by Equation 10 in Chapter 1, written here as

$$\frac{\partial T}{\partial t} = \frac{1}{A}\frac{\partial}{\partial z}\left(AK_v\frac{\partial T}{\partial z}\right) + s \tag{2}$$

where $T$ is temperature; $t$ is time; $z$ is depth measured from lake surface; $A$ is cross-sectional area and varies with the depth, i.e., $A = A(z)$; $K_v$ is vertical turbulent diffusivity; and $s$ is a source term. The diffusivity, $K_v$, can be a function of depth, wind mixing, the Coriolis force, buoyancy effects, internal-wave frequencies, and even molecular diffusion. The following formulation has been tested with Great Lakes data and has produced satisfactory results (Lam et al.).[39] Briefly, the vertical diffusivity is defined over the three layers

$$\text{Epilimnion:} \quad K_v = K_o\left(1 + \sigma R_i\right)^{-1} - \gamma\frac{\partial\rho}{\partial z}\cdot\frac{g}{\rho} \tag{3}$$

$$\text{Mesolimnion:} \quad K_v = K_{TC}\left(\frac{N_{TC}^2}{N^2}\right)^{\frac{1}{2}} \tag{4}$$

$$\text{Hypolimnion:} \quad K_v = K_B \tag{5}$$

where $K_o$ is the air-water diffusion parameter; $K_B$ is the bottom diffusion parameter; $\sigma$, $\gamma$ are model coefficients; $g$ is gravitational acceleration; $\rho$ is water density, $R_i$ is the Richardson number; $N$ is the Brunt-Väisälä frequency; and TC denotes the value at the thermocline, where the maximum temperature gradient occurs. The salient feature of this model is that the definition of vertical diffusivity is divided into three distinct phases, as given in Equations 3 to 5. In the epilimnion, the diffusion is controlled by the thermal-buoyancy effect and the wind-mixing effect. The balance is indicated by the Richardson number $R_i$ (Equation 3), which is basically the ratio of the strength of the buoyancy forcing over the strength of the wind mixing.[70] Thus, when the thermal buoyancy is strong, $R_i$ is positive and large, so that the first term, $K_o(1+\alpha R_i)^{-1}$, becomes small, resulting in a small $K_v$ and a strongly stratified condition. Conversely, when wind mixing is strong, $R_i$ becomes positively small so that $K_v$ becomes large and the epilimnion will entrain downward. The second term, $-(\gamma g/\rho)\,\partial\rho/\partial z$, of Equation 3 becomes important only when $\partial\rho/\partial z$ is large and negative, i.e., for unstable conditions, $R_i < 0$, such as during the fall overturn. The constants $\sigma$ and $\gamma$ in Equation 3 are calibrated constants, and the constant $K_o$ is based on the empirical expression, $K_o = 0.0045\,W$, where $K_o$ is the air-water diffusion parameter (cm²/s) and $W$ is the wind speed (cm/s), as summarized in Table 2.

In the mesolimnion, in Equation 4, $K_{TC}$ denotes the eddy diffusivity at the thermocline (i.e., where $\partial T/\partial z$ is maximum and $K_v$ is minimum). The Brunt-Väisälä frequency squared, $N^2 = \alpha_w g\,\partial T/\partial z$, describes the internal wave mechanisms; the eddy diffusivity will increase with depth from the minimum value of $K_{TC}$ at the top of this layer to a value $K_B$ at the bottom of the same layer. By expressing $K_B$ as $K_B = \beta K_{TC}$, we found that from calibration $\beta = 2.0$ for the central basin and $\beta = 5.0$ for the eastern basin (Table 2).

In the hypolimnion, we assume that $K_v = K_B$ holds for the rest of the water column. Since $K_B$

**TABLE 2**
**Constants Used in the One-Dimensional Thermocline Model**

| Constants | Units |
|---|---|
| **Calibrated Constants** | |
| $\sigma = 0.03$ | — |
| $\gamma = 4.6 \times 10^3$ | $cm^2s$ |
| $\beta_{CB} = 2.0$ | — |
| $\beta_{EB} = 5.0$ | — |
| **Physical Constants** | |
| $g = 981.0$ | $cm/s^2$ |
| $\rho = 1.0$ | $g/cm^2$ |
| $\alpha_w = 13.6 \times 10^{-6}$ | $°C^{-1}$ |
| **Empirical Constant** | |
| $K_o = 0.0045\ W$ | $K_o$ in $cm^2/s$, $W$ in $cm/s$ |

From Lam, D. C. L. and Schertzer, W. M., *J. Great Lakes Res.*, 13(4), 757–769, 1987. With permission.

is related to the daily value of $K_{TC}$, it is indirectly related to the wind conditions at the surface layer. For example, during calm summer periods, $K_{TC}$ and hence $K_B$ is small when thermal buoyancy is dominant in the epilimnion. Storm events, however, would occasionally result in a relatively large $K_{TC}$ and hence a larger $K_B$, which could, in turn, cause the upward entrainment phenomenon.[31] Similarly, toward the end of the stratification, penetrative convection due to rapid cooling at the lake surface invokes an even larger $K_v$ throughout the epilimnetic layer, resulting in a large $K_{TC}$ at the top of the mesolimnion. Thus, the epilimnion grows at the expense of the mesolimnion and hypolimnion, until a uniformly mixed condition is reached for the entire water column. Note that, for the fully mixed condition, only Equation 3 applies, as if there is only one layer throughout the water column.[39]

Since the physical parameters $R_i$, $N^2$, and $\rho$ are functions of the unknown variable (i.e., temperature), Equation 2 is a highly nonlinear, partial differential equation and must be solved by appropriate numerical methods.[34] In particular, iterative procedures are required in which the temperature profile to be solved is first approximated (e.g., by the profile at the previous time step as the initial guess) and then refined by the numerical method. Thus, during the iterative calculation, the maximum temperature gradient is located from the approximated profile. If no maximum occurs, then Equation 3 applies to the whole water column as one fully mixed layer. If the maximum is located, then the diffusivities for all $z$ above this location can be obtained from Equation 3, while the diffusivities for all $z$ below this location can be computed from Equation 4 until $K_v$ attains the value of $K_B$ before the lake bottom is reached, e.g., during deep penetrative convection in the fall; then a two-layer structure is assumed. Otherwise, at the point where $K_v$ computed from Equation 4 attains the value of $K_B$, we switch to Equation 5 to compute the diffusivities for the rest of the water column. During the stratification period, it is this three-layer structure that prevails most often. The next step in the iteration is to compute an updated temperature profile from these approximated diffusivities. The above matching procedure is then repeated, using the updated temperature profile. Usually only two or three iterations are required to achieve consistency between the matched locations and the predicted temperature profile.[39] This iterative procedure is embedded in the numerical integration of Equation 2, which is performed with a time step of 1 day.[34]

To allow for the variation of cross-sectional areas, $A = A(z)$, of the water column or lake basin at different depths, the model accounts for the diffusive fluxes across varying areal interfaces.

This areal dependence sets the model apart from those that assume a uniform cross-sectional area.

The heat source and sink term, $s$ (Equation 2), is a general term that includes the surface heat flux, the solar radiation, and other components. These individual components are presented in Schertzer[58] and Lam et al.[39]

The exchange of thermal energy between the lake surface and the atmosphere involves parameterization of heat sources and sinks. Equation 11 of Chapter 1 can be reexpressed, for use here as

$$\phi_s - \phi_r + \phi_a - \phi_{ar} - \phi_{bs} - \phi_h - \phi_e + \phi_v = \phi_t \qquad (6)$$

where $\phi_s$ = incoming global solar radiation, $\phi_r$ = reflected global solar radiation, $\phi_a$ = incoming long-wave radiation, $\phi_{ar}$ = reflected long-wave radiation, $\phi_{bs}$ = outgoing long-wave radiation, $\phi_h$ = sensible heat flux, $\phi_e$ = evaporative heat flux, $\phi_v$ = net advected heat, and $\phi_t$ = surface heat flux.

Equation 2 disregards minor energy gains or losses, such as from chemical or biological sources, and it is assumed that heat conduction through the bottom is negligible compared to other components. The parameter $\phi_v$ represents the net heat gained or lost by the lake due to exchange of water masses resulting from inflow-outflow balance.[39] Surface heat flux and components are calculated from relationships and coefficients outlined in Tables 3 and 4 (Equations 7 to 17). Wintertime estimates are included from Derecki[19] to complete the seasonal cycle.

If the lateral heat transfer from adjacent water columns or basins can be disregarded, then the surface heat flux, $\phi_t$, can be applied to the air-water interface as part of the surface boundary condition,

$$-K_o \left( \frac{\partial T}{\partial z} \right)_{z=0} = \frac{(\phi_t - R_t)}{\rho c_\rho} \qquad (18)$$

where $c_\rho$ is the specific heat of water and $R_t$ is that part of the incoming solar radiation that is within the photosynthetic wavelengths of 400 to 700 nm. We assume $\phi_t = (\phi_t - R_t) + R_t$ so that the portion of heat flux, $\phi_t - R_t$ would first act on the surface ($z = 0$) and would subsequently be mixed downward by water movements, while the other portion, $R_t$, enters the water column directly through radiative processes. Furthermore, upon entering the water column, $R_t$ is attenuated by the extinction coefficient, $\varepsilon$, such that the radiation $R_z$ at a depth $z$ is given by

$$R_z = R_t \exp(-\varepsilon z) \qquad (19)$$

Thus, the radiation, $R_z$, is distributed in an exponentially decreasing manner over the entire water column. The incorporation of Equation 19 as part of the source term, $s$, in Equation 1 has been discussed in Walters et al.[78] for different numerical schemes. Since not much is known about heat transfer between sediments and overlying waters, the simple no flux condition is used as the water boundary condition:

$$\left( \frac{\partial T}{\partial z} \right)_{bottom} = 0 \qquad (20)$$

## B. NINE-BOX MODEL STRUCTURE

A detailed discussion of the mathematical equations for the nine-box model is given in Lam et al.[39] Briefly, Table 5 summarizes the three basic mass balance equations, Equations 21 to 23, for the three variables, SRP, organic phosphorus (OP), and DO, respectively. For each of the three layers, namely, epilimnion ($i = 1$), mesolimnion ($i = 2$), and hypolimnion ($i = 3$), in each of the three basins, appropriate areas ($AB_i$) and volumes ($V_i$) must be used in these equations (see Table 1). Note that $AB_i$ and $V_i$ are assumed to vary with time. The source term, $F_i$, used in these

## TABLE 3
### Heat Balance Relationships

| Heat balance relationships | Equation number |
|---|---|
| $\phi_s = \phi_o[a + b(S_m/S_p)]$ | 7 |
| $\phi_r = \alpha\phi_s$ | 8 |
| $\phi_a = \sigma T_a^{\,4} - [228.0 + 11.16\,(\sqrt{e_{sa}} - \sqrt{e_a}) - A] \bullet [\phi_s/\phi_{sc}]^n$ | 9 |
| $\phi_{bs} = \varepsilon\,\sigma T_s^{\,4}$ | 10 |
| $\phi_{ar} = (1 - \varepsilon)\,\phi_a$ | 11 |
| $E = 0.0097\,(e_s - e_a \bullet H)_8\,U \bullet R_8$ | 12 |
| $\phi_e = E \bullet L$ | 13 |
| $\phi_h = \beta \bullet \phi_e$ | 14 |
| $\phi* = \phi_s - \phi_r + \phi_a - \phi_{ar} - \phi_{bs}$ | 15 |

$\phi_o$ = extraterrestrial radiation (ly/d)

$\phi*$ = net radiation (ly/d)

$S_m$ = measured sunshine (hours and tenths)

$S_p$ = potential sunshine (hours and tenths)

$T_a$ = air temperature (°K)

$T_s$ = surface water temperature (°K)

$T_d$ = dew point temperature (°C)

$e_s$ = saturation vapor pressure at Ts (mb) (Dilley[20])

$e_a$ = ambient vapor pressure at Td (mb)

$e_{sa}$ = saturation vapor pressure Ta (mb)

A = mean monthly station adjustment term
(Anderson and Baker,[2] Derecki[19]) (ly/d)

$\phi_{sc}$ = clear sky solar radiation (ly/d)

$n$ = exponent of ratio for degree of cloudiness
(approximately 2.0)

$U$ = mean wind speed (m/s)

$\alpha$ = monthly mean surface albedo (Davies and
Schertzer[16,17])

$\sigma$ = Stefan-Boltzmann constant
$(8.1127 \times 10^{11}\ \text{cal/cm}^2/\text{min/K}^4)$

$\varepsilon$ = emissivity at the water surface (0.97)

$\beta$ = Bowen ratio, $\beta = 0.62\,[(T_s - T_a)/(e_s - e_a)]$   (16)

$H$ = lake/land humidity ratio

$R$ = lake/land wind ratio

$L$ = latent heat of vaporization = 595 − 0.52   (17)
(Ts) (cal/cm³)

$a,b$ = radiation constants (see Table 4)

From Schertzer, W. M., *J. Great Lakes Res.*, 13(4), 454–467, 1987. With permission.

## TABLE 4
### Meteorological and Radiation Factors

| | Wind ratio | Humidity ratio | Albedo | Station adjustment | Radiation | Constants |
|---|---|---|---|---|---|---|
| | $R$ | $H$ | $\alpha$ | $A$ | $a$ | $b$ |
| April | 1.81 | 1.14 | 0.08 | 11 | 0.15 | 0.61 |
| May | 1.71 | 0.86 | 0.08 | 8 | 0.22 | 0.56 |
| June | 1.31 | 0.94 | 0.08 | 6 | 0.14 | 0.64 |
| July | 1.16 | 1.09 | 0.08 | 2 | 0.13 | 0.61 |
| August | 1.39 | 1.09 | 0.08 | 4 | 0.17 | 0.59 |
| September | 1.78 | 1.11 | 0.08 | 10 | 0.16 | 0.67 |
| October | 1.99 | 1.15 | 0.09 | 15 | 0.15 | 0.61 |
| November | 2.09 | 1.15 | 0.13 | 17 | 0.12 | 0.58 |

R: Lemire,[43] H: Richards and Fortin,[52] α: Davies and Schertzer,[16] A: Anderson and Baker,[2] Derecki,[19] and a,b: Sanderson[54] Pelee Island data.

Units conversion: 1 ly/d = 0.04186 MJ/m²/d

From Schertzer, W. M., *J. Great Lakes Res.*, 13(4), 454–467, 1987. With permission.

equations refers to the loading as well as the inputs and outputs of the variable into and out of the $i^{th}$ layer as a result of five major physical processes. Again, briefly, Figure 8 shows a schematic description of the five physical processes considered. The hydraulic flow represents the inflow at the Detroit River and the outflow at the Niagara River, resulting in a general west-to-east flow in the lake. We assume that the water transport across the boundary between the western basin and the central basin, as well as that across the boundary between the central basin and the eastern basin, is conservative. Thus, depending on the cross-sectional areas of these

**TABLE 5**
**Equations for the Nine-Box Three-Variable Model**

**A. Mass Balance Equations**

$$\frac{d}{dt}\left(V_i\ \text{SRP}_i\right) = F_i\left(\text{SRP}_i\right) - U_i + R_i + r_p\left(AB_{i-l} - AB_i\right) \tag{21}$$

$$\frac{d}{dt}\left(V_i\ \text{OP}_i\right) = F_i\left(\text{OP}_i\right) + U_i - R_i + r_w\left(AB_{i-l} - AB_i\right) -$$

$$\sigma_i\ AB_{i-l}\left(\text{OP}_i - 0.005\right) + \sigma_{i-l}\ AB_{i-l}\left(\text{OP}_{i-l} - 0.005\right) \tag{22}$$

$$\frac{d}{dt}\left(V_i\ \text{DO}_i\right) = F_i\left(\text{DO}_i\right) + \left[f_{\text{po}}\ U_i - f_{\text{po}}\ R_i - k_s\left(AB_{i-l} - AB_i\right)\right]$$

$$\frac{\text{DO}_i}{\text{DO}_i + K_o} + r_{\text{Ai}}\ AB_{i-l}\left(\text{DO}_s - \text{DO}_i\right) \tag{23}$$

**B. Rate Formulations and Calibrated Constants**

$$U_i = B_i\ \rho_i\ (1.07)^{T_i}\left(\text{OP}_i - 0.005\right) V_i\ \frac{\text{SRP}_i}{\text{SRP}_i + 0.0005},$$

where $B_1 = 0.43$, $B_2 = 0.60$, $B_3 = 0.60$ (d$^{-1}$), and $\rho_1$ = light factor.

$$R_i = Y_i\ (1.07)^{T_i}\left(\text{OP}_i - 0.005\right) V_i,$$

where $Y_1 = 0.02$, $Y_2 = 0.002$, and $Y_3 = 0.001$ (d$^{-1}$).

| | |
|---|---|
| $\sigma$ | Settling velocities in m/d; $\sigma_0 = 0$, $\sigma_1 = 0.2$, $\sigma_2 = 0.4$, $\sigma_3 = 0.4$. |
| $F_i$ | Source term due to loading advection, diffusion, entrainment, and mixing; see Figure 9. |
| $f_{\text{po}}$ | Phosphorus to oxygen ratio in photosynthetic production of chlorophyll: $f_{\text{po}} = 140.0$. |
| $r_w$ | Phosphorus resuspended by wind waves, in g/m$^2$/d = $0.001\ W/W_s$, where $W$ is wind speed in m/day and $W_s = 500,000$ m/day |
| $r_p$ | Phosphorus return per area, in g/m$^2$/d: $r_p = 0.0001$ or $r_p = \text{L(TP)} \times 3 \times 10^{-7}$ if $\text{DO}_3 < 1.5$ mg/l. |
| $R_{\text{Ai}}$ | Reaeration coefficient, in m/d: $r_{\text{A1}} = 2$, $r_{\text{A2}} = 0$, $r_{\text{A3}} = 0$. |
| $\text{DO}_s$ | Saturated oxygen concentration, in mg/l. |
| $K_o$ | Half-saturated coefficient for oxygen: $K_o = 1.4$ mg/l. |
| $k_s$ | Sediment oxygen demand, in g O$_2$/m$^2$/d: $k_s = \text{L(TP)} \times 10^{-5}$, L(TP) = lakewide total phosphorus load in tonnes. |

*Note:* The subscript $i$ denotes the location of the parameter value, with $i = 0$ denoting the surface value and $i = 1,2,3$ denoting the value in the first, second, and third lake model layers. SRP, OP, and DO are concentrations (mg/l) of soluble reactive phosphorus, organic phosphorus, and oxygen, respectively. $AB$ and $V$ are area (m$^2$) and volume (m$^3$); $U$ and $R$ are the algal uptake and release rates (g/d). $T$ is temperature (°C).

From Lam, D. C. L. et al., *J. Great Lakes Res.*, 13(4), 770–781, 1987. With permission.

boundaries in each layer, the velocities due to the hydraulic flow is adjusted to obey mass conservation within each box.

For simplicity, we assume that the interface between the epilimnion and mesolimnion and the interface between the mesolimnion and hypolimnion are at the same depths in all three basins. This assumption facilitates the calculation of vertical entrainment and avoids the complication of further adjusting the hydraulic flows due to mismatched interfaces in different basins. Under this assumption, when one of the interfaces moves up, that part of the water in the upper layer traversed by the interface is mixed into the lower layer. Thus, the concentrations in the new upper layer remain undisturbed. This entrainment mechanism[31] differs from the so-called diffusion process in which the concentrations in both layers are mutually disturbed due to an exchange mechanism. Indeed, the incorporation of the vertical entrainment sets our model apart from other models using fixed thermal interfaces (e.g., DiToro and Connolly[21]) in which only the diffusion process can occur. In our model, both entrainment and diffusion processes are possible.

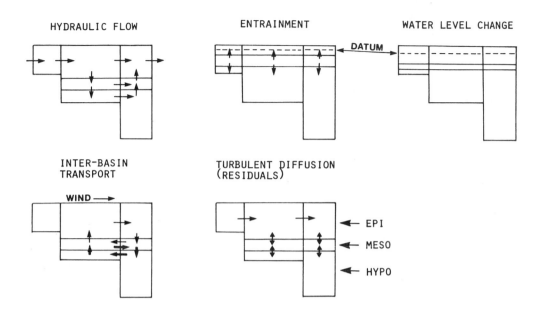

FIGURE 8. Schematic of phosphorus processes incorporated in the nine-box model. (From Lam, D. C. L. et al., *J. Great Lakes Res.*, 13(4), 770-781, 1987. With permission.)

We used daily weather records and a one-dimensional thermocline model[40] to calculate the positions of thermal interfaces *a priori* with reference to the water surface, instead of to the lake bottom. Since the nine-box model uses the lake bottom as the reference point, changes in water level must be taken into account to transfer these positions properly. The water level can affect the heat storage in the lake and hence the thermal-layer structure. As the water level rises, so do the thermal interfaces with respect to the lake bottom and vice versa.[40] Thus, the water level could affect the hypolimnion thickness, which in turn could affect the oxygen depletion. In the nine-box model, this change is conveniently simulated in the same manner as the vertical entrainment process.[39]

Wind-driven circulation can cause interbasin transports (Figure 8) in addition to the hydraulic flow. For simplicity, two vertical gyres are assumed in the nine-box model. The first gyre connects the epilimnion and mesolimnion in the central and eastern basins, and the second connects the mesolimnion and hypolimnion (Figure 8). This circulation pattern is well documented in Boyce et al.[7] Lam et al.[39] derived a simple formula relating the strength of these flows to the wind velocity. Of particular interest is the possibilty[7] that under certain wind conditions there could be interbasin transport from the eastern basin hypolimnion to the central basin hypolimnion, bringing oxygen-rich water from the former to replenish the oxygen depleted in the latter.

We also include turbulent diffusion processes along the three layers in the model (Figure 8), even though the thermal interfaces between them are allowed to move dynamically. The turbulent diffusion coefficient is determined on a daily basis from the heat budget calculated for the nine boxes. Thus, based on the daily computed temperature from the one-dimensional thermocline model[40] and the daily total heat input,[58] we back-calculate the turbulent diffusion coefficients across the interfaces using a diagnostic procedure.[65] In this model, the residual effects resulting from water level changes and from the use of uniform interfacial depths for all basins are included in these balances, so that the scheme remains mass conservative.

## C. PHOSPHORUS-OXYGEN SUBMODEL

The nine-box model structure provides the dynamic framework for defining the boundaries

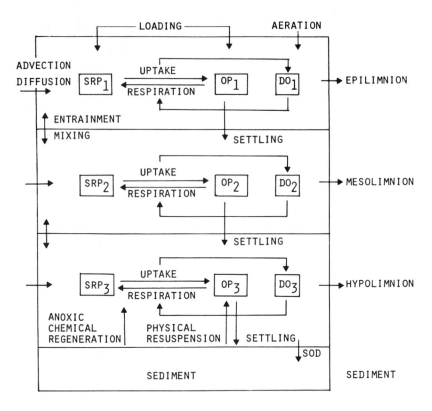

FIGURE 9.   Schematic of biological and chemical processes incorporated for a water column in the nine-box model. The subscripts 1, 2, 3 denote first, second, and third layers, SRP = soluble reactive phosphorus, OP = organic phosphorus, DO = dissolved oxygen. (From Lam, D. C. L. et al., *J. Great Lakes Res.*, 13(4), 770-781, 1987. With permission.)

of the boxes, as well as the movement of substances among the boxes. Within each box, a set of biological and chemical processes also takes place. The Simons and Lam model[65] has been refined by the addition of an oxygen compartment for this purpose, i.e., a three-variable model of SRP, OP, and DO. Figure 9 shows the schematic for the biochemical kinetics of the three variables for a typical basin with a three-layered structure. The detailed equations are given in Lam et al.[39] Again, Table 5 summarizes the governing equations (Equations 21 to 23). In the epilimnion, the oxygen is produced by plankton photosynthesis in the photic zones and by reaeration at the air-water surface. Most of the time, oxygen is saturated or even supersaturated in this layer, with the saturation being a function of the water temperature calculated using the thermocline model.[40] In the mesolimnion and hypolimnion, oxygen can be produced by photosynthesis, since these layers may still be within the photic zone, particularly during the early part of the stratification period. However, as in the case of Lake Ontario,[65] plankton respiration activity is reduced, and hence less oxygen is consumed in the hypolimnion because of lower temperature and smaller cell size.

The SOD is one of the major factors responsible for removing oxygen from the overlying waters in the central basin hypolimnion. The values of SOD measured by, e.g., Lucas and Thomas,[44] Snodgrass,[67] Charlton,[12] and Herdendorf,[28] in this basin range from 0.18 g $O_2$/m²d to 0.88 g $O_2$/m²d, depending on the instrument design and sampling method. Lam et al. reviewed the theoretical background for selection of an appropriate SOD value. It is interesting to note the notion that the SOD changes with the lake trophic status because Lake Erie central basin has passed through the stages (Vollenweider et al.[77]) from eutrophy to near mesotrophy during the period 1967 to 1978. The SOD, therefore, might be expected to decrease over the same time frame.

Snodgrass[67] examined three SOD submodels with varing degrees of complexity and concluded that the simplest one produced essentially the same results as the most complex. In view of the uncertainties, we have adopted the simplest model discussed in Snodgrass.[67] This model uses the Monod kinetic

$$J_o = k_s \frac{DO}{DO + K_o} \tag{24}$$

where $J_o$ is the oxygen flux into the sediment (g $O_2$/m²d), $k_s$ is the sediment oxygen demand rate (g $O_2$/m²d), DO is the oxygen concentration of the overlying water (mg/l), and $K_o$ is the Michaelis constant (mg/l) for oxygen. The Monod kinetic has been found[67] to simulate successfully the biological sediment oxygen demand as well as the water oxygen demand. The chemical sediment oxygen demand is found[67] to be relatively small and can be sufficiently described by first-order kinetics. Snodgrass[67] reported $K_o = 1.2$ mg/l for biological sediment oxygen demand. In the case of Equation 24, with the two Monod expressions combined into one, the Michaelis coefficient $K_o$ is found to be 1.4 mg/l. Lam et al.[42] examined the SOD submodel and found that, as a first-order approximation, $K_s$ could be written as

$$k_s = L(TP) \times 10^{-5} \tag{25}$$

where $L$(TP) is the lake wide total phosphorus load in tonnes. For example, using $L$(TP) = 22,000 t for 1972 and $L$(TP) = 13,000 t for 1979[42], Equation 25 would produce a sediment oxygen demand rate of 0.22 g $O_2$/m²d for 1972 and 0.13 g $O_2$/m²d for 1979. These values are well within the range for SOD in ice-covered lakes with differing trophic levels of Mathias and Barica,[45] who found values of 0.08 g $O_2$/m²d for oligotrophic lakes and 0.23 g $O_2$/m²d for eutrophic lakes. The estimated value for 1972 is also quite consistent with the value of 0.268 g $O_2$/m²d calculated for the year 1970 by Di Toro and Connolly.[21] We must emphasize, however, that Equation 25 is only a first-order approximation. It has a number of shortcomings. For instance, it uses the lakewide load instead of the basin load and uses the current year's load instead of allowing for a time lag. While improvement on this model formulation can be made easily, the lack of accurate and reliable SOD measurements[67] discourages further attempts.

# V. RESULTS

The emphasis in model verification is primarily in the comparison of simulation results with observed data. However, advances can also be derived by investigating situations in which the model provides inaccurate responses. To this end, model results are presented for each variable, each layer, each basin, and each year, with detailed comparisons with observed data. Too often, experiences[65] show that one can easily make a good fit for one part of the model at the expense of another part. The model should be capable of accurately simulating the behavior of all of the variables, including the kinetic rates, and fluxes, as well as the concentrations. Unfortunately, as in most cases, the Lake Erie database contains few measurements of kinetic or flux rates between model compartments, despite the wealth of concentration observations. The following section presents a detailed summary of the results of the modeling framework in which full use of the database is made to establish the credibility of the model. Primary emphasis is given here to the central basin results. For details of the model performance beyond that which is presented here, the reader may refer to the references provided in the introduction (i.e., Lam et al.[39]).

## A. SURFACE HEAT EXCHANGE
A summary of the surface heat flux computations for Lake Erie is given in Schertzer[58] for the period 1967 to 1982, in which the magnitude of the radiative flux and turbulent exchange

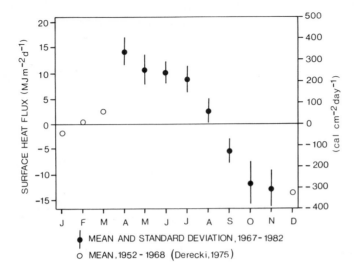

FIGURE 10.    Long-term lakewide monthly mean surface heat flux for Lake Erie. (From Schertzer, W. M., *J. Great Lakes Res.*, 13(4), 454-467, 1987. With permission.)

components are described. Figure 10 illustrates the computed lakewide surface heat flux for Lake Erie expressed as a monthly mean based on daily computations. Detailed computations were conducted over the stratified period of April to November in this study, and wintertime values are shown to complete the annual cycle based on Derecki.[19] As shown, significant heat gains to the lake occur from April to August, with maximum heating in April. Maximum heat storage occurred in August, and thereafter losses from the lake surface to the atmosphere are computed with maximum losses in December. Comparison of the lakewide surface heat flux computations with computations based on midlake buoy observations during intensive surveys (1979 and 1980) in the central basin verify the accuracy of the present approach.[58]

## B. THERMOCLINE SIMULATIONS (1967 TO 1982)
### 1. Central Basin

The daily surface heat fluxes estimated by Schertzer[58] are incorporated within the one-dimensional temperature model to derive daily basin-averaged vertical temperature profiles. Figure 11 illustrates a time series of the central basin thermal structure for the period 1967 to 1982, showing a comparison between measured and computed values. Since the heat fluxes are derived from daily weather records, the variability in the surface conditions strongly influence the computed daily profiles. In each year, simulations are introduced using either an estimated vertically mixed temperature profile or an observed profile during the early spring period. Figure 11 shows simulations over the stratified period, generally considered to be from May to October.

Figure 11 demonstrates the daily fluctuations in the mean basin temperature and shows seasonal and annual changes in the temperature profiles. Included in Figure 11 is the computed position of the interface between the mesolimnion and the hypolimnion.[39,47] Using a simple nine-box model,[42] average temperatures for each of the thermal layers can be computed. Comparison of observed and computed temperatures is accomplished by computing the median relative error (MRE).[39,71] Table 6 shows MRE in percent for each of the basins and layers. For the central basin, the MRE statistics are 2.7%, 5.7%, and 5.3% for the epilimnion, mesolimnion, and hypolimnion layers, indicating particularly good results, especially in the mesolimnion.

Changes in the thermal structure from year to year can be demonstrated by considering the position of the interface depths relative to the 21-m depth line, which is superimposed on each year (Figure 11). This depth is the maximum depth at the boundary, i.e., the Pennsylvania Ridge, between the central basin and eastern basin. Its significance is such that whenever the interface between the hypolimnion and the mesolimnion is deeper than this depth in the central basin, the hypolimnion is completely blocked from the lateral transport with the eastern basin. Under such conditions, and in the absence of occasional storm-induced internal waves, which may facilitate some east-west transport of hypolimnion waters, the central basin hypolimnion waters are constrained to interact physically only with its mesolimnion. This configuration has significant consequences for the oxygen budget of the central basin hypolimnion, because the only major oxygen supply under restricted exchange with the eastern basin is mesolimnetic diffusion or entrainment, which are comparatively slow processes.

The thermal structure depicted in Figure 11 can be described by considering the dynamics of the thermocline positions. Five years (1970, 1977, 1978, 1980, and 1982), in particular, show prolonged periods of shallow hypolimnion, which coincide with observed anoxic occurrences.[42] Four other years (1969, 1972, 1974, and 1979) are cases in which springtime thermocline development is interrupted by strong winds and reversal entrainment episodes,[31] resulting in a thicker hypolimnion in midsummer (July). In 1972, in particular, the developing thermal structure in June was perturbed by severe storms. The thermocline model predicted the collapse of the three-layer structure into a one-layer structure in October 1972. The transition from stratified to fully mixed conditions were supported by measured temperature profiles, thus confirming the success of the matching procedures (Equations 3 to 5) to cover such an event. Five other years (1967, 1968, 1971, 1973, and 1976) do not have periods of prolonged shallow hypolimnion, but show a gradual reduction of the hypolimnion thickness. These years are not characterized by prolonged periods of intense surface heating or strong wind episodes during the critical summertime period and are considered more normal conditions. Such categorizations are general according to the physical conditions, however, detailed examination of two contrasting years, 1972 and 1979,[41] support such a scheme for describing the one-dimensional thermal characteristics of the central basin.

The model has been calibrated with the observed data of 1970 and then verified with the 11 years of data of 1967 to 1969 and 1971 to 1978, inclusively,[39] and postaudited using the same coefficients for the period 1979 to 1982.[42] The one-dimensional lake temperature model is regarded as validated for Lake Erie.[15]

## 2. Western and Eastern Basins

As indicated in Figure 5, the shallow western basin is nearly isothermal throughout the whole year. Conversely, the deeper (maximum depth, 64 m) eastern basin stratifies much like the deeper temperate lakes.[14] Lam and Schertzer[40] detail the results of the thermocline model in the eastern basin. Since we are more concerned with the central basin in this analysis, only an example of the model performance in the eastern basin is shown in Figure 12 for 1970 and 1980, a decade apart.

Morphological and topographic differences between the central and eastern basins cause different responses in the bottom turbulence to wind-driven circulations and internal waves. Consequently, a change of the coefficient $h$, which relates bottom turbulence to the internal wave oscillation, is increased from $h = 2$ in the central basin to $h = 5$ for the eastern basin. With one change in this coefficient, the model that was originally developed for the central basin is applied to the eastern basin. Median relative errors between computed and observed temperature results in the eastern basin are similar to values for the central basin, being 3.4%, 5.0%, and 9.3% for the eplimnion, mesolimnion, and hypolimnion, respectively.

The thermodynamic regime of the eastern basin is important to the central basin oxygen balance. Because of its greater depth, anoxia does not occur within the hypolimnion, and although oxygen is depleted to below the saturation level, east-west exchange with the central basin can replenish oxygen-deficient water in the central basin. Restrictions occur, however, in

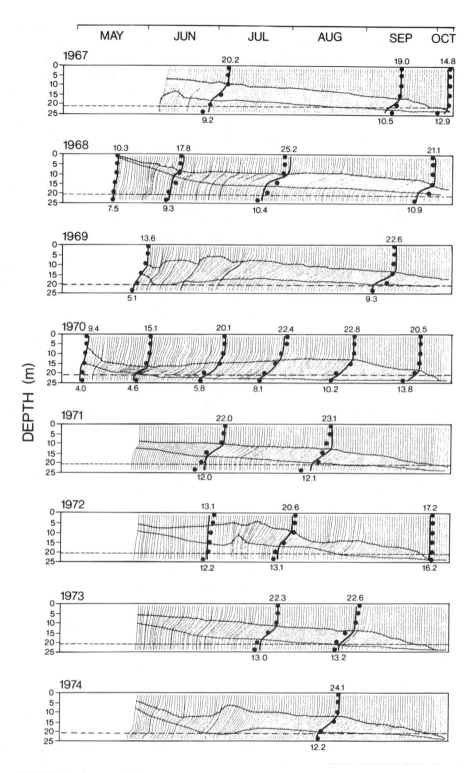

FIGURE 11.   Computed daily vertical temperature profiles (thin lines) for the central basin of Lake Erie showing a comparison between computed (heavy line) and observed (dots) temperatures at the cruise midpoint dates from 1967 to 1982. Calculated depths of the bottom of the epilimnion layer and the top of the hypolimnion are also illustrated. (From Lam, D. C. L. and Schertzer, W. M., *J. Great Lakes Res.*, 13(4), 757–769, 1987. With permission.)

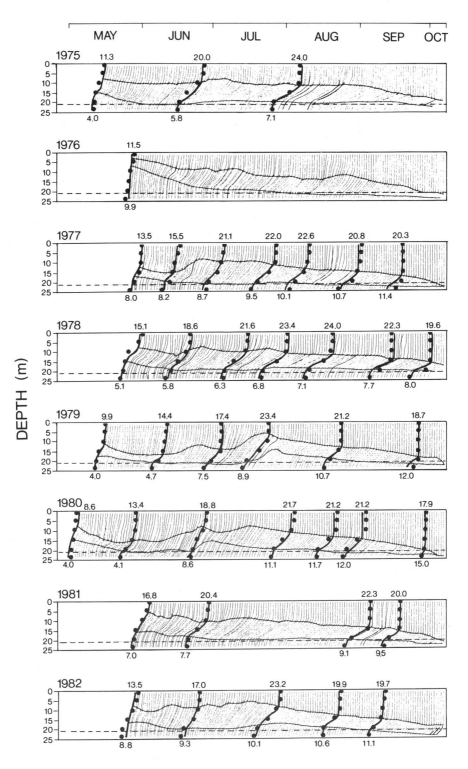

FIGURE 11 (continued).

EASTERN BASIN

1970    TEMPERATURE (°C)

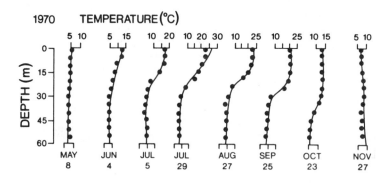

1980    TEMPERATURE (°C)

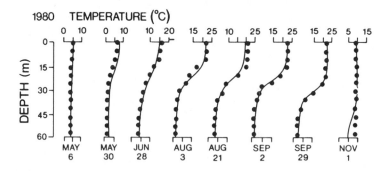

FIGURE 12.   Computed (line) and observed (dot) vertical temperature profiles at the cruise midpoint dates for the eastern basin of Lake Erie from 1970 and 1980.

the case of the central basin hypolimnion when the interface between the mesolimnion and hypolimnion in the central basin is below the Pennsylvania Ridge.

## C. WATER QUALITY SIMULATIONS

The nine-box water quality model described in the previous section is used to simulate basin and layer concentrations of TP, SRP, and DO. Simulation results are compared to concentrations derived from measurements conducted over a grid of locations within each basin.

### 1. Western Basin

Computed and observed concentrations of the three variables are shown for the western basin in Figure 13. Only the epilimnetic concentrations are shown, since the basin is virtually completely vertically mixed throughout the year. The shallow western basin is influenced to a high degree by the large flow-through resulting from the Detroit River inflow, as well as influences from surrounding tributary inflows. The irregular seasonal patterns observed in the TP and SRP computed and observed concentrations result from the influence of loading inputs (e.g., spring pulse in 1976, see Lam et al.[39]) and the influence of wind wave resuspension (e.g., winter pulse in 1970, see Lam[35]). Figure 13 also demonstrates a clear response to loading reduction efforts, as the TP and SRP concentrations show significant reductions in response to the decreases in inputs, largely from the Detroit River. In contrast, DO concentrations show no obvious general trend in the basin, and computed concentrations compare consistently very well with observed values in all seasons.

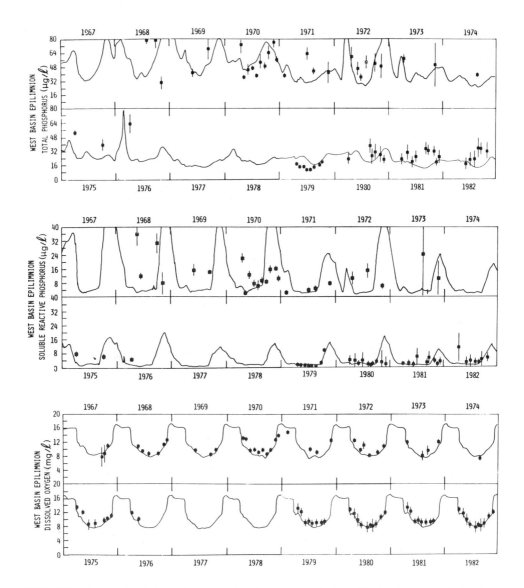

FIGURE 13. Computed and observed concentrations in the western basin epilimnion for (top) total phosphorus, (middle) soluble reactive phosphorus, and (bottom) dissolved oxygen. (From Lam, D. C. L., et al., *J. Great Lakes Res.*, 13(4), 782-800, 1987. With permission.)

## 2. Central Basin

Simulations for TP, SRP, and DO for the central basin are shown in Figures 14, 15, and 16. In contrast to the conditions observed in the western basin, the central basin shows only a gradual decrease in concentrations of TP and SRP over the period 1967 to 1982. In addition, seasonal fluctuations in the epilimnion are not as prominent. This is expected, since the central basin volume is larger than the western basin and the direct effects of the Detroit River are not as apparent.

Mesolimnion concentrations of TP show relatively high concentrations, which are related to the release of phosphorus from the sediments (e.g., 1969, 1970, 1977, and 1980). In these years, the peaks are particularly more pronounced in the hypolimnion, with very high concentrations in 1969, 1970, and 1977.

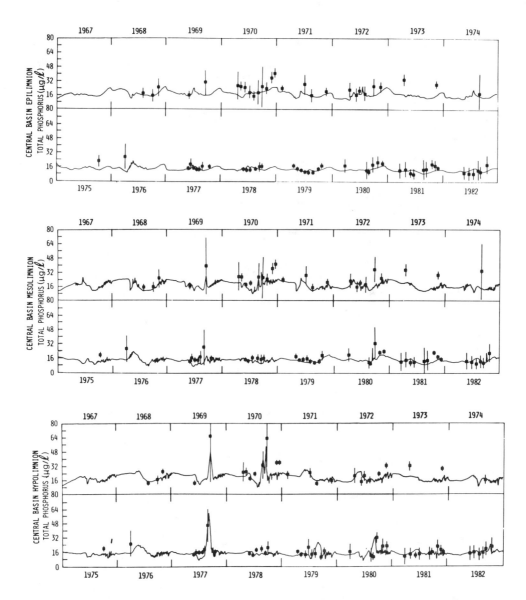

FIGURE 14.    Computed and observed total phosphorus concentration for the central basin. (From Lam, D. C. L., et al., *J. Great Lakes Res.*, 13(4), 782-800, 1987. With permission.)

In the case of SRP (Figure 15), the influence of chemical regeneration from the hypolimnion is evident in the mesolimnion, especially in 1969, 1970, and 1977. Releases from the sediments to the hypolimnion are particularly noticeable in these years, with significant anoxia occurrence. Other years, such as 1979, 1980, and 1982, appear to have conditions that may be conducive to anoxia occurrence.[40] However, the chemically regenerated SRP appears not to be as abundant as in 1969 or 1970. This may be attributable to a decrease in the phosphorus available in the sediment for chemical regeneration resulting from reductions in phosphorus loading. Lam et al.[41,42] noted a significant reduction in the SRP regeneration between that observed in 1970 by Burns and Ross[10] at 7.5 mg/m²/d and model calibrations using 1978 data of 4.5 mg/m²/d.

Simulation of dissolved oxygen concentrations in the three layers of the central basin is shown in Figure 16. The saturation levels of oxygen within the epilimnion, as well as oxygen production due to photosynthesis, is strongly influenced by water temperature. In this modeling

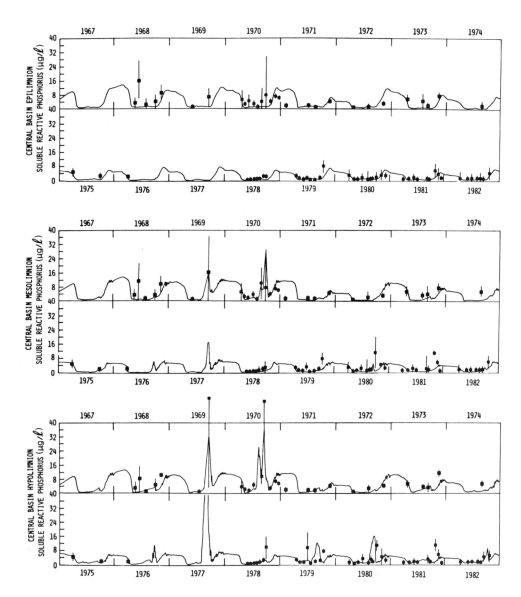

FIGURE 15. Computed and observed soluble reactive phosphorus concentration for the central basin. (From Lam, D. C. L., et al., *J. Great Lakes Res.*, 13(4), 782-800, 1987. With permission.)

framework, the water temperatures are derived from the thermocline model[40] and the accuracy of the temperature results are reflected in the excellent agreement between observed and computed epilimnion DO concentrations. Mesolimnetic DO concentrations display a larger seasonal variability, and in years with severe anoxia occurrence (e.g., 1970 and 1977), extremely low concentrations can develop. The period of 1979 to 1982 shows a more stable seasonal cycle compared to previous years.

The central basin hypolimnion shows large seasonal fluctuations in the DO concentration. During isothermal periods, concentrations are high near saturated conditions, while the stratified period shows general declines in DO concentration until the late summer minimum is reached, usually in September, followed by reoxygenation during fall overturn. Anoxic conditions are assumed to prevail if oxygen concentrations are below 1.5 mg/l, and such conditions occurred in the years 1970, 1977, 1978, 1980, and 1982. Lam et al.[41] discussed the controlling influences

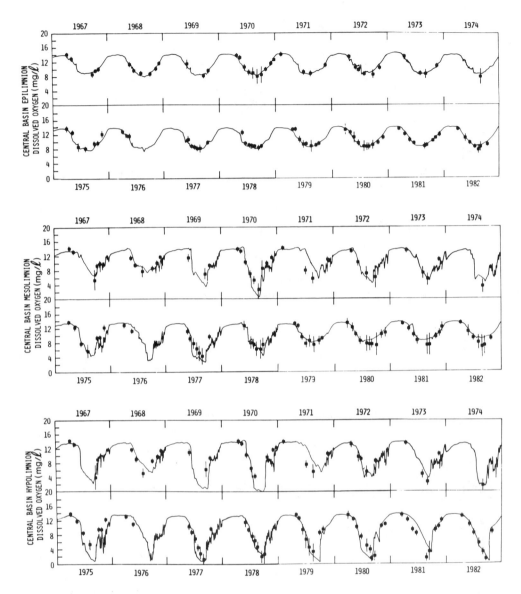

FIGURE 16.   Computed and observed dissolved oxygen concentration for the central basin. (From Lam, D. C. L., et al., *J. Great Lakes Res.*, 13(4), 782-800, 1987. With permission.)

on the development of central basin anoxia occurrence. Considering the controversy regarding the influences of loading reductions and meteorological influences as represented by the thermocline development,[4] it was shown that both effects are present. Loading reductions influence the long-term trend of the occurrence of hypolimnetic anoxia, however, the thermocline position exerts a dominant influence on the annual variability of anoxic occurrence. For example, in years with a significant occurrence of anoxic regions, the hypolimnion thickness is small and the vertical diffusion is small. Conversely, under conditions of a thick hypolimnion with large vertical diffusion, anoxia is unlikely to occur in general. Exceptions to this observation are noted in Lam et al.,[42] related to the long-term reduction in phosphorus loading and consequent effects on the sediment oxygen demand rate.

Table 6 shows the median relative error for basin and layers for TP, SRP, and DO. Error statistics between measured and computed concentrations for DO are generally low and within

**TABLE 6**
**Median Relative Errors in Percentage of the Variables in the Nine-Box**
**Model Based on Computed and Observed Results for 1967 to 1978**

|  | Total phosphorus | Soluble reactive phosphorus | Dissolved oxygen | Temperature |
|---|---|---|---|---|
| Western basin |  |  |  |  |
| Epilimnion | 13.2 | 48.2 | 4.8 | 2.2 |
| Central basin |  |  |  |  |
| Epilimnion | 11.7 | 32.1 | 1.2 | 2.7 |
| Hypolimnion | 16.4 | 41.1 | 5.8 | 5.7 |
| Mesolimnion | 12.2 | 37.4 | 7.5 | 5.3 |
| Eastern basin |  |  |  |  |
| Epilimnion | 9.6 | 28.0 | 1.9 | 2.9 |
| Hypolimnion | 9.8 | 37.0 | 4.6 | 3.1 |
| Mesolimnion | 7.1 | 42.8 | 3.8 | 10.1 |
| Lakewide mean | 11.4 | 38.1 | 4.2 | 4.6 |

the range shown for temperature. The median relative error for TP is approximately 10% for the lakewide case, while errors are considerably larger for SRP.

**3. Eastern Basin**

The eastern basin, located at the downstream end of the lake, is not directly affected by the TP loadings reductions, which have occurred primarily in the Detroit River, and consequently its response is diminished compared to the western and central basins. In contrast to the central basin, where the thermocline position played a significant role in the development of anoxia, the hypolimnion in the eastern basin is relatively deep (Figure 12) for the entire stratified period. Consequently, the volume of oxygen in the hypolimnion is too great for the SOD to deplete to the level of anoxia before the onset of fall overturn.

Detailed results for TP, SRP, and DO for all layers in the eastern basin are shown in Lam et al.,[42] and only the hypolimnion values are shown here in Figure 17. Compared to the western and central basins, the TP concentrations in the eastern basin are relatively invariant (Figure 17A). Considering the absence of anoxia, Figure 17B shows virtually no chemical regeneration of SRP from the sediment. Seasonal cycles observed in the SRP are the primary result of plankton uptake and respiration processes. Definite seasonal cycles in the DO concentrations are observed in all layers[42] (Figure 17C). Sediment and water oxygen demand have the effect of reducing DO concentrations in this basin to 8 to 10 mg/l for the mesolimnion and 6 to 8 mg/l in the hypolimnion, well above the anoxic criterion of 1.5 mg/l.

# VI. DISCUSSION

Having established the accuracy of the various models, we can now examine applications, particularly as research tools for scenario testing of management strategies. Inasmuch as the prediction of water quality responses is important, the quantification of uncertainties is equally essential. In the case of Lake Erie, there are many environmental factors that can affect the oxygen depletion in the central basin. The influences of these factors are best discussed in the context of probabilities and frequency distributions.

## A. HEAT STORAGE

The results achieved in the simulation of the thermal structure through the one-dimensional thermocline model[40] were dependent on the accuracy achieved in specifying the

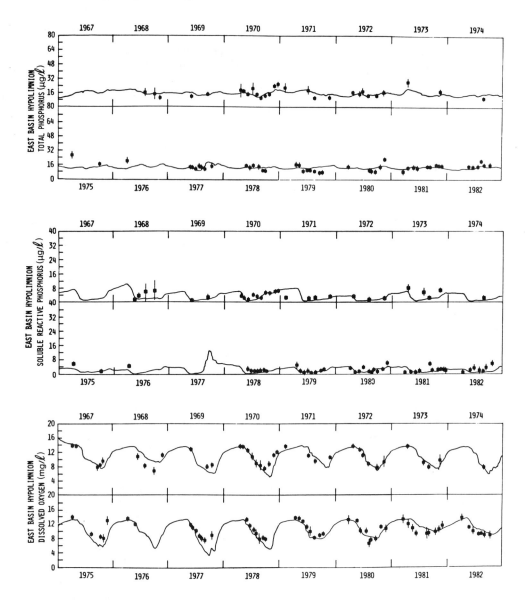

FIGURE 17.    Computed and observed total phosphorus, soluble reactive phosphorus, and dissolved oxygen concentration for the eastern basin hypolimnion. (From Lam, D. C. L., et al., *J. Great Lakes Res.*, 13(4), 782-800, 1987. With permission.)

hydrometeorological conditions at the air-water interface. As shown previously, the components of the energy budget terms are combined to describe the surface heat flux. A measure of the accuracy in the surface heat flux computations is demonstrated by deriving the lake heat storage. Figure 18 illustrates a comparison between computed and observed heat storage for Lake Erie. Measured heat storage is based on observed temperature profiles at a grid of stations sampled during lakewide monitoring by the NWRI.[58] Computed heat storage represents the summation of the calculated daily surface heat flux initialized in each year at the initial cruise date or back-calculated to the springtime period. Considering the approximations in the heat balance formulations, comparison between observed and computed heat storage for Lake Erie is very good. Correlations between measured and computed values[58] show r ~ 0.95 for both the heating and cooling phases.

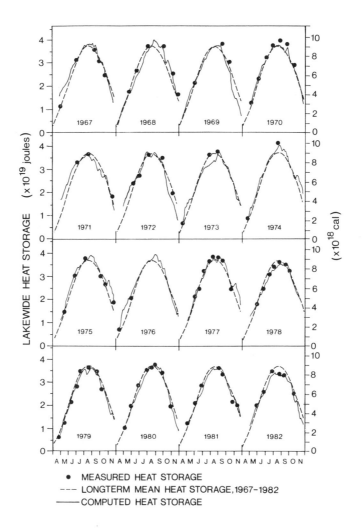

FIGURE 18. Comparison of computed lakewide heat storage with measured heat storage and long-term mean heat storage for the 18 months April to November from 1967 to 1982. (From Schertzer, W. M., *J. Great Lakes Res.*, 13(4), 454-467, 1987. With permission.)

## B. FREQUENCY DISTRIBUTIONS IN RELATION TO ANOXIC OCCURRENCE
### 1. Thermal Layer Positions

The success of the heat flux one-dimensional thermocline and the nine-box water quality models allows us the opportunity to examine the relative importance of stratification and loading effects on the occurrence of anoxia in the central basin. One of our main hypothesis was that the uncertainties in the controversy[4] result from weather variabilities. To address this problem, we use the computed thermal layer positions over the 16 years as a surrogate to provide some insight into the effect of weather variability. Lam et al.[42] discuss the rationale for the statistical procedures and limitations on studies of uncertainty using environmental data.

The thermocline data are used in three approaches to examine the variabilities. Figures 11 and 12 illustrated the daily vertical temperature responses for each basin. In this section, trends and variations over a longer term are examined by (1) analyzing the frequency distribution of these thermal layer depths for each month during the stratification period and (2) selecting any one day in order to compare the computed thermal layer depths for that day in each of the 16 years, i.e., to compare the year-to-year variabilities. In all cases, we assume statistical randomness.[42]

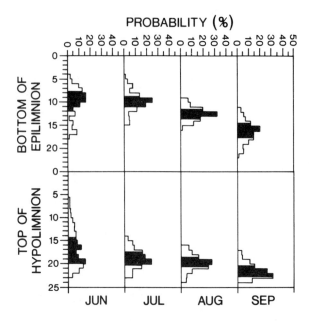

FIGURE 19.   Frequency distributions for the central basin depths of the bottom of the epilimnion and the top of the hypolimnion for the period June to September based on data from 1967 to 1982. The shaded area represents the interquartile range, i.e., between the 75% and 25% quartile. (From Lam, D. C. L. and Schertzer, W. M., *J. Great Lakes Res.*, 13(4), 757-769, 1987. With permission.)

Figure 19 shows the frequency distributions of the bottom of the epilimnion and the top of the hypolimnion for the stratified period from June through September 1967 to 1982. Both layer depths show a similar seasonal response to the weather influences. The distributions of the layer positions are wider earlier in the stratification cycle and become more narrowly defined in September. A measure of the central tendency is indicated by the shaded area, which represents the interquartile range (i.e., between the 75% and 25% quartile. The other distinctive feature illustrated in Figure 19 is that both interfaces show a downward trend in time, as expected in the seasonal development of the thermocline. For example, for half of the time (the interquartile range), the top of the hypolimnion would be found at 14.2 to 20 m, 17.5 to 20 m, 18.5 to 20 m, and 20.5 to 23 m depth in June, July, August, and September, respectively. In general, these results compare well with observations on the thermal cycle of Lake Erie[59] and provide valuable information for surveying the lake temperature and biochemical parameters in this basin related to water quality problems.

## 2. Hypolimnion Thickness and Diffusion vs. Dissolved Oxygen Concentration

As shown earlier, central basin anoxia is likely to occur when the top of the hypolimnion is below the depth of the Pennsylvania Ridge (21 m depth) and when the turbulent diffusion is small. Based on these two criterion (i.e., the hypolimnion thickness $HD \leq 4$ m and a vertical turbulent diffusivity $(K_v)_L \leq 1$ cm$^2$/s), an estimate of the probability of this condition occurring can be derived for each daily profile. A 5-d mean of the derived daily probabilities is computed for each year, starting with the period Julian day 150 to 154. Further averaging over each 5-d period over the 16 years provides a reliable mean probability distribution, describing the likely occurrence of the condition.

Figure 20a shows the estimated probability presented as 5-d means. The most likely (66% probability) date for such an occurrence is between Julian days 255 and 259, i.e., about mid-

September. The analysis indicates that there is over 50% probability that the condition (HD ≤ 4 m; $(K_v)_L$ ≤ 1 cm²/s) would occur within the first 25 d of September. The probability of occurrence at other times of the year is substantially lower. A secondary mode in late June (i.e., 19%) is associated with years of very shallow hypolimnia.

A comparison between the two probability distributions[40] results in a correlation of r = 0.83. This correspondence confirms that in a lake such as the central basin of Lake Erie, the physical influence on the occurrence of anoxia can be summarized by considering the criteria HD ≤ 4 m; $(K_v)_L$ ≤ 1 cm²/s, which strongly supports the hypothesis that the weather influences on lake stratification can have a profound effect on the occurrence of central basin anoxia.

### 3. Dissolved Oxygen (Observed vs. Computed)

Considering the success of the nine-box water quality model in computing the seasonal DO concentrations for each layer within the central basin, the daily DO layer concentrations can be formed into probability distributions in the manner described above for the layer interface depths.

Figure 20b illustrates the distribution at 5-d intervals over the 16-year period from June to October. Also included in Figure 20b are the DO probabilities derived from observations as above. Correspondence between the two distributions[42] is good, with correlations of r = 0.73. Both distributions indicate that the most probable dates for the occurrence of anoxia in the central basin hypolimnion is between Julian day 240 and 270 (approximately from the end of August to the end of September).

### 4. Water Level vs. Dissolved Oxygen Concentration

Over the study period, a mean water level is computed as 174.2 m above sea level (ASL) (Figure 20c). Considering that the central basin is 25 m deep and that the summertime hypolimnion thickness is 4 m or less, it is instructive to investigate the correspondence between anoxic occurrence and water level changes, since a change in water level could influence the level of the top of the hypolimnion interface. Figure 20c shows the probability distribution for water level occurrence below the mean level of 174.2 m (ASL) determined over a 5-d mean for the 16-year period as above. A comparison between the water level distribution and the probability distribution for DO < 1.5 mg/l shows reasonably good correspondence. Correlation between the two distributions yields r = 0.60 or even higher (r = 0.82) if the last point (days 270 to 274) is removed, considering that stratification is nearly complete at this time.[42] This correspondence serves to indicate that in the central basin there is a connection between water levels and anoxic occurrences.

### C. EQUILIBRIUM RESPONSE CURVES

In the foregoing, dynamic changes in lake concentrations were described under the influence of phosphorus loading and thermal stratification. Lam et al.[42] produced a series of equilibrium response curves, which are useful as control curves for planning and assessing the phosphorus removal program, especially when the loading influence is masked by the stratification effects. Details of the procedures applied to estimate lake concentrations at equilibrium with the loadings and the approach used to quantify concentrations in the annual stratification cycle are given in Lam et al.[39]

In general, three cases of lake equilibrium concentrations are established by running the nine-box water quality model for a 12-year period of 1967 to 1978 for a given phosphorus load. Case 1 uses the concentration of the 12th year representing the lake concentration at equilibrium with the given load under long-term average weather conditions. Case 2 considers the computed concentration of the fourth year (1970) as at equilibrium with the same loading and as under a shallow hypolimnion influence. Case 3 uses the concentration of the sixth year (1972) so that the computed equilibrium concentration will be directly under the influence of a thick hypo-limnion.[40]

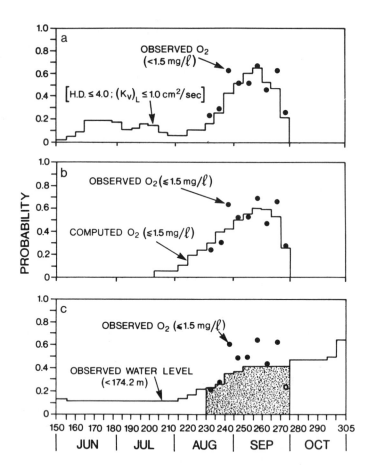

FIGURE 20. Comparison of probability distributions for (a) the computed depth of the top of the hypolimnion (HD ≤ 4 m and ($K$v)L ≤ 1 cm²/s) and observed dissolved oxygen concentrations (DO < 1.5 mg/l), (b) anoxic occurrences estimated from observed and computed $O_2$ concentrations, and (c) observed oxygen concentration less than 1.5 mg/l with water level under the mean level (174.2 m). (From Lam, D. C. L. and Schertzer, W. M., *J. Great Lakes Res.*, 13(4), 757-769, 1987; Lam, D. C. L., et al., *J. Great Lakes Res.*, 13(4), 782-800, 1987. With permission.)

Figure 21 (top) shows equilibrium response curves generated by varying the phosphorus loading from 28000 t to 8000 t. As indicated, anoxia tends to occur under case 2 with a shallow hypolimnion. Although there is a slight increase in the equilibrium oxygen concentration in the hypolimnion at decreased loading, concentrations are generally below 1.5 mg/l. Conversely, under conditions of thick hypolimnion, case 3, anoxia is not expected to occur. Under the scenario of case 1, the tendency is for anoxia to occur only under very high loading conditions. As a point of illustration, Figure 21 (top) includes observed hypolimnion DO concentrations for those years in which measurements were conducted close to but before turnover. Lam et al.[42] discuss some of the inherent limitations of the observed concentrations with respect to the response curves, i.e., the measured data may not necessarily be in equilibrium with the loadings and the measurements may not necessarily be at the lowest depletion value before the fall overturn, as in the cases of 1968, 1969, 1971, 1975, and 1976, which are excluded in Figure 21. Nevertheless, despite these uncertainties, the data depicted in Figure 21 (top) support the equilibrium response curves and hypotheses regarding the stratification effects on anoxic occurrence. Detailed discussion is presented in Lam et al.[42] In general, those years that have shallow

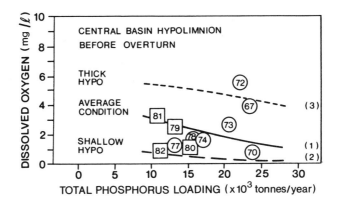

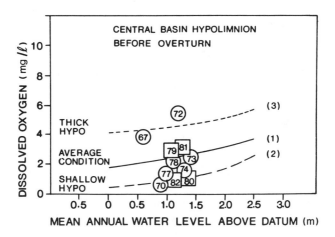

FIGURE 21.   Estimated response curves for central basin hypolimnion dissolved oxygen equilibrium concentration just before overturn in response to (top) lakewide phosphorus loading and (bottom) water level changes. Curves are derived for three contrasting cases being (1) an average condition based on a 12-year simulation, (2) a shallow hypolimnion, and (3) a thick hypolimnion. Actual loadings and water levels are used in the computations. The curves indicate dynamic data observed in the years shown by the numbers. (From Lam, D. C. L., et al., *J. Great Lakes Res.*, 13(4), 782-800, 1987. With permission.)

hypolimnia (1970, 1977, 1978, 1980, and 1982) adhere closely to curve 2. Of the years depicted as having a thick hypolimnion (1969, 1972, 1974, and 1979), only 1972 strongly adheres to curve 3, which is expected because of the extreme storm events that occurred in the early summer period. As mentioned earlier, 1969 data were not measured close enough to the period of lowest oxygen depletion and were excluded. The years 1974 and 1979, which experienced "reversal entrainment" episodes, did not have a hypolimnion as thick as in 1972, and hence they adhered to a more normal condition curve (1). Other years that were included in this analysis generally compared with the expected responses between the two extreme conditions.

An alternate control strategy for the alleviation of anoxic conditions in the central basin involves manipulation of lake water levels. As shown previously, hypolimnion DO concentrations showed some good correspondence with water level changes. Equilibrium response curves are produced following the procedures outlined above. In this case, the water level is kept constant in each of the three cases in the 12-year simulation runs, while the phosphorus loading is allowed to vary as observed. As above, the response curves (Figure 21 [bottom]) are derived by plotting equilibrium DO concentrations reached at the fourth (curve 2), sixth (curve 3), and twelfth (curve 1) years, thus providing results corresponding to shallow hypolimnion, thick hypolimnion, and average conditions. All the curves show that an increase in water level serves to improve the hypolimnion DO content. The addition of observed cruise data in Figure 21 (bottom) shows good correspondence with expected responses based on results from the thermal structure and mass balance analyses discussed earlier.

## VII. SUMMARY

A case study on the anoxia problem in central basin, Lake Erie was presented by using 16 years of observed data and results from a new water quality model developed at the NWRI in Canada. Recurrent hypolimnetic anoxia, even under conditions of reduced phosphorus loadings, led to the question of whether thermal processes or loading reductions were responsible for the oxygen depletion. This analysis demonstrated that both factors must be considered. The sediment oxygen demand is related to the level of phosphorus loading and the long-term effect of loading decreases (i.e., 28,000 to 11,000 t) over the study period helped to alleviate the anoxic conditions in the central basin hypolimnion. However, the model results strongly demonstrated the importance of the weather-induced thermal stratification response, which exerted a dominant influence over the frequency and duration of the central basin hypolimnetic anoxia. The formation of a shallow hypolimnion of long duration over the stratified period has a high probability of hypolimnion anoxic occurrence.

Complications in the understanding of the water quality (anoxia) problem in Lake Erie arose because the effects of short-term processes, such as photosynthesis, and long-term processes, such as SOD, were interrupted and masked by the dominating effects of the thermal stratification. Solution to the problem was possible only by use of a long-term and consistent surveillance and meteorological database, which allowed the formulation of a complex modeling framework that integrated physical, chemical, and biological components into a system considering the heat flux, thermocline, and a nine-box water quality model. The new feature in this model is that it incorporates the dynamic changes in the thermal layers. Comparisons between observations and simulation results for temperature, thermocline position, TP, SRP, and DO provided verification of the accuracy of the model framework. The predictive capability of the model framework provided the confidence from which probability distributions and response curves could be derived for understanding process interrelationships and for management of the lake water quality problems.

The heat fluxes combined with the thermocline model showed that there were three main weather-induced thermal responses, namely, the "normal", the "shallow", and the "entrainment reversal" types. At the present rate of phosphorus loading reduction, anoxia is very probable under the shallow hypolimnion case, but not as likely in the other two categorizations. Because of the high accuracy demonstrated in the thermal stratification model, the uncertainties in the anoxic occurrences could be directly related to the weather records, so that meaningful probabilities of such occurrences could be derived. It was established that there was more than a 50% probability that anoxia would occur over the period Julian day 240 to 270 each year. A reliable indicator of the occurrence can be derived by considering the condition (HD $\leq$ 4.0; $(K_v)_L \leq 1.0$ cm$^2$/s), where HD is the hypolimnetic depth and $(K_v)_L$ is the vertical turbulent diffusivity.

Equilibrium response curves, while demonstrating the correspondence between the central basin hypolimnetic DO concentration and TP loading, also showed that manipulation of the mean annual water level could be used as a management strategy in the shallow basin. Recognizing that manipulation of the water levels with current control structures on the Great Lakes is a difficult task at best, the correspondence between hypolimnion DO concentration and physical factors, such as the weather, thermocline depth, and water level, indicating the susceptibility of the central basin to water quality problems in response to the weather influences. This underscores the concern for deleterious effects on the lake water quality from potential global and regional scale climate warming, which may result from increases in atmospheric "greenhouse gases" (see also discussion in Chapter 9). Such effects have the potential to alter the surface heat fluxes and hydrological characteristics of Lake Erie significantly and to impact on the water quality problems.

## ACKNOWLEDGMENTS

This chapter presents the essential features of the water quality modeling study conducted on Lake Erie as part of the environmental simulation investigations conducted by the National Water Research Institute of Canada Center for Inland Waters. We appreciate the many individuals within institutions in Canada and the United States who generously provided data, useful comments, and discussions on the physical and biochemical aspects of the study. In particular we are grateful to R. A. Vollenweider, T. J. Simons, C. R. Murthy, F. M. Boyce, and J. Barica for their critical reviews and scientific advice. In addition, we wish to thank G. K. Rodgers and F. E. Elder for their support of the project.

## REFERENCES AND BIBLIOGRAPHY

1. **AES**, Surface Water Temperatures for Lake Erie by Infra-red Thermometer Technique, Hydrometeorology and Applications Division, Atmospheric Environment Service, Downsview, Ontario, 1967 to 1978.
2. **Anderson, E. A. and Baker, D. R.,** Estimating incident terrestrial radiation under all atmospheric conditions, *Water Resour. Res.*, 3, 975–988, 1967.
3. **Barica, J. and Mathias, J. A.,** Oxygen depletion and winterkill risk in small prairie lakes under extended ice cover, *J. Fish. Res. Board Can.*, 36, 980–986, 1979.
4. **Barica, J.,** 1982, Lake Erie oxygen controversy, *J. Great Lakes Res.*, 8, 719–722, 1982.
5. **Beale, E. M. L.,** Some uses of computers in operational research, *Industrielle Organisation*, 31, 51–52, 1962.
6. **Bedford, K. W. and Babajimopoulos, C.,** Vertical diffusivities in areally averaged models, *J. Environ. Eng. Div.*, ASCE, 103, 113–125, 1977.
7. **Boyce, F. M., Chiocchio, F., Eid, B., Rosa, F., and Penicka, F.,** Hypolimnion flow between the Central and East basins of Lake Erie 1977, *J. Great Lakes Res.*, 6, 290–306, 1980.
8. **Burns, N. M.,** Nutrient budgets for Lake Erie, *J. Fish. Res. Board Can.*, 33, 520–536, 1976.
9. **Burns, N. M.,** Oxygen depletion in the Central and Eastern basins of Lake Erie, 1970, *J. Fish. Res. Board Can.*, 33, 512–519, 1976.
10. **Burns, N. M. and Ross, C.,** Project Hypo, Canada Center for Inland Waters, Burlington, Ontario, Paper No. 6, 1972.
11. Coordinating Committee on Great Lakes Basic Hydraulic and Hydrologic Data, Lake Erie Outflow 1860–1964 with addendum 1965–1975, Chicago, Illinois and Cornwall, Ontario, June 1976.
12. **Charlton, M. N.,** Oxygen depletion in Lake Erie: Has there been any change? *Can. J. Fish. Aquat. Sci.*, 37, 72–81, 1980.
13. **Chapra, S. C.,** Total phosphorus model of the Great Lakes, *J. Environ. Eng. Div. Am. Soc. Civil. Eng.*, 103(EE2), 147–161, 1977.
14. **Chen, C. W. and Orlob, G. T.,** Ecological Simulation for Aquatic Environments, Report to Office of Water Resources Research, OWRR C-2044, Water Resources Engineers, Walnut Creek, CA, 1972.
15. **de Broissia, M.,** Review of the Current and Potential Application of Modelling and Simulation in Environmental Assessment in Canada, Contract Report No. 1656917, Federal Environmental Assessment Review Office, Environment Canada, Ottawa, 1984.
16. **Davies, J. A. and Schertzer, W. M.,** Canadian radiation measurements and surface radiation balance estimates for Lake Ontario during IFYGL, IFYGL Project Nos. 71EB and 80EB, Contract No. OSP3-0017, Canada Center for Inland Waters, 1974.
17. **Davies, J. A. and Schertzer, W. M.,** Estimating global solar radiation, *Boundary-Layer Meteorology*, 9, 33–52, 1975.
18. **Delorme, L. D.,** Lake Erie oxygen: the prehistoric record, *Can. J. Fish. Aquat. Sci.*, 39, 1021–1029, 1982.
19. **Derecki, J. A.,** Evaporation from Lake Erie, TOAA Technical Report ERL342-GLERL3, Ann Arbor, MI, 1975.

20. **Dilley, A. C.,** On the computer calculation of vapor pressure and specific humidity gradients from psychrometric data, *J. Appl. Meteorol.*, 7, 717–719, 1968.

21. **DiToro, D. M. and Connolly, J. P.,** Mathematical models of water quality in large lakes, Part 2: Lake Erie, EPA-600/3-80-065 Report, Duluth, MN, 1980.

22. **Dobson, H. H. and Gilbertson, M.,** Oxygen depletion in the hypolimnion of the Central Basin of Lake Erie, 1929–1970, in *Proc. 12th Conf. Great Lakes Res.*, 14, 743–748, 1971.

23. **Dolan, D. M., Yui, A. M., and Feist, R. D.,** Evaluation of river load estimation methods for total phosphorus, *J. Great Lakes Res.*, 7, 207–1214, 1981.

24. **Elder, F. C., Andren, A., Eisenrich, T. J., Murphy, T. J., Sanderson, M., and Vet, R. J.,** Atmospheric Loadings to the Great Lakes, unpublished manuscript, University of Windsor, Ontario, 1977.

25. **Fraser, A. S. and Willson, K. E.,** Loading Estimates to Lake Erie, 1967–1976, Environment Canada, Inland Waters Directorate, Science Series No. 129, Ottawa, Ontario, 1981.

26. **Fraser, A. S.,** Tributary and point source total phosphorus loadings to Lake Erie, *J. Great Lakes Res.*, 13(4), 659–666, 1987.

27. **Henderson-Sellers, B. and Lam, D. C. L.,** Reformulation of Eddy Diffusion Thermocline Models, Unpublished manuscript, University of Salford, England, 1982.

28. **Herdendorf, C. E.,** Lake Erie Nutrient Control Program – An Assessment of its Effectiveness in Controlling Lake Eutrophication, EPA-600/3-80-062, Duluth, MN, 1980.

29. **International Joint Commission,** Report to the International Joint Commission on the Pollution of Lake Erie, Lake Ontario and the International Section of the St. Lawrence River, Vol. 2, 1969.

30. **International Joint Commission,** Great Lakes Water Quality Agreement Between the United States of America and Canada, IJC, Windsor, Ontario, 1978.

31. **Ivey, G. N. and Boyce, F. M.,** Entrainment by bottom currents in Lake Erie, *Limnol. Oceanogr.*, 27, 1029–1028, 1982.

32. **Kraus, E. B. and Turner, J. S.,** A one-dimensional model of the seasonal thermocline: II. The general theory and its consequences, *Tellus*, 19, 98–105, 1967.

33. **Lam, D. C. L.,** Temporal and spatial constraints in data estimation problems, *Appl. Math. Notes*, 6, 20–32, 1981.

34. **Lam, D. C. L.,** Finite Element Analysis of Water Quality in Lake Erie, NWRI unpublished manuscript, Burlington, Ontario, 1982.

35. **Lam, D. C. L. and Jaquet, J. M.,** Computations of physical transport and regeneration of phosphorus in Lake Erie, fall 1970, *J. Fish. Res. Board Can.*, 33, 550–563, 1976.

36. **Lam, D. C. L. and Halfon, E.,** Model of primary production, including circulation influences in Lake Superior, *Appl. Math. Modelling*, 2, 30–40, 1978.

37. **Lam, D. C. L. and Schertzer, W. M.,** Modelling the interaction of climate and aquatic regimes of large lakes, invited paper, in *Proc. Canadian Climate/Water Workshop*, University of Alberta, Feb. 28–29, 1980, AES, Downsview, Ontario, 1980, 157–160.

38. **Lam, D. C. L., Schertzer, W. M., and Fraser, A. S.,** Mass balance models of phosphorus in sediments and water, *Hydrobiologia*, 91, 217–225, 1982.

39. **Lam, D. C. L., Schertzer, W. M., and Fraser, A. S.,** Simulation of Lake Erie Water Quality Responses to Loading and Weather Variations, Environment Canada, Inland Waters Directorate, Science Series No. 134, NWRI, Burlington, Ontario, 1983.

40. **Lam, D. C. L. and Schertzer, W. M.,** Lake Erie thermocline model results: comparison with 1967–1982 data and relation to anoxic occurrences, *J. Great Lakes Res.*, 13(4), 757–769, 1987.

41. **Lam, D. C. L., Schertzer, W. M., and Fraser, A. S.,** Oxygen depletion in Lake Erie: modelling the physical, chemical and biological interactions, 1972 and 1979, *J. Great Lakes Res.*, 13(4), 770–781, 1987.

42. **Lam, D. C. L., Schertzer, W. M., and Fraser, A. S.,** A post-audit analysis of the NWRI nine-box water quality model for Lake Erie, *J. Great Lakes Res.*, 13(4), 782–800, 1987.

43. **Lemire, F.,** Winds on the Great Lakes, CIR-3560, TEC-380, Canada Department of Transport, Metropolitan Branch, 1961.

44. **Lucas, A. M. and Thomas, N. A.,** Sediment oxygen demand in Lake Erie's Central Basin, 1970, in *Project Hypo*, Burns and Ross, Eds., Canada Center for Inland Waters Paper No. 6, Burlington, Ontario, 1972, 45–48.

45. **Mathias, J. A. and Barica, J.,** Factors controlling oxygen depletion in ice-covered lakes, *Can. J. Fish. Aquat Sci.*, 37, 185–194, 1980.

46. **Mellor, G. L. and Durbin, P. A.,** The structure and dynamics of the ocean surface mixed layer, *J. Phys. Oceanogr.*, 5, 718–728, 1975.

47. **Papadakis, J. E.,** Determination of the Oceanic Wind Mixed Layer Depth by an Extension of Newton's Method, Pacific Marine Science Report, 81–9, I.O.S. Sidney, B.C., 1981.

48. **Phillips, D. W. and Irbe, J. G.,** Lake to Land Comparison of Wind, Temperature, and Humidity on Lake Ontario During the International Field Year for the Great Lakes, Environment Canada, Report Number CL1-2–77, 1978.

49. **Phillips, D. W. and McCulloch, J. A. W.,** The Climate of the Great Lakes Basin. Environment Canada, Climatological Studies, Number 20, AES, Downsview, Ontario, 1972.

50. **Platt, T., Denman, K. L., and Jassby, A. D.,** The Mathematical Representation and Prediction of Phytoplankton Productivity. Environment Canada, Fish. Mar. Serv., Tech. Rep. 523, 1975.

51. **Resio, D. T. and Vincent, C. L.,** Estimation of winds over the Great Lakes, Amer. Soc. Civil Eng. Waterway, Port and Coast, *Ocean Div. J.*, 102, 265–283, 1977.

52. **Richards, T. L. and Fortin, J. P.,** An Evaluation of the Land-Lake Vapor Pressure Relationships for the Great Lakes, Pub. No. 9, University of Michigan, Great Lakes Research Division, Ann Arbor, MI, 1962, 103-110.

53. **Robertson, D. G. and Jordan, D. E.,** Digital Bathymetry of Lakes Ontario, Erie, Huron, Superior and Georgian Bay, Canada Center for Inland Waters, Basin Investigation and Modelling Section, Burlington, Ontario, 1978.

54. **Sanderson, M. E.,** The Climate of the Essex Region of Canada's Southland, Department of Geography, University of Windsor, Windsor, Ontario, 1980.

55. **Sawchuk, A. M. and Schertzer, W. M.,** Climatology of the Lake Ontario and Lake Erie Surface Heat Exchanges 1953–1983, NWRI Contribution No. 88–37, National Water Research Institute, Canada Centre for Inland Waters, Burlington, Ontario, 1988.

56. **Schertzer, W. M., Elder, F. C., and Jerome, J.,** Water transparency of Lake Superior in 1973, *J. Great Lakes Res.*, 4, 350–358, 1978.

57. **Schertzer, W. M.,** Energy budget and monthly evaporation estimates for Lake Superior, 1973, *J. Great Lakes Res.,* 4, 320–330, 1978.

58. **Schertzer, W. M.,** Heat balance and heat storage estimates for Lake Erie 1967 to 1982, *J. Great Lakes Res.*, 13(4), 454–467, 1987.

59. **Schertzer, W. M., Saylor, J. H., Boyce, F. M., Robertson, D. G., and Rosa, F.,** Seasonal thermal cycle of Lake Erie, *J. Great Lakes Res.*, 13(4), 468–486, 1987.

60. **Schwab, D. J. and Morton, J. A.,** Estimation of over-lake wind speed from over-land wind speed: a comparison of three methods, *J. Great Lakes Res.*, 10, 68–72, 1984.

61. **Simons, T. J.,** Continuous dynamical computations of water transports in Lake Erie for 1970, *J. Fish. Res. Board Can.*, 33, 371–384, 1976.

62. **Simons, T. J.,** Analysis and simulation of spatial variations of physical and biochemical processes in Lake Ontario, *J. Great Lakes Res.*, 2, 215–233, 1976.

63. **Simons, T. J.,** Verification of Seasonal Stratification Models, Department Report, Institute for Meteorology and Oceanography, University of Utrecht, Netherlands, 1980.

64. **Simons, T. J. and Lam, D. C. L.,** Nutrient exchange between volume elements of large lakes, in *Proc. Symp. Modelling the Water Quality of the Hydrological Cycle*, Baden, Austria, September 1978, IAHS-AISH Pub. No. 125, 185–191, Washington, D.C., 1978.

65. **Simons, T. J. and Lam, D. C. L.,** Some limitations of water quality models for large lakes, a case study of Lake Ontario, *Water Resour. Res.*, 16, 105–116, 1980.

66. **Simons, T. J., Boyce, F. M., Fraser, A. S., Halfon, E., Hyde, D., Lam, D.C.L., Schertzer, W.M., El-Shaarawi, A. H., Willson, K., and Warry, D.,** Assessment of Water Quality Simulation Capability for Lake Ontario, Environment Canada, Inland Waters Directorate, Science Series No. 111, Ottawa, Ontario, 1979.

67. **Snodgrass, W. J.,** Analysis of models and measurements for sediment oxygen demand in Lake Erie, *J. Great Lakes Res.*, 13(4), 738–756, 1987.

68. **Stadelmann, P., Moore, J., and Pickett, E.,** Primary production in relation to temperature structure, biomass concentration and light conditions at an inshore and offshore station in Lake Ontario, *J. Fish. Res. Board Can.*, 31, 1215–1232, 1974.

69. **Steele, J. H.,** Notes on some theoretical problems in production ecology, *Mem. Ist. Ital. Idrobiol. Dott Marco de Marchi Pallanza Italy*, 18 (Suppl.), 383–398, 1965.

70. **Sundaram, T. R. and Rehm, R. G.,** The seasonal thermal structure of deep temperature lakes, *Tellus*, 25, 157–167, 1973.

71. **Thomann, R. V. and Segna, J. S.,** Dynamic phytoplankton-phosphorus model of Lake Ontario: ten-year verification and simulations, in *Proc. Conf. Phosphorus Management Strategies for Lakes*, Ann Arbor Sci., Michigan, 1980, 153–205.

72. **Vollenweider, R. A.,** Scientific Fundamentals of the Eutrophication of Lakes and Flowing Waters, with Particular Reference to Nitrogen and Phosphorus as Factors in Eutrophication, OECD, PAS/CSI/68.27, 1968.

73. **Vollenweider, R. A.,** Models for calculating integral photosynthesis and some implications regarding structural properties of the community metabolism of aquatic systems, in *Proc. IBP/PP Technical Meeting*, Trebon, 14–21 September, 1969, 455–472, 1970.

74. **Vollenweider, R. A. and Dillon, P. J.,** The application of the phosphorus loading concept to eutrophication research, Nat. Res. Council Canada Rep. NRCC 13690, 1974.

75. **Vollenweider, R. A.,** Advances in defining critical loading levels for phosphorus in lake eutrophication, *Mem. Ist. Ital. Idrobiol. Dott Marco de Marchi Pallanza Italy*, 33, 53–85, 1976.

76. **Vollenweider, R. A. and Janus, L. L.,** Statistical models for predicting hypolimnetic oxygen depletion rates. National Water Research Institute Report, Burlington, Ontario, 1981.

77. **Vollenweider, R. A., Rast, W., and Kerekes, J.,** The phosphorus loading concept and Great Lakes eutrophication, in *Proc. Conf. Phosphorus Management Strategies for Lakes*, Ann Arbor Sci., Michigan, 1980, 207–234.

78. **Walters, R. Z., Carey, G. F., and Winter, D. F.,** Temperature computation for temperate lakes, *Appl. Math. Modelling*, 2, 41–48, 1978.

79. WPM, Water level and flow data, Water Planning and Management Branch, Ontario Region, Environment Canada, Burlington, Ontario (personal communication with Peter Yee), 1983.

Chapter 3

# WATER QUALITY MODELING: APPLICATION TO LAKES AND RESERVOIRS

## CASE STUDY OF LAKE BALATON, HUNGARY

**Peter Shanahan, Richard A. Luettich, Jr., and Donald R. F. Harleman**

## TABLE OF CONTENTS

# I. INTRODUCTION

## A. OVERVIEW
### 1. The Lake Balaton Case Study

From 1979 through 1987, The Ralph M. Parsons Laboratory at the Massachusetts Institute of Technology (MIT) conducted a program of research organized around a case study of Lake Balaton in Hungary. The research was conducted in cooperation with concurrent research programs at the International Institute for Applied Systems Analysis (IIASA) in Laxenburg, Austria and the Institute for Water Pollution Control (VITUKI) of the Hungarian National Water Authority. Several previous publications have reported on the entire Lake Balaton case study, drawing on the results achieved by these and other institutions. Our purpose here is not to report on the full breadth of the Lake Balaton case study as conducted at these several institutions. For that the reader is referred to the comprehensive description in the book, *Modeling and Managing Shallow Lake Eutrophication with Application to Lake Balaton*, edited by L. Somlyódy and G. van Straten.[1] Rather, our aim in this chapter is to focus on the particular interest of the MIT research: the formulation of deterministic lake water quality models with specific attention to the interaction between hydrodynamics and water-quality processes.

Lake Balaton is a unique case study owing to its extreme shallowness. Balaton is the largest lake in Central Europe. It is approximately 75 km long and 8 km wide. Its shallowness is remarkable: the average depth of the lake is only 3.2 m and at its deepest point Balaton is but 11.6 m deep. The shallow lake responds vigorously to wind events, preventing the seasonal thermal stratification seen in deeper lakes. The wind creates complex patterns of horizontal circulation and a strong seiche. Another effect of the wind is the frequent resuspension of the fine carbonate sediments that cover most of the lake's bottom. The sediments impart a milky appearance to the lake's water, so regular is their resuspension.

The suspended sediment does not interfere with the water quality and attractiveness of the lake. It is a popular summer resort and is vitally important to the Hungarian economy as a source of foreign currency. However, with its popularity and other pressures from land development, has come cultural eutrophication. The correction and reversal of the lake's eutrophication was the goal of the Lake Balaton case study, of which the MIT program was a part.

This chapter summarizes the findings on three major issues addressed during the MIT research program. First is the representation of nutrient uptake and algal growth kinetics in lake eutrophication models. The research, previously reported by Luettich and Harleman,[2] compares results from a phosphorus-based model using a kinetic formulation based on a variable cell quota with an otherwise similar model that assumes a constant cell quota. Model experiments using Lake Balaton data investigate the significance of the cell quota model in the shallow lake environment.

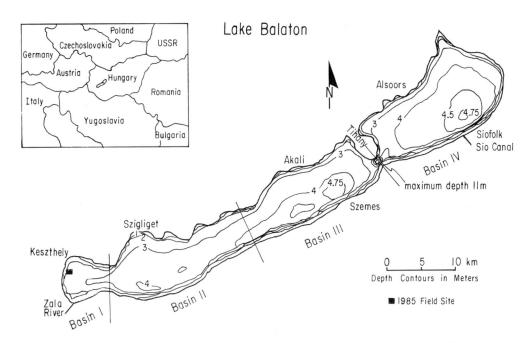

FIGURE 1. Lake Balaton, Hungary.

The second major issue is the appropriate representation of hydrodynamic influences in a model of shallow lake eutrophication. This work, published as an MIT report by Shanahan and Harleman[3] and in other publications,[4–7] is based upon investigations in which a hydrodynamic model of circulation in Lake Balaton is linked with a eutrophication model. The linked model results illustrate the fundamental importance of the model spatial structure and hydrodynamic transport functions.

The third major issue addressed during the MIT research program is another manifestation of the interaction between hydrodynamics and the water quality of Lake Balaton: the resuspension of lake bottom sediments by wind-induced water movement. This recently completed work is described in detail.[8,52,53] The modeling effort drew upon field program results that illustrated the dominance of wind-induced wave action in sediment resuspension. A simple model of the depth-averaged suspended sediment resuspension was developed, calibrated, and verified using data obtained in Lake Balaton.

## B. DESCRIPTION OF LAKE BALATON
### 1. Geometry
The unique geometry of Lake Balaton, introduced in the preceding section, is illustrated in Figure 1. The surface area of the lake is approximately 600 km². The lake has been segmented for analysis purposes into four basins: I. Keszthely Basin, II. Szigliget Basin, III. Szemes Basin, and IV. Siófok Basin.

The lake is nearly divided by the Peninsula of Tihany. In the narrow straits at Tihany, the lake is less than 2 km wide. It is at this point that the lake is deepest, 11.6 m. Outside of this small area, the lake is everywhere less than 5 m deep.

### 2. Climate
The climate of the Balaton region is temperate, with an average annual air temperature of 10.7°C. The average monthly temperature drops below freezing in winter (–1.0°C in January) and ice covers the lake for roughly 2 months each year. The maximum mean monthly air

temperature of 21.4°C occurs in July. Water temperatures vary similarly, from a low monthly average water temperature of 0.7°C in January to a high of 24.1°C in July. The shallowness of the lake causes it to warm rapidly in the summer sun, leading to the high July water temperature.

Wind is an important meteorological factor in the behavior of shallow lakes. At Balaton, strong winds occur on several stormy days each month in a roughly uniform annual distribution. The prevailing wind direction is from the northwest, passing across the width of the lake.

### 3. Circulation

Two water courses are major hydrologic influences on the circulation in the lake. The lake's only outlet is the Sió Canal, located at the eastern end of the lake, while the largest inflow to the lake, the Zala River, is located at the extreme opposite end of the lake. The Zala drains roughly one-half of the lake's drainage basin, and additional local drainage is concentrated in the western part of the lake. There is thus created a net hydrological flow, albeit at low velocities, through the lake from west to east.

The average annual flows associated with hydrological influences on the lake are 18 m$^3$/s due to stream flow and runoff, 12 m$^3$/s from direct precipitation, 17 m$^3$/s removed by evaporation, and 13 m$^3$/s discharged at the Sió Canal. The total lake volume is approximately 1900 m$^3$, so that the mean hydraulic residence time of Lake Balaton is roughly 4 years.

The influence of the wind overwhelms the slow hydrological flow in establishing the pattern of flow in Lake Balaton. The shallowness of the lake leads to a circulation response to even mild winds, producing currents that are one to two orders of magnitude greater than those associated with the hydrological through-flow. Uneven wind patterns, some formed by hills north of the lake that block and deflect the wind, lead to a complex pattern of flow in the lake. Flow occurs in a series of horizontal gyres along the lake, with little discernible vertical velocity gradient.

The most striking circulation phenomenon in Lake Balaton is its seiche. The shallowness of the lake enhances wind set-up, leading to observed longitudinal denivellations as great as 1 m. Both transverse and longitudinal seiches are observed in the lake, with multiple seiche periods created by the response of the entire lake and the two lake basins. Muszkalay[9] collected water-level measurements throughout the lake for a period of nearly a decade. The measurements show the lake to be in nearly constant oscillation. Only a few hours of strong wind will cause an observable seiche; a typical month-long record from Muszkalay's observations shows frequent events, with both longitudinal and transverse modes evident.

### 4. Water Quality

Lake Balaton is a hardwater lake, owing to the dolomite geology of the region. The lake is characterized by high concentrations of calcium and magnesium (approximately 40 mg/l each), a high alkalinity (200 – 250 mg/l as CaCO$_3$) and a high pH (8.4). The shallow waters are typically well mixed throughout the water column, with rare intermittent thermal stratification.

Although the water quality of the lake is generally good, signs of rapidly developing eutrophication became evident in the early 1970s. Since that time the lake's water quality has degraded from a uniformly mesotrophic state to one in which eutrophication is at times extreme in Keszthely Bay, at the western end of the lake. Accompanying this has been the development of a gradient in almost all other water quality parameters (including primary productivity, chlorophyll-*a* [chl*a*], biomass, and total phosphorus) along the axis of the lake. For example, Figure 2 shows field measurements of total phosphorus and chl*a* for 1977. The western end of the lake is hypertrophic, while the eastern end remains mesotrophic.

As is true in many temperate lakes, phosphorus is the principal limiting nutrient in Lake Balaton, although there is evidence that nitrogen alone or jointly with phosphorus also limits algal growth in some situations.[10] Phosphorus in the lake is predominantly in particulate form, and the spatial distribution of phosphorus reflects the distribution of historical nutrient loads to

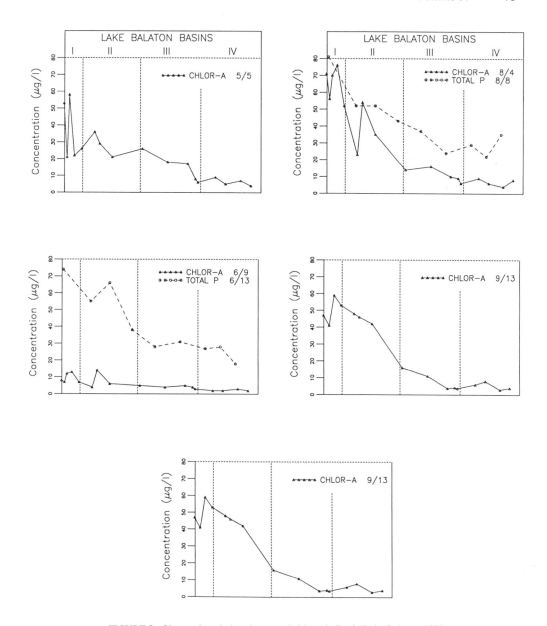

FIGURE 2.  Observed total phosphorus and chlorophyll-*a* in Lake Balaton, 1977.

the lake. Somlyódy and Jolánkai[11] estimate the total phosphorus load to the lake from external sources as 865 kg/d, of which approximately one-half is in the biologically available forms of orthophosphate and soluble organic phosphorus. The available phosphorus load comes as four roughly equal parts from the Zala River, other tributaries, sewage discharges, and other sources (direct runoff and atmospheric loads). The total nitrogen load is approximately 8600 kg/d, with approximately 55% coming from the Zala River and tributary inflows, 10% from sewage, 20% from atmospheric deposition, and 15% from direct runoff. The nutrient inflows to the lake show significant variation through the year. Sewage loads peak during the summer tourist season, at three to four times the off-season rate.

In the 1980s, the Hungarian government took the first steps towards restoring Lake Balaton's water quality by reducing nutrient inputs. In 1985, the Zala River was impounded so as to be

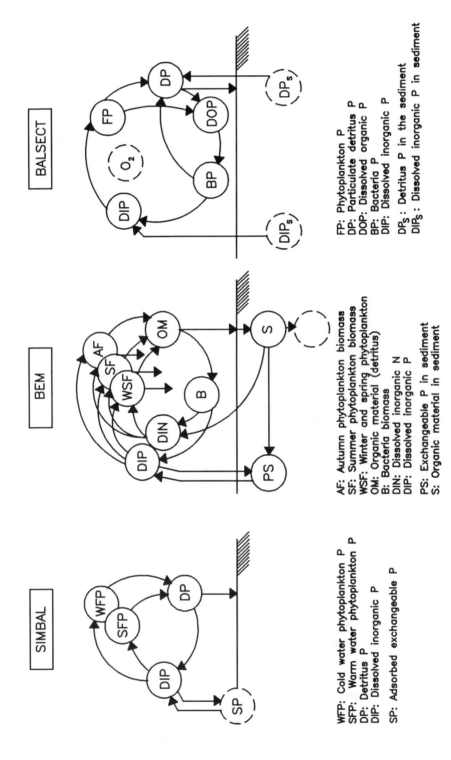

WFP: Cold water phytoplankton P
SFP: Warm water phytoplankton P
DP: Detritus P
DIP: Dissolved inorganic P

SP: Adsorbed exchangeable P

AF: Autumn phytoplankton biomass
SF: Summer phytoplankton biomass
WSF: Winter and spring phytoplankton
OM: Organic material (detritus)
B: Bacteria biomass
DIN: Dissolved inorganic N
DIP: Dissolved inorganic P

PS: Exchangeable P in sediment
S: Organic material in sediment

FP: Phytoplankton P
DP: Particulate detritus P
DOP: Dissolved organic P
BP: Bacteria P
DIP: Dissolved inorganic P

DP$_s$ : Detritus P in the sediment
DIP$_s$ : Dissolved inorganic P in sediment

FIGURE 3. Schematic drawing of Lake Balaton eutrophication models. (From van Straten, G., *Modeling and Managing Shallow Lake Eutrophication with Application to Lake Balaton*, Somlyódy, L. and van Straten, G., Eds., Springer-Verlag, Berlin, 1986. With permission.)

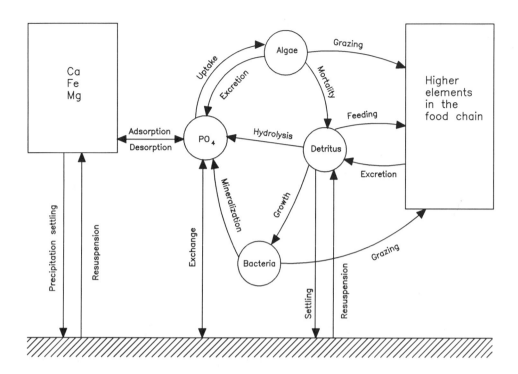

FIGURE 4. The lake phosphorus cycle. (From van Straten, G., *Modeling and Managing Shallow Lake Eutrophication with Application to Lake Balaton,* Somlyódy, L. and van Straten, G., Eds., Springer-Verlag, Berlin, 1986. With permission.)

rerouted through the Kis Balaton, a marshland near its mouth expected to act as a settling basin and nutrient removal system. At present an extensive sanitary sewer system is being constructed. At its completion, interceptor sewers will ring the lake, with the treated sewage diverted outside the Balaton basin. Attempts are also being made to reduce nutrient runoff from agricultural lands by altering land-use strategies and by using more efficient fertilizer application techniques.

## II. WATER QUALITY MODEL FORMULATION

### A. OVERVIEW OF PHOSPHORUS MODEL CONSTRUCTION

Three eutrophication models, SIMBAL, BEM, and BALSECT, have been developed for Lake Balaton and are shown schematically in Figure 3. For complete details of each, the reader is referred to the original publications[12-15] and an overview by van Straten.[16] In each case the models attempt to represent the life cycle of phytoplankton in the lake via temperature-, light-, and nutrient-limited growth functions, and population- and temperature-dependent mortality. Zooplankton were not included, as they have been shown to be relatively ineffective grazers in Lake Balaton.[17] All of the models assume that a constant ratio exists between phytoplankton biomass and the nutrients required for growth, i.e., phosphorus in SIMBAL and BALSECT, and phosphorus and nitrogen in BEM. As a result, it is only necessary to model explicitly the biological cycling of the limiting nutrient(s). Since phosphorus is the most common limiting nutrient in Lake Balaton, it is reasonable to use it as the basis for a water quality model, as was done in SIMBAL and BALSECT.

Figure 4, adapted from van Straten,[16] illustrates the dominant pathways for phosphorus cycling as it is believed to occur in Lake Balaton. The growth of algae is fueled by the uptake of dissolved inorganic phosphorus (DIP). In lakes in which phosphorus limits algal growth, this is a rapid and efficient process, as DIP concentrations are typically below chemically detectable limits (~1 µg/l). This has been shown to be true in Lake Balaton as well.[10] Some of the phosphorus contained in the phytoplankton is returned to inorganic form by algal excretion,[19] while mortality converts the rest into particulate detrital phosphorus. Detrital phosphorus may be lost from the water column due to settling or is converted back into DIP via lysis and bacterially mediated mineralization. The phosphorus cycle is influenced by interaction with the sediments through several pathways, although the details of this interaction are only beginning to be understood (see Section IV.A.1).

Operationally, the biological cycling of phosphorus is represented as a set of model compartments linked by mathematical representations of the reactions that transform phosphorus from one form (model compartment) to another. There is extensive literature addressing how best to formulate such models. Key issues include the appropriate level of model complexity, as expressed by the number of model compartments, the representation of sediment-water interaction, the use of multiple algal species or time-dependent parameters to capture seasonal effects, the trophic level included in the model, and the procedure used to calibrate the model.

The work reported in the remainder of this section addresses the additional question of how to translate the model compartment that represents algal phosphorus into the quantity directly related to eutrophication, algal biomass. As mentioned above, the three Lake Balaton models use the commonly held assumption that phosphorus comprises a constant fraction of the cell biomass. Such models are known as constant cell quota (CCQ) models. An alternative favored by laboratory experimental evidence is that the phosphorus to biomass ratio is not constant. This type of model is called a variable cell quota (VCQ) model and considers algal growth as a two-step process, in which algal growth — the creation of biomass — is modeled as distinct from phosphorus uptake.

## B. AN INVESTIGATION OF THE EFFECT OF NUTRIENT KINETICS ON WATER QUALITY MODEL BEHAVIOR

The purpose of the work described below was twofold: (1) to examine the effect that a VCQ model formulation had on conclusions about significant processes in the eutrophication of Lake Balaton, which were based on a CCQ model, and (2) to compare the predicted lake behavior using the two model formulations in a management application, i.e., to reductions in the external nutrient loading. To avoid masking the effect of the cell quota formulation, the least complex of the existing Lake Balaton eutrophication models, SIMBAL, was chosen to be the basis of our study. We first briefly review the original model formulation and then present the modifications that were made for our study.

### 1. The Development of the SIMBAL Model for Lake Balaton

SIMBAL was originally developed for the purpose of testing assumptions about the major modes of phosphorus cycling in Lake Balaton.[14] As shown in Figure 3, SIMBAL state variables include phytoplankton phosphorus (PP) in winter and summer algal groupings, detrital phosphorus (DP), and DIP. Growth is multiplicatively limited by light, temperature, and DIP, the latter being formulated using the CCQ assumption with different parameters for winter and summer algal populations. Mortality is a function of population size and temperature. The remineralization of DP to DIP is represented without explicitly including bacteria by making this process temperature dependent. Exchange with the sediments is modeled using a temperature-dependent release of DIP to simulate regeneration by bacteria at the sediment surface, and via a two-way bulk exchange process to simulate the adsorption/desorption of DIP to the sediments. DIP can also be lost from the water column by coprecipitation with biogenic lime.

The model was applied simultaneously in each of the four basins shown in Figure 1, assuming that each basin was fully mixed and that the basins interacted via calibrated exchange flows. Data from 1977 were used to define acceptable ranges for total phosphorus, total dissolved phosphorus, and DIP in each basin. Monte Carlo simulations were used to define parameter sets that produced model behaviors within the acceptable ranges.

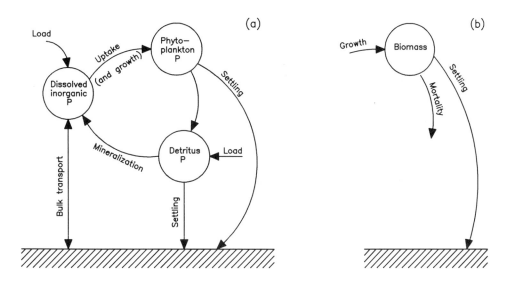

FIGURE 5. Schematic drawing of water quality models. (a) Constant cell quota (CCQ) model; (b) additional compartment added to the CCQ model to create a variable cell quota (VCQ) model.

Initial model runs were made without the terms representing the sorption/desorption of phosphorus to sediments. However, no parameter combinations could be found that produced acceptable model results. In particular, unreasonably high DIP concentrations occurred during and after the die-off of the summer plankton bloom. Upon inclusion of the sediment exchange process, it was possible to calibrate the model, and from this it was concluded that phosphorus exchange with the sediment was a significant process in Lake Balaton.

## 2. Modifications Made for Kinetics Study

To focus the modeling application on the VCQ/CCQ issue, several simplifications were made to SIMBAL. The model was applied only to the summer phytoplankton bloom (which was the period for which van Straten found the sediment interaction to be most important[14]) and only in basin I, which is the smallest and most eutrophic basin. The application to basin I eliminates the possibility of any interbasin exchange flows. Also, recent experimental evidence has shown that the coprecipitation of DIP with biogenic lime[20] and the temperature-dependent release of DIP from the sediment[21] are unimportant in Lake Balaton, and therefore these processes were dropped from the model. (van Straten[14] found both of these processes to have little effect on the behavior of SIMBAL using realistic parameter values.) Finally, PP was allowed to settle at the same rate as DP. (PP did not settle at all in SIMBAL.)

The model that resulted (hereafter called the CCQ model) is shown schematically in Figure 5A, and its equations are presented in Table 1. A second model (VCQ model) was then constructed that was identical to model I, with the exception that a VCQ formulation was used (Figure 5 and Table 2).

Both models were applied for the period from June 14, 1977 to November 25, 1977. The starting date corresponded to a low point in the chl*a* data which was presumed to separate the winter and summer blooms. Initial conditions were determined as follows. For the CCQ model, initial PP was determined from chl*a* data using a chl*a*/PP ratio of 2.[22,23] For the VCQ model, the initial PB was determined from chl*a* data using a chl*a*/PB ratio of 100,[22,24] and PP was determined from PB assuming the phytoplankton were at the maximum cell quota. For both models initial DIP values were arbitrarily set to 1 µg/l (the models were reasonably insensitive to this value), while DP was set by subtracting PP and DIP from total phosphorus data.

**TABLE 1**
**CCQ Model Equations**

### State Equations

$$\dot{PP} = -O_{PP} + PPGR - PPMORR - PPSR \tag{1}$$

$$\dot{DP} = -O_{DP} + PPMORR - MINR - DPSR + L_{DP} \tag{2}$$

$$\dot{DIP} = -O_{DIP} - PPGR + MINR + SEXCR + L_{DIP} \tag{3}$$

$\dot{}$ (superscripted dot) = indicates time derivative
PP = phytoplankton phosphorus
DP = detrital phosphorus
DIP = dissolved inorganic phosphorus

### Data Requirements

#### External Loadings

$L_{DP}$ = volumetric external detrital phosphorus loading
$L_{DIP}$ = volumetric external dissolved inorganic phosphorus loading

#### Outflows

$$O_{PP} = PP \cdot Q/V \tag{4}$$

$$O_{DP} = DP \cdot Q/V \tag{5}$$

$$O_{DIP} = DIP \cdot Q/V \tag{6}$$

$Q$ = hydrologic outflow from basin I
$V$ = volume of basin I

#### Water Temperature, $T$

#### Incident Solar Radiation, $I$

### Rates

#### Growth rate

$$PPGR = k_g \cdot f_n \cdot f_l \cdot f_T \cdot PP \tag{7}$$
$k_g$ = maximum specific DIP uptake and PP growth rate
$f_n$ = CCQ nutrient limitation factor

$$f_n = \frac{DIP}{DIP_g + DIP} \tag{8}$$

$DIP_g$ = half saturation constant for CCQ uptake/growth
$f_l$ = light limitation factor (depth-averaged Steele equation)

$$f_l = \frac{\varepsilon}{\varepsilon H}\left\{\exp\left[-\frac{I}{I_s}\exp\left(-\varepsilon H\right)\right] - \exp\left[\frac{I}{I_s}\right]\right\} \tag{9}$$

$H$ = water depth
$\varepsilon$ = total extinction coefficient

## TABLE 1 (continued)
## CCQ Model Equations

$$\varepsilon = \varepsilon_o + \alpha \cdot PP \tag{10}$$

$\varepsilon_o$ = extinction coefficent without algae
$\alpha$ = phytoplankton self-shading coefficient

$I_s$ = optimal light intensity for algal growth

$$I_s = I_{sm} + I_{se} \cdot T \tag{11}$$

$I_{sm}$ = base optimal light intensity
$I_{se}$ = temperature correction coefficient

$f_T$ = temperature limitation factor

$$f_T = \begin{cases} \dfrac{T_c - T}{T_c - T_o} \exp\left[ 1 - \dfrac{T_c - T}{T_c - T_o} \right] & \text{if } T \le T_c \\ 0 & \text{if } T > T_c \end{cases} \tag{12}$$

$T_c$ = critical water temperature
$T_o$ = optimal water temperature

### Mortality Rate

$$PPMORR = k_d \cdot \theta_d^{T-20} \cdot PP \tag{13}$$

$k_d$ = phytoplankton mortality rate coefficient at 20°C
$\theta_d$ = mortality temperature coefficient

### Settling Rate

$$PPSR = v_s \cdot PP/H \tag{14}$$
$$DPSR = v_s (1 - \gamma) \, DP/H \tag{15}$$

$v_s$ = settling velocity
$\gamma$ = fraction of detritus that is dissolved

### Mineralization Rate

$$MINR = k_m \theta_m^{T-20} \cdot DP \tag{16}$$

$k_m$ = mineralization rate coefficient at 20°C
$\theta_m$ = mineralization temperature coefficient

### Sediment Exchange Rate

$$SEXCR = k_{ex} \left( DIP_{eq} - DIP \right) \tag{17}$$

$k_{ex}$ = transport coefficient for exchange between sediment and water
$DIP_{eq}$ = sediment equilibrium DIP concentration

Model calibrations were accomplished by systematically varying parameter values by hand within reasonable bounds. Whenever possible, the original SIMBAL parameter values were maintained so that as close a comparison as possible could be made to conclusions reached by van Straten.[14] The models were calibrated to chl*a* measurements that were made in basin I during

## TABLE 2
## VCQ Model Equations

### State Equations

$$\dot{PP} = -O_{PP} + DIPUR - PPMORR - PPSR \tag{18}$$

$$\dot{PB} = -O_{PB} + PBGR - PBMORR - PBSR \tag{19}$$

$$\dot{DP} = -O_{DP} + PPMORR - MINR - DPSR + L_{DP} \tag{20}$$

$$\dot{DIP} = -O_{DIP} - DIPUR + MINR + SEXCR + L_{DIP} \tag{21}$$

$\dot{}$ (superscripted dot) = indicates time derivative
PP = phytoplankton phosphorus
PB = phytoplankton biomass
DP = detrital phosphorus
DIP = dissolved inorganic phosphorus

### Data Requirements

#### External Loadings - Same As CCQ Model

#### Outflows

$$O_{PB} = PB \cdot Q/V \tag{22}$$

All others the same as CCQ model

#### Water Temperature - Same As CCQ Model

#### Incident Solar Radiation - Same As CCQ Model

### Rates

#### DIP Uptake Rate

$$DIPUR = k_u \, \frac{DIP}{DIP_u + DIP} \, \frac{q_{max} - q}{q_{max} - q_{min}} \, PB \tag{23}$$

$q$ = cell quota; $q$ = PP/PB
$k_u$ = maximum DIP uptake rate
$DIP_u$ = half-saturation constant for VCQ uptake
$q_{max}$ = maximum cell quota
$q_{min}$ = minimum cell quota

#### Growth Rate

$$PBGR = k_g \cdot f_q \cdot f_I \cdot f_T \cdot PB \tag{24}$$

$k_g$ = maximum specific growth rate
$f_q$ = VCQ growth reduction factor

$$f_q = \frac{q - q_{min}}{q} \tag{25}$$

Light and temperature limitation are the same as CCQ model

## TABLE 2 (continued)
## VCQ Model Equations

### Mortality Rate

$$PBMORR = k_d \cdot \theta_d^{T-20} \cdot PB \tag{26}$$

PPMOR, $k_d$, and $\theta_d$ are the same as CCQ model

### Settling Rate

$$PBSR = v_s \cdot PB/H \tag{27}$$

All other variables are the same as CCQ model

### Mineralization Rate - Same As CCQ Model

### Sediment Exchange Rate - Same As CCQ Model

this time. Total phosphorus data was also available for this period, however, the measurements were made at intervals of more than a month, and therefore this information was compared to the model predictions only qualitatively to eliminate completely unreasonable model calibrations. Measurements of dissolved reactive phosphorus were not used, as these chemically determined values have been shown to not be representative of DIP.[10,18]

### 3. Results from Kinetics Study

Initial model runs were performed to examine the conclusion reached by van Straten[14] that sediment adsorption/desorption is critically important in the phosphorus cycle in Lake Balaton. First, an acceptable calibration was obtained for the VCQ model. This model is called VCQ-1 and the predicted chl*a* concentrations are shown in Figure 6A. (The parameter values for this and all other runs are listed in Table 3.) Then, keeping all other parameters constant, the sediment exchange was disabled. This model, called VCQ-1ns, predicts chl*a* concentrations as shown in Figure 6B. The results show enhanced growth early in the summer and a large increase in the predicted maximum chl*a* level. This suggests that the sediments acted as a significant phosphorus sink during the critical early summer growth period and supports the earlier conclusion about their importance in the phosphorus cycle.[14]

An alternative calibration for the VCQ model to the same data, called model VCQ-2, is shown in Figure 6C. If all model parameters are again kept constant and sediment exchange is disabled with this alternative calibration (model VCQ-2ns), an entirely different conclusion is reached, Figure 6D. In this case, the predicted chl*a* concentrations decrease by a small amount in comparison with model VCQ-2. Given the scatter in the data, this could arguably be considered a reasonable model calibration. Therefore we must conclude that the behavior attributed to sediment interaction in van Straten's earlier model study can be equally well attributed to VCQ nutrient uptake kinetics.

The VCQ/CCQ issue was examined further by attempting to calibrate the CCQ model to the 1977 data. To isolate the uptake/growth effects, all recurring parameter values were kept equal to those in VCQ-1. Figure 7A shows the model output (CCQ-nnl) if $f_n = 1$ in Equation 7, which corresponds to no nutrient limitation. Surprisingly, chl*a* predictions are quite close to the observed data, suggesting that nutrients may not have limited growth during much of the study period. To obtain this type of behavior using the CCQ nutrient-limitation equation (Table 1, Equation 8), the DIP level must always remain much larger than the half-saturation constant, $DIP_g$. Assuming that DIP concentrations are in the 1 µg/l or less range, we found it impossible with the CCQ model, using reasonable parameter values, to maintain DIP at sufficient

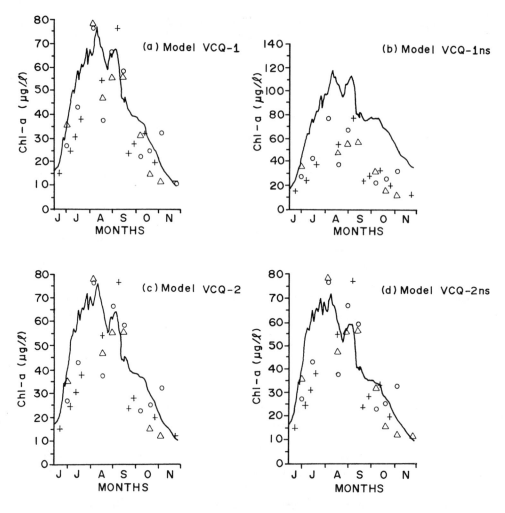

FIGURE 6.  VCQ model calibrations and responses to the elimination of sediment interaction. Chl*a* data were collected in basin I at three different spatial locations: + center; Δ western end; and ○ eastern end.

concentrations to provide $f_n$ ~1 during the entire summer. The best calibration that could be obtained using the CCQ model is shown in Figure 7B. In order to attain the maximum measured chl*a* values, it was necessary to use a higher maximum growth rate (Table 3). As a result, at the beginning of the modeling period when the dying off of the spring bloom released a significant amount of DIP into the water, growth was much more rapid than the data would indicate. After about 2 weeks, nutrient limitation occurred frequently, thereby reducing the growth. A comparison with the field data shows that this resulted in predicted algal levels that increased too quickly and decreased too slowly. A comparison of Figures 6 and 7 shows that the VCQ model experiences much less nutrient limitation during July than the CCQ model, because a high cell quota is able to sustain high growth rates, even during times of low DIP.

Figure 7C shows the result of running the CCQ model (CCQ-ns) without sediment interaction. This run suggests that the sediments act as a DIP source during the beginning of the model period and a sink during the later stages, which is similar to the conclusions of van Straten.[14] However, as shown in Figure 6, when a VCQ formulation is introduced into the model, the importance of this sediment interaction becomes questionable.

From a practical standpoint, eutrophication models are often used to predict changes in water quality for various management strategies. In the present study, the model behavior was

**TABLE 3**
**Model Parameter Values**

| Parameter | SIMBAL | VCQ-1 | VCQ-1ns | VCQ-2 | VCQ-2ns | CCQ-nnl | CCQ | CCQ-ns |
|---|---|---|---|---|---|---|---|---|
| $k_g$ (d⁻¹) | 6 | 2.13 | 2.13 | 2.22 | 2.22 | 1.72 | 3.0 | 3.0 |
| $T_c$ (°C) | 30 | 27.5 | 27.5 | 27.5 | 27.5 | 27.5 | 27.5 | 27.5 |
| $T_o$ (°C) | 26 | 23.5 | 23.5 | 23.5 | 23.5 | 23.5 | 23.5 | 23.5 |
| $\varepsilon_o$ (m⁻¹) | 3.2 | 3.2 | 3.2 | 3.2 | 3.2 | 3.2 | 3.2 | 3.2 |
| $\alpha$ (m⁻¹/µg PP/l) | 0.015 | 0.015 | 0.015 | 0.015 | 0.015 | 0.015 | 0.015 | 0.015 |
| $I_{sm}$ (cal/cm²) | 96 | 96 | 96 | 96 | 96 | 96 | 96 | 96 |
| $I_{se}$ (cal/cm²/°C) | 9.6 | 9.6 | 9.6 | 9.6 | 9.6 | 9.6 | 9.6 | 9.6 |
| $DIP_g$ (µg/l) | 10.2 | – | – | – | – | – | 0.001 | 0.001 |
| $k_d$ (d⁻¹) | 0.13 | 0.13 | 0.13 | 0.13 | 0.13 | 0.13 | 0.13 | 0.13 |
| $\theta_d$ | 1.14 | 1.2 | 1.2 | 1.2 | 1.2 | 1.2 | 1.2 | 1.2 |
| $v_s$ (m/d) | 0.036 | 0.036 | 0.036 | 0.036 | 0.036 | 0.036 | 0.036 | 0.036 |
| $\gamma$ | 0.4 | 0.4 | 0.4 | 0.4 | 0.4 | 0.4 | 0.4 | 0.4 |
| $k_m$ (d⁻¹) | 0.035 | 0.05 | 0.05 | 0.05 | 0.05 | 0.05 | 0.05 | 0.05 |
| $\theta_m$ | 1.18 | 1.18 | 1.18 | 1.18 | 1.18 | 1.18 | 1.18 | 1.18 |
| $k_{ex}$ (d⁻¹) | 0.16 | 0.25 | 0 | 0.16 | 0 | 0.25 | 0.25 | 0 |
| $DIP_{eq}$ (µg/l) | 5.8 | 2 | 2 | 2 | 2 | 2 | 2 | 2 |
| $k_u$ (µg DIP/µg PB/d) | – | 0.01 | 0.01 | 0.01 | 0.01 | – | – | – |
| $DIP_u$ (µg/l) | – | 2.0 | 2.0 | 6.0 | 6.0 | – | – | – |
| $q_{max}$ (µg PP/µg PB) | – | 0.02 | 0.02 | 0.02 | 0.02 | – | – | – |
| $q_{min}$ (µg PP/µg PB) | – | 0.001 | 0.001 | 0.001 | 0.001 | – | – | – |
| Chla/PP | 2 | – | – | – | – | 2 | 2 | 2 |
| PB/chla | – | 100 | 100 | 100 | 100 | – | – | – |

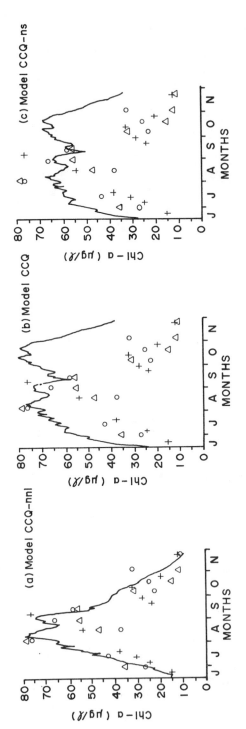

FIGURE 7.   CCQ model calibration and response to the elimination of sediment interaction and nutrient elimination. Chl*a* data were collected in basin I at three different spatial locations: + center; △ western end; ○ eastern end.

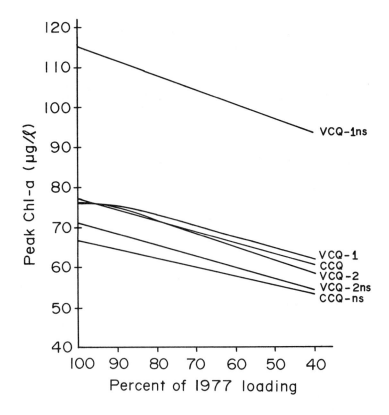

FIGURE 8.  Predicted peak chl*a* concentrations for reduced external phosphorus loading.

examined for reductions in the external nutrient loading. The peak chl*a* value was taken as a measure of the effect on water quality. Figure 8 shows the predicted responses. Each model predicts a very similar behavior, which is quite linear over the range of reductions examined. This is somewhat surprising, considering all the nonlinear dynamics included in the model equations, and since the VCQ models matched the dynamics of the chl*a* data much more closely than the CCQ models. In fact, a greater difference occurs in the responses of the two alternate calibrations of the VCQ model, VCQ-1 and VCQ-2, than between the VCQ and CCQ models. This suggests that both the CCQ and VCQ models may predict similar results for peak algal concentrations under load reductions, as long as they are calibrated to the same peak values. The inclusion or exclusion of sediments also seemed to have little effect.

### 4. Conclusions about the Application of a Water Quality Model in Lake Balaton

The work presented above suggests that a variable cell quota (VCQ) model formulation may give more reasonable results than a constant cell quota (CCQ) formulation when DIP concentrations are below 1 µg/l, as is typically observed in lakes. It is important to maintain a reasonable representation of the DIP concentration in the water if there is to be any hope of properly modeling a sediment interaction that is dependent on this DIP level. The use of a CCQ model under such circumstances requires an elevated maximum growth rate to produce significant growth at times of low phosphate concentrations, although this may cause too much growth at higher concentrations. The inclusion of VCQ expressions into a modified version of SIMBAL overshadowed the importance of the sediments and, in fact, made a sediment subroutine difficult or impossible to calibrate.

Load reduction runs for each of the models showed very similar, linear responses for peak chl*a* levels, regardless of the uptake/growth or sediment interaction formulation, as long as the peak value was accurately reproduced by the calibration. We must view the lack of significance of sediment interaction with some skepticism, however, as this result may be due to inadequacies in the formulation or calibration of this part of the model.

## III. LINKAGE OF WATER QUALITY AND HYDRODYNAMICS

### A. BACKGROUND

All lake water quality models are based upon the mass transport equation. This equation expresses in its three-dimensional form the conservation of mass for a dissolved or suspended substance in lake water as:

$$\frac{\partial c}{\partial t} = -u\frac{\partial c}{\partial x} - v\frac{\partial c}{\partial y} - w\frac{\partial c}{\partial z} + \frac{\partial}{\partial x}\left(e_x\frac{\partial c}{\partial x}\right) +$$

$$\frac{\partial}{\partial y}\left(e_y\frac{\partial c}{\partial y}\right) + \frac{\partial}{\partial z}\left(e_z\frac{\partial c}{\partial z}\right) + s \tag{28}$$

where $x, y$ are the horizontal direction coordinates; $z$ is the vertical direction coordinate; $t$ is time; $u, v, w$ are the fluid velocity components in the $x, y,$ and $z$ directions, respectively; $e$ is the turbulent diffusion coefficient in the direction indicated by its subscript; and $s$ represents the rate of net addition of mass per unit volume due to internal sources and sinks.

Equation 28 states that the change in concentration with time is equal to the change due to advective transport, turbulent diffusive transport, and addition or removal of mass by sources and sinks. Internal sources and sinks include biological, chemical, and physical reactions. External sources and sinks, which are incorporated in the boundary conditions to Equation 28, include the mass added by inflows or removed by outflows.

Two types of transport are considered in Equation 28: advective transport due to organized large-scale motion and diffusive transport due to small-scale turbulent fluctuation in accordance with the usual Fickian assumption. The separation of transport into advective and diffusive components is arbitrary, to the extent that it depends upon the time scale assumed to represent advection. Formally, turbulent diffusion is the residual transport that remains after averaging the extremely transient field of velocity and concentration over a short but finite time period. The time period implicit in Equation 28 is no longer than a few minutes.

If Equation 28 is averaged over one or more spatial dimensions, there is created another form of apparent transport, dispersion. Dispersion arises from spatial nonuniformities in velocity and concentration over the dimension or dimensions of averaging, and as such varies with the spatial structure of the lake model. Typically, dispersion is assumed to adhere to a Fickian-type proportionality with the coefficient of proportionality being the dispersion coefficient, $D$.

How the lake water quality model accounts for transport processes is intrinsically dependent upon the model's spatial structure. All lake water quality models must incorporate a statement of the conservation of mass. Formally, that statement arises when Equation 28 is averaged over time or space or both, with differences in the period of averaging (in time or space) leading to different separations between advective transport and diffusive or dispersive transport.

Shanahan and Harleman[5] have investigated the formulation of transport, as it depends upon the spatial structure of the lake water quality model. They identify two fundamental approaches to model construction: the multiple-box model, in which the lake is divided into a set of completely mixed volume elements, and the finite-difference model, in which the transport equation is solved as an approximately continuous function in one or more dimensions. The multiple-box model is implicitly a zero-dimensional formulation: each model box is assumed to be fully mixed and thus without dimensional structure. In contrast, the finite-difference model seeks to represent concentration as a continuous function over one or more spatial dimensions.

A specific outcome of the MIT research was a critical comparison of the multiple-box and finite-difference model formulations, leading to recommendations for their application. Shanahan and Harleman[5] present a theoretical analysis that compares these alternative model structures. They conclude that multiple-box models carry within their formulation a substantial implicit dispersion that is related to the number of boxes in the model. Implicit dispersion is similar, but not identical, to numerical dispersion as defined by Bella and Grenney.[25] As a consequence of this implicit dispersion, model dispersive transport cannot be determined directly from field measurements or other information on the actual lake. Rather, the modeler must treat interbox exchange flow as an empirical model parameter to be determined by calibration. Most importantly, if the model is constructed of box elements that are simply too large, the model will be intrinsically overdispersive and impossible to calibrate accurately. In contrast, with the finite-difference approach, numerical dispersion can be computed explicitly and controlled through the choice of the numerical solution technique and by varying the time step and spatial discretization of the model.

This section of the chapter presents the results of investigations of alternative model formulations for Lake Balaton as a case study on model spatial structure. Two models are contrasted: (1) a linked model in which hydrodynamic information from a two-dimensional depth-averaged flow model is used to specify transient flow and dispersion in a one-dimensional finite-difference water quality model, and (2) a box model employing previously calibrated exchange flows. The models both incorporate the SIMBAL phosphorus model described in Section II above. Results from the two models are compared with water quality observations from Lake Balaton and assessed in terms of the theoretical considerations developed by Shanahan and Harleman.[5]

## B. LINKED HYDRODYNAMIC-BIOGEOCHEMICAL MODEL
### 1. Model Construction

The linked hydrodynamic-biogeochemical model of Lake Balaton was constructed to solve Equation 28 in one dimension along the long axis of the lake. A one-dimensional model was employed for several reasons: it conforms well with the long and narrow geometry of the lake; nutrient sources are concentrated at one end of the lake, leading to pronounced longitudinal gradients in water quality constituent concentrations; and the one-dimensional framework is directly comparable with results from a previously developed four-box model. Moreover, insufficient field data exist to test a model's predictions of water quality variations across the lake. In the final analysis, the one-dimensional model was determined to satisfy all objectives in modeling the lake, but with much less computational expense than higher dimensional models.

To construct a one-dimensional model, Equation 28 must be averaged over the lake's cross-section. The resulting equation is

$$\frac{\partial \mathbf{C}}{\partial t} = -\frac{1}{A}\frac{\partial(Q\mathbf{C})}{\partial x} + \frac{1}{A}\frac{\partial}{\partial x}\left(DA\frac{\partial \mathbf{C}}{\partial x}\right) + \mathbf{N} + \mathbf{L} \tag{29}$$

where, $\mathbf{C}$ is a vector of cross-sectionally averaged phosphorus component concentrations, $A$ is the cross-sectional area, $Q$ is the cross-sectional average flow, $D$ is the longitudinal dispersion coefficient, $\mathbf{N}$ is a biogeochemical reaction vector, and $\mathbf{L}$ is a vector of loading per unit volume. This equation computes the change in phosphorus component concentration over time due to transport by advection and dispersion, and mass addition or removal by biogeochemical reaction and mass loading.

The linked water quality model of Lake Balaton brings together three main model components to solve Equation 29. The model components and their forcing functions are shown schematically in Figure 9. The model components are

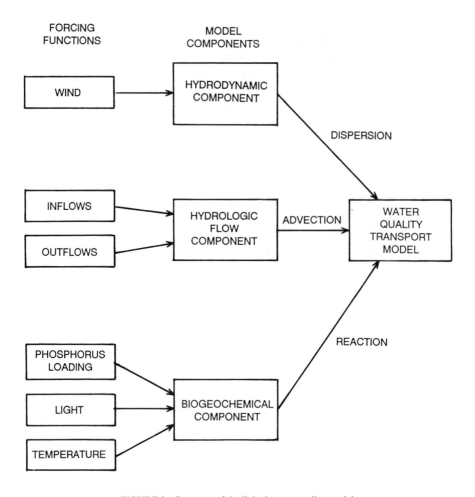

FIGURE 9.  Structure of the linked water quality model.

*Hydrodynamic Component* — The hydrodynamic component is a transient, depth-averaged circulation model that determines the two-dimensional horizontal flow pattern in the lake. This model component accounts for circulation variation on the time scale of hours and does not consider the much slower process of hydrologic flow.

*Biogeochemical Component* — The biogeochemical component is the SIMBAL model developed by van Straten[14] to simulate the phosphorus cycle in Lake Balaton (see Figure 3).

*Hydrologic Flow Component* — The hydrologic flow component is a water balance calculation that calculates monthly average flows through the lake.

Each of these three components furnishes a particular input to the water quality transport model: dispersion is computed in the hydrodynamic component, advection in the hydrologic flow component, and sources and sinks in the biogeochemical component. The following subsections describe each component in greater detail.

## 2. Hydrodynamic Component

The hydrodynamic model component is a two-dimensional depth-averaged lake circulation model developed by Shanahan and Harleman.[3,6] It is a fully transient model based on the linearized equations of motion, with a nonlinear boundary condition for bottom friction. Input to the model is a time-varying wind field. The model solves for displacement of the water surface and for flow in the two horizontal coordinate directions as functions of time and horizontal space. An explicit finite-difference technique is used to solve the equations. Full details on the hydrodynamic model may be found in Shanahan and Harleman.[3]

The hydrodynamic model was applied to Lake Balaton on a square-mesh finite difference grid with grid spacing $\Delta x = \Delta y = 1900$ m. This relatively coarse spacing was tested against results from a finer mesh grid and found to be adequate. Two model parameters, friction factors for the surface wind stress and bottom friction, required calibration. Calibration was accomplished in simulations of historical seiche events recorded by Muszkalay[9] and achieved similar parameter values as an independent one-dimensional model of the lake.[6] After calibration, the model was verified against an independent set of historical events.

The geometry of Lake Balaton led to an interesting application of the hydrodynamic model results in the linked water-quality model. Longitudinal advection predicted by the model was found to occur only as oscillatory seiche motion, with a very minor influence on the net mass transport. However, the oscillatory motion was found to be a significant factor in dispersion, owing to cross-lake differences in the flow velocity. Thus, the seiche creates negligible net transport along the lake, but significant cross-lake shear currents. These shear currents, together with cross-sectional mixing due to wind-induced secondary currents, give rise to a significant dispersive transport. The situation is complicated, however, by the intermittency of the seiche. Significant dispersion occurs on a sporadic basis following wind events.

The character of dispersive transport in Lake Balaton necessitated that the dispersion coefficient be computed as variable function of time. The coefficient was computed from the time history of horizontal flow determined by the hydrodynamic model. The method of calculation is an adaption of formulae developed by Fischer.[26] Fischer computed dispersion caused by transverse variations in velocity with cross-sectional transport by transverse turbulent diffusivity as

$$D_h \propto \frac{W^2 u_w^2}{\varepsilon_t} \tag{30}$$

where $D_h$ is the dispersion coefficient due to transverse velocity variations, $W$ is the width, $u_w$ is the maximum deviation of the velocity from the transverse mean velocity, and $\varepsilon_t$ is the transverse turbulent diffusivity. Dispersion in Lake Balaton was found to be dominated by lateral velocity variations with cross-sectional mixing by secondary (lateral) currents. Accordingly, the following proportionality was derived[3] following Fischer's approach:

$$D_w \propto \frac{W u_w^2}{v} \tag{31}$$

where, $D_w$ is the dispersion coefficient due to lateral variations and $v$ is the lateral velocity. Full details on the derivation of Equation 31 are given in Shanahan and Harleman.[3]

In applying Equation 31 in the Lake Balaton case study, the output time history from the hydrodynamic component directly furnished $v$ and could be processed to compute $u_w$. With these data and the known width of the lake, the dispersion coefficient was computed as a time-varying function at regular intervals along the lake corresponding to the model grid spacing $\Delta x$.

### 3. Biogeochemical Component

The biogeochemical component used in the linked model is the unmodified version of

SIMBAL (the CCQ model in Section II). The model is applied within the transport model finite-difference solution scheme to determine the time-varying distribution of phosphorus components as approximately continuous functions along the lake. In addition, the biogeochemical component provides for the import and export of nutrient loads due to inflow and outflows from the lake. Nutrient loadings employed in the model were based on estimates developed by Somlyódy and Jolankai.[11] The model inputs accounted for the longitudinal distribution of nutrient inflows along the lake, as well as the significant seasonal variation.

## 4. Hydrological Flow Component

The hydrodynamic model considers only the short-term (hourly) transient motion associated with wind-induced circulation; it does not include inflows and outflows to the lake. Thus there is neither net through-flow nor net variation in lake volume. In fact, the annual discharge at the Sió Canal equals roughly one fifth of the lake's volume, and the lake is typically 30 cm lower before spring runoff than it is after. The effects of these changes on transport in the lake were evaluated in a hydrological flow computation. The computation was simply a water balance, computed on a monthly basis, that accounted for average inflows and outflows, and direct precipitation and evaporation for each month. Flow through the lake was computed by apportioning flow to the model segments so as to achieve an equal change in elevation in each segment. The resulting net hydrological flows, computed for each segment for each month of the year, were used in the transport model as the advection term.

## 5. Numerical Solution

The one-dimensional mass transport equation, Equation 29, was solved in the linked model on a one-dimensional finite-difference grid with a grid spacing of 1900 m. The selected grid spacing corresponds to the longitudinal spacing of the hydrodynamic model and provides 40 computation points along the lake.

The linkage of the hydrodynamic, biogeochemical, and hydrological flow components combines processes occurring at radically different characteristics time scales. Each, in turn, gives rise to a limiting time step for solution of the equations by finite difference techniques. Unfortunately, conventional solution approaches would require that transient simulation be done at the most restrictive of these time steps — with severe cost implications for long-term simulations. In order to avoid these problems, a mixed-time step solution scheme based on the fractional step method was developed.[3]

The fractional step method, or time splitting, is the basis for the well-known alternating direction implicit (ADI) scheme that solves two-dimensional finite-difference problems in separate steps for each coordinate direction.[27-29] Verboom[30] developed a fractional step solution for the advection dispersion equation in which the solution was split not by coordinate directions, but simply into conveniently separated solution steps for advection and dispersion. Shanahan and Harleman[3] extended this approach to include different time steps for each of three solution steps for advection, dispersion, and reaction. The method is strictly accurate only for linear operations, but there is only minor approximation introduced in solving the nonlinear reaction step. The mixed time steps afford significant computational efficiencies by allowing much longer time steps for reaction than for advection and dispersion.

## 6. Multiple-Box Model

An alternative model structure to the linked model is the multiple-box model. In the application to Lake Balaton, model simulations were performed in parallel on both the linked model and the previously developed four-box model of the lake (Figure 10). Both models employed the same loading and hydrological flow data, and identical reaction components based on the previously developed SIMBAL model. Thus, the only differences between the models was their spatial structure and consequent formulation of mass transport.

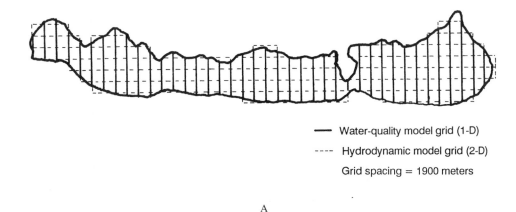

FIGURE 10.   Water quality model structure. (A) Linked water quality model; (B) four-box model.

## C. MODEL APPLICATION TO LAKE BALATON
### 1. Simulation Conditions

Evaluation of modeling approaches was accomplished through a series of simulations of Lake Balaton water quality for the year 1977. This particular year was selected because available field data satisfied the data input needs of the model and provided real-world information against which model results could be evaluated. The simulations were approached with the philosophy that they were to be experiments to evaluate the influence of hydrodynamics and model spatial structure on water-quality model predictions. Insufficient data and resources precluded detailed calibration (or reformulation) of the biogeochemical model in an attempt to duplicate field water-quality data. The approach taken was thus to produce a plausible model of the lake's behavior and concentrate on experiments that tested model spatial structure and hydrodynamic transport.

The hydrodynamic model was run using a continuous record of wind speed and direction averaged over 3-hour periods. To reduce computation costs, we took advantage of the nearly uniform pattern of wind over the year and simulated the circulation during the months of July and August only. This hydrodynamic information was then used repeatedly to create an artificial longer record for input to the water quality model.

The water quality model was run for the period beginning late February and ending in most cases at the end of October. The simulations were started with initial concentration distributions based on field data. Input to the model included the following major items:

- An artificial time history of the dispersion coefficient developed from the hydrodynamic model simulation described above,

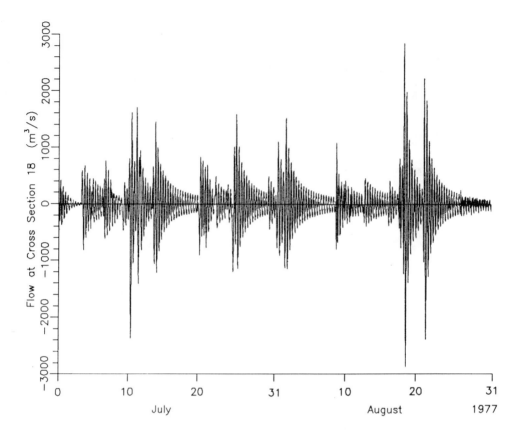

FIGURE 11.    Advective flow at midlake.

- Monthly average hydrologic flows computed from stream gauge records and tributary runoff data furnished by the IIASA Lake Balaton project, and

- The estimated 1977 phosphorus loads developed by Somlyódy and Jolankai,[11] which included a continuous record for the Zala River and monthly average inputs for other sources.

Finally, the model parameters used in the biogeochemical component were those determined previously by van Straten[14] during development of the SIMBAL model.

### 2. Hydrodynamic Model Results

Figure 11 shows the longitudinal water motion predicted by the model at a cross-section roughly midway along the lake. The oscillatory behavior of the advective flow is striking, but it should be clear that it creates no net transport and is not used as an input to the water quality model. Nonetheless, the higher advective velocities created during the larger seiche events lead to greater shear currents and more dispersion. Longitudinal dispersion is thus episodic, with negligible dispersion during calmer periods interspersed with occasional dispersion events.

The dispersion coefficient is computed according to the modified Fischer procedure (Equation 31). An example time history of dispersion is shown in Figure 12, while the longitudinal distribution of the average dispersion coefficient is shown in Figure 13. Dispersion is highest at sections of greatest change in shoreline geometry due to the larger secondary currents created at those locations. Zero dispersion is computed in the Strait of Tihany, because the hydrodynamic model is only one grid wide at that section.

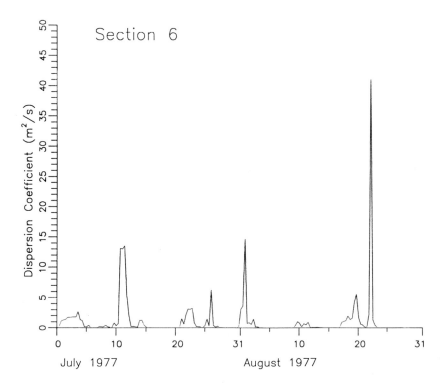

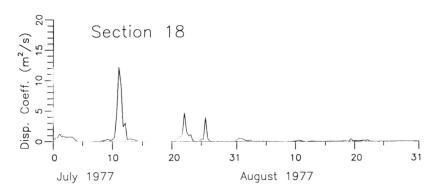

FIGURE 12.  Computed dispersion coefficient.

## 3. Water Quality Model Results

The influence of the hydrodynamic component on the linked water-quality model was evaluated in a series of simulations that employed different representations of dispersion. The large quantity of output generated by simulating 40 model grids over a period of 1 year forces us to be selective in the material we present. Results are generally presented as the spatial concentration profile of total and algal phosphorus along the lake predicted for August 4, a period of low flow when phytoplankton is near the peak summer concentration. In addition, some time histories of concentration at selected locations in the lake are also shown.

Figure 14a shows the base-case simulation results. The base case is the model application employing the time histories of dispersion computed as described in Section A.2 above. These results may be compared with measured concentrations in the lake, shown in Figure 2. The predicted total phosphorus concentrations are high compared with the field data, but the

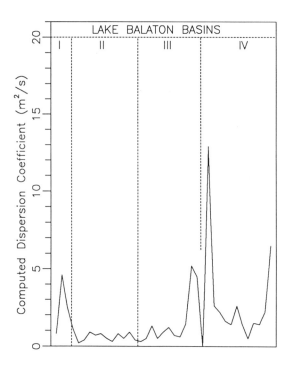

FIGURE 13.    Spatial distribution of computed dispersion coefficient.

character of the concentration profiles is generally similar. The model-field data agreement is adequate to satisfy the modeling program goal to be able to compare different model approaches, without necessarily achieving a full calibration.

The importance of the hydrodynamic influence is clearly illustrated in Figure 14b, a comparison of the base case with a simulation using only hydrological flow (that is, without dispersion). The latter simulation is the equivalent of a plug-flow reactor and is clearly unrealistic. Without the effects of dispersion, the hydrological flow-only simulation produces a jagged concentration profile, with sharp peaks in the locality of each nutrient source. This is particularly striking at the lake's eastern end, where there is a large sewage discharge.

Figure 14c shows results employing spatially and temporally constant dispersion coefficients. Constant dispersion coefficients are often used in water quality models in the absence of detailed field data. The simulation employing a constant dispersion coefficient of 1 m$^2$/s produces fair agreement with the base case. In contrast, the higher dispersion of 10 m$^2$/s is overly dispersive, removing all local concentration peaks and causing substantially greater mixing in basin IV.

Figure 14d gives a perspective on the differences associated with the dispersion coefficients. This simulation assesses the potential impact of reducing the Zala River nutrient load by one half. The simulation shows significant reductions in total phosphorus and algal phosphorus concentrations in Keszthely Bay. However, the predicted concentrations are approximately equal to the concentrations in Keszthely Bay predicted by the overly dispersive simulation using a dispersion coefficient of 10 m$^2$/s. The implications of this are that mixing may be as effective as load reduction in ameliorating the eutrophication in Keszthely Bay, but that an improper dispersion coefficient in the water quality model may mask the predicted effects of water quality improvements.

From these experiments with the linked model, we can conclude that the dispersion coefficient computed by the hydrodynamic model furnishes the most detailed predictions, but that a constant dispersion coefficient is a valid and useful approximation so long as it is a

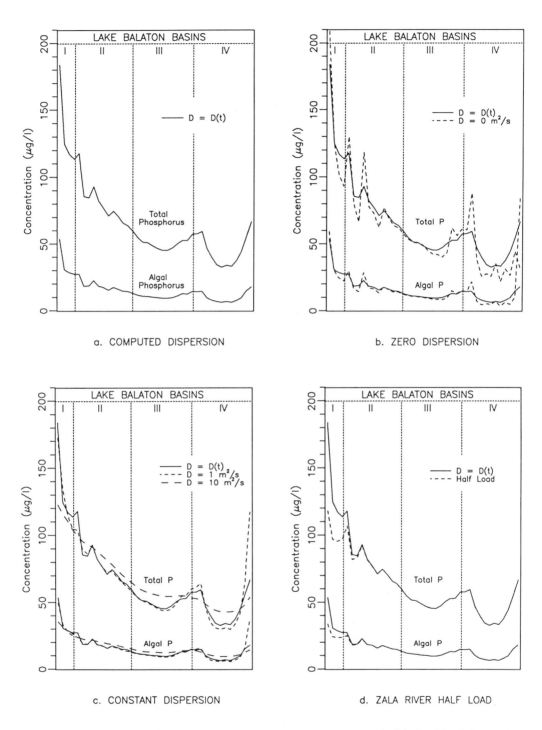

FIGURE 14.   Comparison of predicted spatial distribution of total and algal phosphorus by linked model variations. (a) Base case (computed dispersion coefficient); (b) zero dispersion; (c) constant dispersion; and (d) effect of 50% reduction in Zala River phosphorus load.

reasonable value. The chief value in using hydrodynamic information to compute a dispersion coefficient directly is that it ensures a value reasonably representative of the actual transport on the lake. In contrast, a reasonable dispersion coeffient is not guaranteed when it is determined by calibration along with other water-quality model parameters.

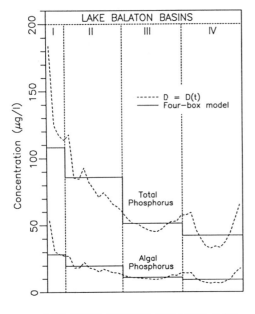

a. BOX MODEL WITHOUT EXCHANGE FLOW

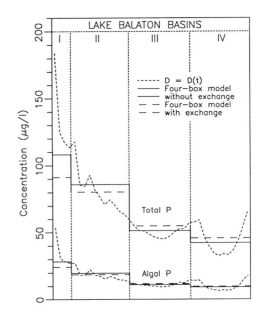

b. BOX MODEL WITH EXCHANGE FLOW

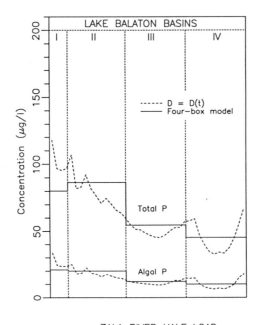

c. ZALA RIVER HALF LOAD

FIGURE 15.   Comparsion of predicted spatial distribution of total and algal phosphorus by linked and four-box models. (a) Four-box model without exchange flow; (b) four-box model with exchange flow; and (c) effect of 50% reduction in Zala River phosphorus load.

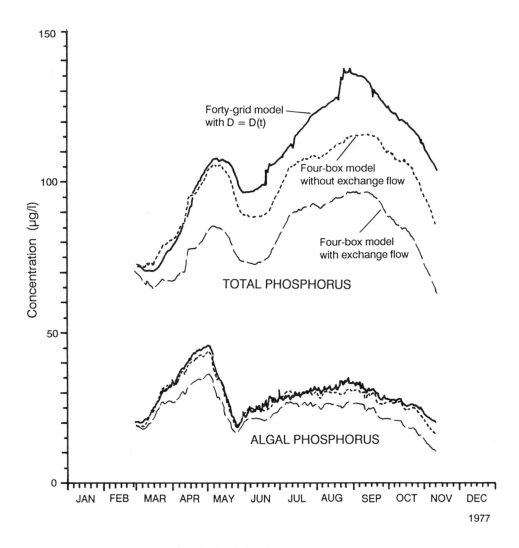

FIGURE 16. Predicted phosphorus concentrations vs. time.

## 4. Comparison with Box-Model Results

In Section A above, we discuss the potential limitations of multiple-box models owing to the problem of implicit dispersion. This hypothesis was further explored by comparing results from the 40-grid linked model with those from the previously developed four-box model. The spatial profile predicted by the four-box model is compared with the linked model base case in Figure 15a. Of course, the four-box model is unable to capture the spatial detail of the linked model. It also underpredicts total phosphorus concentrations in Keszthely Bay, a critical location from the standpoint of perceptible water quality problems. Nonetheless, the four-box model satisfactorily predicts the general character of the lakewide longitudinal gradients. Figure 15b contrasts the model predictions incorporating interbox exchange flows in the four-box model. The exchange flows were developed by van Straten[14] in an early calibration of the SIMBAL model. Inclusion of exchange flows smooths the concentration profile along the lake still more, substantially reducing the predicted peak concentration in Keszthely Bay.

Figure 16 addresses the temporal character of the model predictions by comparing concentration histories in Keszthely Bay at the western end of the lake. The linked model results are from grid 2, roughly the middle of Keszthely Bay. The four-box model without exchange flow predicts lower concentrations through the summer, owing to the greater longitudinal mixing caused by implicit dispersion. Nonetheless, this version of the four-box model adequately represents the concentration dynamics, particularly of algal phosphorus, the constituent of greatest interest. The same cannot be said of the four-box model with exchange flows: the peak algal concentration is underpredicted by about 20% and total phosphorus is always low.

The final analysis of the models' capabilities is their predictive abilities. Figure 15c compares the predicted phosphorus concentration profiles after reduction of the Zala River load to one half. The linked model predictions are the same as those in Figure 14d. The relative comparison is essentially the same as for the full load: there is a loss of spatial detail and generally lower predicted concentrations with the four-box model. The loss of spatial detail becomes important in evaluating future water quality. As the Keszthely Bay situation improves, other areas become relatively more important. For example, Szigliget Bay, the second peak in the linked model results, already has poor water quality, and predictions of its response to load reductions is important. That information is unavailable from the four-box model, however, and the four-box model is generally ineffective for examining water quality problems other than in Keszthely Bay.

## D. CONCLUSIONS

The results of this investigation show that multiple-box models can be useful and accurate tools for the evaluation of lake water quality. Box models are limited in their ability to consider spatial variations in water quality, but for many problems their predictive ability is adequate. However, if box models are constructed of inappropriately large volume elements, or with incorrect dispersive exchange flows, then they can be greatly in error and unsuitable for predictive purposes. Thus, there remains a substantial impetus for the modeler to consider the spatial structure of the multiple-box model carefully and to develop a correct representation of dispersive transport carefully. Unfortunately, multiple-box models constructed of large volume elements carry large implicit dispersion. This precludes direct determination of dispersive exchange flow from hydrodynamic information. Rather, exchange must be determined by calibration with a conservative tracer, such as chloride. For many situations, inadequate supporting data or the effort required to calibrate transport mitigate against a multiple-box formulation and favor a finite-difference approach, in which dispersion can be directly computed from the hydrodynamic properties of the lake.

# IV. MODELING SUSPENDED SEDIMENT

## A. INTRODUCTION
### 1. Suspended Sediments in Lake Balaton

During the course of our early studies in Lake Balaton, it became increasingly evident that the bottom sediments have a large role in determining the lake's water quality and in its potential for long-term recovery in the aftermath of reduced external nutrient loads. Three areas in which the sediments are particularly important are summarized below.

First, the presence of suspended sediment in the water column decreases light penetration so much that Secchi disk depths in the lake are frequently as little as 20 cm and rarely include the entire water column. In areas of high nutrient loading, e.g., the western end of the lake, it is likely that light often limits phytoplankton growth (see also Section II).

Second, in areas of high productivity, the sediments have a high organic content and are strongly reducing in nature, thereby consuming significant amounts of oxygen from the water column. This oxygen depletion results in episodic fish kills, again mainly at the eastern end of the lake. In this regard, it is quite fortunate that the wind is frequently strong enough to mix the entire water column, thereby supplying the sediment oxygen demand.

Third, the sediments are capable of acting as a source of internal nutrient loading. As much as 95% of the external phosphorus load to the lake is retained in the bottom sediments. Detrital phosphorus that settles to the lake bottom is mineralized at the sediment surface, converting it from biologically unavailable forms to forms that can be readily assimilated by algae. The mineralized phosphorus is released to the sediment pore water, where it is found at concentrations much greater than in the overlying water column. Lijklema et al.[31] report orthophosphorus concentrations in Lake Balaton sediments to be two orders of magnitude greater than in the overlying lake water. The expected diffusive flux of orthophosphorus into the overlying water is drastically reduced by adsorption of phosphorus onto solid phases — particularly iron, aluminum, and calcium minerals — which act as a trap, sealing off the sediments from active exchange with the water column.

This trap may be broken when the sediment is resuspended into the water column. Experimental work by Gelenscér et al.[31,32] has shown that phosphorus adsorption and desorption onto calcite and ferric hydroxide occurs rapidly and is highly dependent on the phosphorus concentration of the surrounding water and is somewhat dependent on pH. This suggests that if bottom sediment is resuspended during a storm event, a change in pH as well as ambient phosphorus concentration would cause phosphorus to be desorbed into the water. Gelenscér et al.[32] calculated that the desorption of phosphorus from sediments resuspended by even a moderate storm in Lake Balaton would be comparable in magnitude to the daily average external phosphorus load.

As a complementary effort to ongoing work into the effects of bottom sediment on water quality when it is introduced into the water column, our efforts focused on predicting the amount of sediment that was resuspended by typical storm events on the lake.

Recently the bottom sediments in Lake Balaton have been mapped extensively by Máté,[33] who has analyzed over 6000 sediment samples. He concludes that approximately half of the total bed material consists of fine-grained carbonate precipitates, while the other half enters the lake as a suspended load in the tributary waters. Particle size distributions show that, in general, the sediments fall into the clay and silt category, except near the lake's southern shore and the straits of Tihany, where coarser sediments prevail. Therefore, except in the areas noted above, it is expected that the bottom sediments are cohesive in nature.

## 2. Models of Cohesive Sediment Transport

The starting point for a model of suspended sediment is a three-dimensional mass transport statement that is analogous to that presented in Equation 28. In conservative form this is written as

$$\frac{\partial \bar{c}}{\partial t} + \frac{\partial}{\partial x}\left[\overline{uc}\right] + \frac{\partial}{\partial y}\left[\overline{vc}\right] + \frac{\partial}{\partial z}\left[(\bar{w} - w_s)\bar{c}\right] = -\frac{\partial}{\partial x}\left[\overline{u'c'}\right] - \frac{\partial}{\partial y}\left[\overline{v'c'}\right] - \frac{\partial}{\partial z}\left[\overline{w'c'}\right] + s \quad (32)$$

In Equation 32, overbars have been used to emphasize time-averaged quantities and primes to denote turbulent fluctuations from the time-averaged values. All other variables are as presented in Section III, with the addition of $w_s$, which is a particle settling velocity. If Equation 32 is applied to the total suspended sediment concentration, $w_s$ should be at least a weak function of time, since large particles settle faster than small ones, causing the particle size distribution to vary with time. Alternatively, Equation 32 can be applied to a particular sediment size class for which $w_s$ is the representative settling velocity. This requires writing a separate equation for each size class and, if the equations are coupled by particles coagulating or breaking apart to form different size classes, solving the resulting set of equations simultaneously.

Boundary conditions for Equation 32 at lateral boundaries are that the horizontal fluxes of material, $\left[\overline{uc} + \overline{u'c'}\right]$ and $\left[\overline{vc} + \overline{v'c'}\right]$, are equal to the external loading minus any outflow, while at the free surface the vertical flux, $\left[(\bar{w} - w_s)\bar{c} + \overline{w'c'}\right]$, is equal to the atmospheric loading, which is often taken to be zero. The major source of uncertainty for sediment transport studies is the boundary condition at the bottom of the water column. If the vertical turbulent flux is written in terms of the mean suspended sediment concentration gradient, i.e.,

$$\overline{w'c'} = -e_z \frac{\partial \overline{c}}{\partial z} \qquad (33)$$

the bottom boundary condition can be specified as a reference concentration at the bottom.[34,35] This is typically done in models of cohesionless sediment transport, where the reference concentration is taken to be the concentration in a bed load layer. There is evidence, however, that bed load transport does not exist over cohesive sediment beds.[36] Therefore, a relationship for the sediment flux between the water column and the bottom is used as the boundary condition. The bottom flux is defined as

$$\Phi \equiv -w_s \overline{c}_o + \overline{w'c'}\Big|_o \qquad (34)$$

where the subscript $o$ is used to denote quantities that are evaluated at the bottom. We note that, although it is not typically used, a flux boundary condition is also theoretically valid for the modeling of cohesionless suspended sediment transport.

Expressions for $\Phi$ must take into consideration the hydrodynamic forcing at the sediment bed, as well as the cohesive properties of the sediments. Unfortunately, it is extremely difficult to measure $\Phi$ directly, either in the laboratory or the field. Therefore, it is usually necessary to measure $\overline{c}$, use an assumed form for $\Phi$ to solve Equation 32, and then calibrate coefficient values by matching model predictions for $\overline{c}$ with measured values. Mechanistically, much of what has been learned about the erosion and deposition of cohesive sediments has come from the experimental investigation of E. Partheniades, A. Mehta, and associates at the University of Florida. They have concluded that, while erosion and deposition of cohesionless sediments are simultaneous processes, erosion and deposition of cohesive sediments are exclusive processes.[36] Given sufficient time, a cohesive sediment bed will erode to the depth where the bed strength (often expressed in terms of a critical stress, $\tau_c$) is equal to the erosive force (or the stress, $\tau$) applied to the bed.[37] They have found that $\tau_c$ increases with distance into the bed for a flow-deposited cohesive sediment bed and that the profile is affected by the time since the bed was created; the make-up of the bed, including mineralogy, particle size, and organic content; the temperature; the chemistry of the overlying water and the pore fluid; bioturbation by benthic organisms; and the presence of cementing agents such as iron oxide or mucus excreted by benthic organisms.[38] Deposition does not occur if $\tau > \tau_c$. When it does occur, deposition is a function of the initial concentration of sediment in suspension, the physicochemical properties of the sediment in suspension, and the ratio of the applied stress to a minimum stress required to maintain any sediments in suspension.[39]

Two consequences of the fact that the erosion and deposition of cohesive sediments are separate and very different processes are important. First, the steady-state suspended sediment concentration, $c_{ss}$, reached after a period of erosion at a constant applied shear stress will not be equal to $c_{ss}$ if an initial suspension of the same sediment is allowed to deposit under the same applied stress.[40] Second, when a steady state exists in the suspended sediment concentration, it indicates both the erosive flux and the depositional flux are equal to zero. Both of these conclusions contradict the behavior of cohesionless sediments for which $c_{ss}$ is the same regardless of whether it is preceded by net deposition or erosion and at steady-state deposition exactly balances erosion resulting in a zero net flux.

An expression for $\Phi$ can be obtained by a simple manipulation of Equation 34. Defining

$$c_{ss} \equiv \frac{\overline{w'c'}\big|_o}{\beta} \tag{35a}$$

$$\beta \equiv w_s \frac{\bar{c}_o}{\tilde{c}} \tag{35b}$$

$$\tilde{c} \equiv \frac{1}{h} \int_o^h \bar{c}\, dz \tag{35c}$$

where $\tilde{c}$ is the depth-averaged suspended sediment concentration over the water depth $h$, allows Equation 34 to be written as

$$\Phi = \beta\big(c_{ss} - \tilde{c}\big) \tag{36}$$

Equation 35b indicates that $\beta$ is equal to the settling velocity multiplied by a factor that depends on the vertical distribution of suspended sediment in the water column. (For very small particles, the effects of Brownian motion should also be included in $\beta$.[41]) Laboratory evidence suggests that in fresh water at the relatively dilute suspended sediment concentrations representative of many field conditions (e.g., less than 300 to 500 mg/l), the particle size distribution, and therefore $w_s$, are not affected either by the suspended sediment concentration or by the hydrodynamic forcing.[40,42] When mixing in the water column is fast enough for the suspended sediment concentration to be uniform with depth, $\beta$ is equal to $w_s$. $\beta$ increases in magnitude as the mixing rate decreases.

The results of the laboratory erosion tests using deposited cohesive sediment beds suggest that it is reasonable to relate $c_{ss}$ and $\tau$ by a power law of the form:[8]

$$c_{ss} = K\big(\tau - \tau_{co}\big)^n \tag{37}$$

where $K$ and $n$ are empirical parameters and $\tau_{co}$ is a minimum critical shear stress required to erode any sediments. Assuming that the bed has a critical stress that increases with depth, $\tau_{co}$ must correspond to the critical shear stress at the surface of the bed. Since $K$, $n$, and $\tau_{co}$ define the strength profile of the bed, there is no reason to believe that they are universal constants. Rather, they will vary in time as the bed consolidates and from bed to bed as the physicochemical and biological conditions change.[38,40]

Laboratory deposition tests[39,43] have suggested that a rather complicated and highly empirical log-normal expression relates $c_{ss}$ with the initial concentration, a minimum stress, and the applied stress. The use of this expression for $c_{ss}$ in place of Equation 37 during deposition requires the specification of several additional parameters, along with a criterion for when to switch from one expression to the other. Had we not obtained success using a simpler assumption (discussed below), we would have pursued this approach further.

There are few field data sets available with enough detail that changes in the suspended sediment concentration can be used to make judgements about the validity of specific expressions for $\Phi$. Krone[44] collected suspended sediment concentrations at seven positions in the vertical over a little more than a tidal cycle in the Savannah River. His plots of concentration vs. time are reasonably symmetric about the times of maximum and minimum velocity, and therefore do not show the result of any characteristic difference in $\Phi$ during erosion or deposition. Lavelle et al.[45] collected 10 days of velocity measurements and transmissometer-derived suspended sediment concentrations 5 m above the bottom of Puget Sound. They were reasonably successful in modeling these data using a one-dimensional (in the vertical) transport equation, together with a flux boundary condition equivalent to Equations 36 and 37, with $\tau_{co} = 0$ and the parameters $K$, $n$, and $\beta$ held constant over the entire period. Both of these results contradict the behavior expected from cohesive sediments.

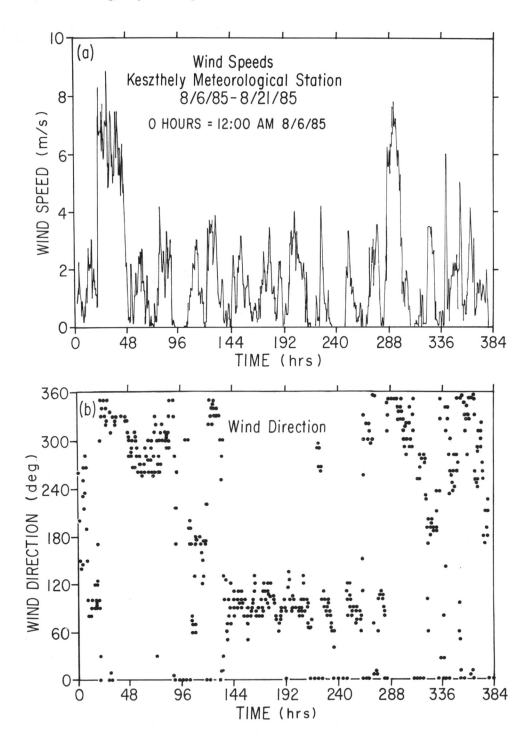

FIGURE 17.   30-min average wind speed and direction measured at the Keszthely meteorological station.

## B. LAKE BALATON FIELD STUDY

A field study was conducted in 2-m deep water about 300 m from the western end of the lake (see Figure 1) from August 6 through August 21, 1985. Bottom material collected at the site showed the sediments to consist principally of clay and fine silt, with 61% of the particle size

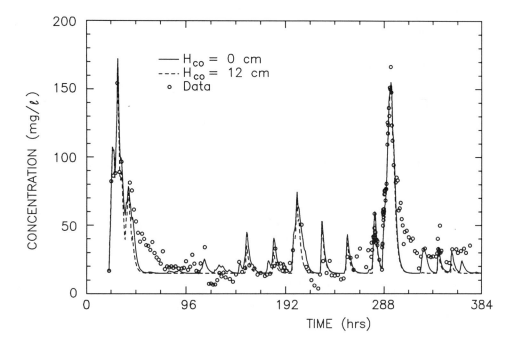

FIGURE 18.    Suspended sediment concentrations at the Keszthely field site.

distribution falling within the range of 8 to 18 μm. The suspended sediment concentration was determined gravitimetrically from bottle samples collected at mid-depth, typically every 2 h, although at times samples were collected as often as every 10 min. Wind speed and direction were recorded continuously at a land-based meteorological station about 500 m from the site. For part of the time, additional data were collected from instruments mounted on a tripod that was deployed at the study site. These data included wind velocities 2 m above the water surface and three-dimensional water velocities 24 and 84 cm above the bottom, using two BASS current meters.[46] A complete description of the instrumentation system can be found in Luettich.[8]

Plots of wind speed and direction from the meteorological station (Figure 17) show that two major storm events (beginning at 24 h and 280 h, respectively) occurred during the study period, each having northeast winds with 30-min averaged speeds of 7 to 9 m/s. These events were responsible for increases in suspended sediment from a background level of about 15 mg/l to maximum concentrations greater than 150 mg/l (Figure 18).

The tripod-based system recorded data for a little more than 60 h from 20:00 on August 15 through 8:00 on August 18, (hours 235 to 296 in Figures 17 and 18), the final 10 h of which corresponded to the latter storm event. Current meter measurements were decomposed into mean water velocities and wave-induced water velocities.[8] Linear wave theory was used to transfer the vertical velocity spectrum from the uppermost measurement height to a wave height spectrum at the surface, from which the significant wave height and the equivalent velocity period were computed. The equivalent velocity period is defined to be the period that, when combined with the significant wave height, gives the same root mean square bottom orbital velocity as would be measured by a current meter mounted at the bottom. In general, the equivalent velocity period was about 10% less than the mean period. The results are presented in Figure 19 and show three intervals during the 60 h that wave activity was significant — two small events when the wind was blowing along the long axis of the lake and the big storm when the winds came from across the lake. In each case the wave period was remarkably constant, with a value of about 2 s.

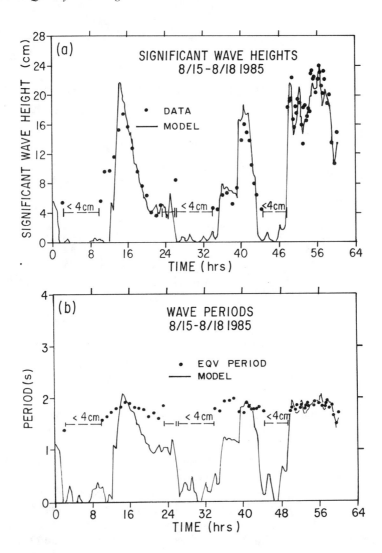

FIGURE 19.   Significant wave heights and equivalent velocity periods at the Keszthely field site.

Water velocities associated with the mean water motion (not shown) reached a maximum speed at the lower BASS of 12 cm/s during the storm event and were directed parallel to the shoreline during virtually the entire 60 h of measurement.

For modeling purposes, it is illustrative to compare the shear stresses exerted on the sediment bed by the mean current and the surface waves. The bottom stress associated with the mean current is related to the current friction velocity, $U_{*\text{curr}}$, using

$$\tau_{\text{curr}} = \rho U^2_{*\text{curr}} \tag{38a}$$

The current friction velocity can be determined from the mean current velocity measured at the lower current meter and the assumption that a logarithmic boundary layer extends to this elevation.

$$\frac{U_{\text{curr}}(z)}{U_{*\text{curr}}} = \frac{1}{0.4} \ln\left[\frac{zU_{*\text{curr}}}{\nu}\right] + 5.5 \tag{38b}$$

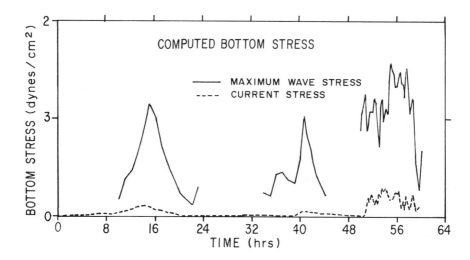

FIGURE 20. Comparison between the computed current-induced bottom stress and the computed maximum wave-induced bottom stress at the Keszthely field site. Gaps in the wave stress indicate intervals of negligible waves.

where $v$ is the kinematic viscosity. Equation 38b assumes a hydrodynamically smooth bottom, since bedforms were not observed during any of the dives made during the study period.

The maximum bottom stress during a wave cycle can be computed from

$$\tau_{\text{wave}} = \frac{f_w}{2} \rho U^2_{\text{wave}} \tag{39}$$

where $U_{\text{wave}}$ is the maximum bottom orbital velocity associated with the waves and $f_w$ is the wave friction factor as found in Jonsson.[47]

Figure 20 compares $\tau_{\text{wave}}$ and $\tau_{\text{curr}}$ for the 60 h of data and shows clearly that $\tau_{\text{wave}}$ dominates $\tau_{\text{curr}}$, typically by more than an order of magnitude. This occurs because bottom shear is generated by the velocity gradient in the boundary layer. The mean current boundary layer has a characteristic period that is on the order of hours, and therefore it can grow to a thickness comparable to the depth of the water column. However, because the surface waves have periods of only a few seconds, the wave boundary layer does not have a chance to grow to more than a few millimeters in depth. As a result, the same bottom stress can be generated by wave orbital velocities that are much smaller than the mean current velocity. Since the wave orbital velocity and the mean current velocity are typically of the same order of magnitude in Lake Balaton, it is reasonable to neglect the stress due to the mean current, in comparison to that due to the waves when specifying the forcing responsible for eroding bottom sediments.

## C. COMPONENTS OF THE SUSPENDED SEDIMENT MODEL

An examination of the time and length scales associated with the hydrodynamic forcing and the sediment properties[8] suggest that horizontal transport can be neglected at the Keszthely field site, except possibly during prolonged periods of deposition (see below). Vertical profiles collected in Keszthely Bay,[48] near Szemes,[49] and near Tihany[8] all indicate that suspended sediment concentrations are relatively uniform vertically in the water column. Physically, this is due to the small particle sizes in suspension and the lake's shallowness, which allows turbulence to be nearly uniform over the depth. The lack of significant vertical concentration gradients makes it convenient to model the depth-averaged suspended sediment concentration.

Equation 32 can be integrated over the depth, assuming no horizontal transport, no sediment flux at the free surface, and the unspecified flux $\Phi$ at the bottom, to give

$$\frac{d\tilde{c}}{dt} = \frac{\Phi}{h} \tag{40}$$

Due to a lack of data regarding temporal changes in the size distribution of suspended particles, multiple size classes were not explicitly included in the present model.

## 1. Formulation and Calibration of Sedimentation Flux

In this study Equations 31 and 32 were used with constant parameter values to specify $\Phi$ both during erosion and deposition. Based on the laboratory results reviewed above, this form of a relationship for $\Phi$ implies that the sediment bed is noncohesive. However, there is a lack of evidence from the few existing field data sets that cohesive effects strongly influence suspended sediment concentrations. Moreover, the proposed formulation limited the quantity of parameters in the model to a number that might reasonably be calibrated and verified using the available data.

As shown in Figure 18, the suspended sediment concentration rarely dropped below 15 mg/l. This "background" concentration is attributed mostly to very small inorganic particles, as well as various plankton species that were at major bloom levels during the course of the measurement period. Introducing both a background concentration and Equations 36 and 37 into Equation 40 results in the following equation:

$$\frac{d\tilde{c}}{dt} = -\frac{\beta}{h}\left(\tilde{c} - \tilde{c}_{bak}\right) + \frac{\beta K}{h}\left(\tau - \tau_{co}\right)^n \tag{41}$$

where $\tilde{c}_{bak} = 15$ mg/l.

The comparison presented in Figure 20 suggests that wave-induced bottom stress is most appropriate for inclusion in Equation 41. Using linear wave theory and the assumption of a laminar wave boundary layer, the maximum wave-induced bottom stress is linearly related to the wave height, $H$, as follows:

$$\tau = H\left[\rho \frac{\sqrt{v\omega^3}}{2\sinh kh}\right] \tag{42}$$

where $\omega$ is the radian wave frequency and $k$ the radian wave number. Equation 42 can be used to simplify Equation 41, yielding

$$\frac{d\tilde{c}}{dt} = -\frac{\beta}{h}\left(\tilde{c} - \tilde{c}_{bak}\right) + \frac{\beta\theta}{h}\left(H - H_{co}\right)^n \tag{43}$$

where

$$\theta = K\left[\rho \frac{\sqrt{v\omega^3}}{2\sinh kh}\right]^n \tag{44}$$

and $H_{co}$ is a minimum critical wave height to erode any sediments. Since our observations suggest that the wave period, and therefore $\omega$ and $k$, were relatively constant in time, assuming that $K$ and $n$ in Equation 41 are constant in time is equivalent to assuming that $\theta$ and $n$ in Equation 43 are constant in time. In the present study, Equation 43 was used to model suspended sediment.

## TABLE 4
### Acceptable Parameter Combinations from Model Calibration

| Mean square error | | 2559—3327 | $(mg/l)^2$ | 30%[a] |
|---|---|---|---|---|
| $c_{bak}$ | = | 15.0 | mg/l | |
| $H_{co}$ | = | 0.0—16.75 | cm | |
| $\beta$ | = | 0.015—0.031 | cm/s | |
| $n$ | = | 0.15—3.95 | | |
| $\theta$ | = | 0.00086—124.6 | | |

[a]Variation from lowest mean square error.

*Note:* Specific parameter combinations within the above listed ranges yield mean square errors between 2559 and 3327.

The model was calibrated using the observed significant wave heights and suspended sediment concentration values obtained during the second major storm event. To do this it was assumed that the mid-depth measurements were representative of the depth-averaged concentrations.

Calibration was begun by systematically varying the parameters $\theta$, $H_{co}$, $\beta$, and $n$ over a wide range of possible values and recording the sum of the mean square error (MSE) between the model predictions and the observed suspended sediment concentrations. Plots of the MSE vs. each parameter value were U-shaped, indicating the model behaved optimally over a single range of values for each parameter. Using narrower parameter ranges that were determined from the initial calibration step, a second set of calibration runs was conducted, systematically varying each parameter value using a more refined increment size. While it is tempting to select the single parameter set that gave the lowest MSE from these runs and call it the optimal model calibration, that would not allow for the possibility that measurement errors exist in the observed data values. To allow for this possibility, all parameter sets giving MSE values within 30% of the lowest MSE were considered to be equally acceptable.

Table 4 lists the range in parameter values that were found to give acceptable values of MSE. If we assume a vertically well-mixed water column, $\beta \approx w_s$. The values of $\beta$ listed in Table 4 correspond to equivalent particle diameters, assuming Stoke's settling, of 14 to 19 μm and are in excellent agreement with the particle sizes measured in bottom sediment samples.

The wide range in acceptable parameter values that resulted from the calibration runs indicates that either the model is insensitive to variations in one or more of the parameters or that too many degrees of freedom exist in the model. It was possible to shed some light onto these issues by examining covariances between different pairs of parameters. As a result of this excercise, it was found that

1. $n$ and $\theta$ were closely correlated through the relationship

$$n = -0.67 \log \theta + 1.8; \qquad R^2 = 0.987 \tag{45}$$

2. $\beta$ was not correlated with any of the other parameter values and therefore was assigned an average value of 0.022 cm/s, and
3. $n$ and $H_{co}$ were well correlated through the relationship

$$\log n = -0.040 H_{co} + 0.48; \qquad R^2 = 0.996 \tag{46}$$

(To reduce the scatter in Equation 46 this equation is based only on parameter sets in which $\beta = 0.022$ and $n$ and $\theta$ were related via Equation 45.)

Substituting $\beta = 0.022$ and Equations 45 and 46 into the model Equation 44 gives

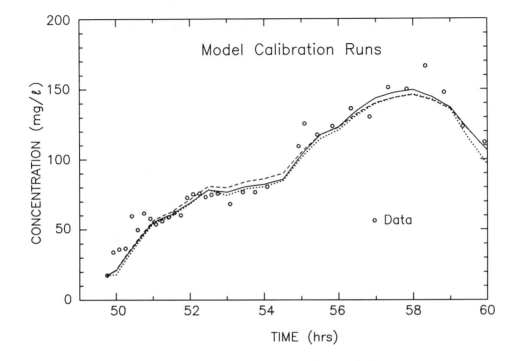

FIGURE 21.   Comparison between the model and calibration data for three parameter sets:

$n = 3;$    $\theta = 0.0151;$    $\beta = 0.022$ cm/s;   $H_{co} = 0$ cm;    MSE = 2560

$n = 1.75;$   $\theta = 1.19;$    $\beta = 0.022$ cm/s;   $H_{co} = 5.92$ cm;   MSE = 2820

$n = 0.88;$   $\theta = 23.5;$    $\beta = 0.022$ cm/s;   $H_{co} = 13.4$ cm;   MSE = 3330

$$\frac{d\tilde{c}}{dt} = -\frac{0.022}{h}(\tilde{c}-15) + \frac{11}{h}\left[\frac{H-H_{co}}{31}\right]^{3\times10^{-0.04\,H_{co}}} \tag{47}$$

where $H_{co}$ is in centimeters. Equation 47 contains one free parameter, $H_{co}$. Unfortunately, based on the data available for calibration, it was not possible to select a unique value of $H_{co}$. Rather, the model remains calibrated for a range of values, as shown in Figure 21 and Table 5.

### 2. Formulation and Calibration of the Wave Model

The use of the suspended sediment model, either in its more general form Equation 41 or in the simplified form of Equation 43 requires information about the surface waves. This information was generated in the present study using the shallow-water modifications made to the SMB method.[50] The significant wave height and period are given for fetch-limited waves by the empirical equations:

$$\frac{gH_s}{W_a^2} = 0.283 \tanh \alpha \cdot \tanh\left[\frac{\gamma}{\tanh \alpha}\right] \tag{48}$$

$$\frac{gT}{W_a} = 2.8\pi \tanh \zeta \cdot \tanh\left[\frac{\delta}{\tanh \zeta}\right] \tag{49}$$

$$\alpha = 0.530\left(gh/W_{10}^2\right)^{0.75} \tag{50a}$$

$$\zeta = 0.833\left(gh/W_{10}^2\right)^{0.375} \tag{50b}$$

**TABLE 5**
**Summary of Results for Varying $H_{co}$ with Calibration Data**

| $H_{co}$ (cm) | $c_{bak}$ (mg/l) | $\beta$ (cm/s) | $n$, Equation 46 | $\theta$, Equation 45 | MSE (mg/l)$^2$ |
|---|---|---|---|---|---|
| 0.0 | 15 | 0.022 | 3.02 | 0.01479 | 2650 |
| 1.0 | 15 | 0.022 | 2.75 | 0.03704 | 2563 |
| 2.0 | 15 | 0.022 | 2.51 | 0.08554 | 2621 |
| 3.0 | 15 | 0.022 | 2.29 | 0.1835 | 2704 |
| 4.0 | 15 | 0.022 | 2.09 | 0.3682 | 2767 |
| 5.0 | 15 | 0.022 | 1.90 | 0.6947 | 2802 |
| 6.0 | 15 | 0.022 | 1.74 | 1.240 | 2816 |
| 7.0 | 15 | 0.022 | 1.59 | 2.102 | 2823 |
| 8.0 | 15 | 0.022 | 1.45 | 3.403 | 2831 |
| 9.0 | 15 | 0.022 | 1.32 | 5.280 | 2841 |
| 10.0 | 15 | 0.022 | 1.20 | 7.881 | 2849 |
| 11.0 | 15 | 0.022 | 1.10 | 11.36 | 2851 |
| 12.0 | 15 | 0.022 | 1.00 | 15.85 | 2888 |
| 13.0 | 15 | 0.022 | 0.91 | 21.48 | 3113 |
| 14.0 | 15 | 0.022 | 0.83 | 28.34 | 3756[a] |
| 15.0 | 15 | 0.022 | 0.76 | 36.49 | 4904[a] |
| 16.0 | 15 | 0.022 | 0.69 | 45.95 | 8183[a] |
| 17.0 | 15 | 0.022 | 0.63 | 56.70 | 18271[a] |

[a] MSE error greater than 30% range in Table 4.

$$\gamma = 0.0125\left(gF/W_{10}^2\right)^{0.42} \tag{50c}$$

$$\delta = 0.077\left(gF/W_{10}^2\right)^{0.25} \tag{50d}$$

where $g$ is the acceleration of gravity, $F$ is the effective fetch,[50] $h$ is the water depth, and $W_{10}$ is the wind speed measured 10 m above the water surface. In applying Equations 48 to 50, it was assumed that the waves were in local equilibrium with the wind, i.e., that they travel in the direction of the wind and that the local depth is appropriate for use in Equations 50a and 50b. This assumption is reasonable throughout much of Lake Balaton because of the very gradual changes in water depth. Wind speeds were adjusted up to the required 10 m using a logarithmic velocity profile, together with the drag coefficient formula suggested by Wu.[51]

The model results for the 60 h of data are shown in Figure 19. The observed wave heights are reproduced quite well, both during the two early periods when the winds were blowing along the lake's long axis and during the storm when the winds were oriented across the lake. The slight overprediction near the beginning of the two events aligned with the lake's axis suggests that the waves were initially duration rather than fetch limited. Attempts were made to include duration limitation in the model.[8] However, no consistent improvement was found in model predictions, and therefore the fetch-limited equations were retained. During the storm the nonsteady nature of the observed wave heights is reproduced very well by the model due to the short cross-lake fetch and therefore the short time required to reach fetch-limited conditions.

Unfortunately, the wave periods were not as well predicted. In order to make the predicted period match the observed period during the main part of each wind event, the leading coefficient in Equation 49 was adjusted from its value of 2.4 in the original publication[50] to 2.8. A more noticeable discrepancy arises because the observations tended to maintain a nearly constant period of about 2.0 s, while the model showed a much more dynamic behavior. From this standpoint, a more accurate representation of the observations results from assuming a constant period equal to the average value predicted by the model during the main part of each wind event.

**TABLE 6**
**Summary of Results for Varying $H_{co}$ with Verification Data**

| $H_{co}$ (cm) | $c_{bak}$ (mg/l) | $\beta$ (cm/s) | $n$, Equation 46 | $\theta$, Equation 45 | MSE (mg/l)$^2$ |
|---|---|---|---|---|---|
| 0.0 | 15 | 0.022 | 3.02 | 0.0148 | 34400 |
| 2.0 | 15 | 0.022 | 2.51 | 0.0855 | 34400 |
| 4.0 | 15 | 0.022 | 2.09 | 0.368 | 34300 |
| 6.0 | 15 | 0.022 | 1.74 | 1.24 | 34300 |
| 8.0 | 15 | 0.022 | 1.45 | 3.40 | 34800 |
| 10.0 | 15 | 0.022 | 1.20 | 7.89 | 37000 |
| 12.0 | 15 | 0.022 | 1.00 | 15.9 | 40500 |
| 13.0 | 15 | 0.022 | 0.91 | 21.5 | 43800 |

This behavior was the basis of our use of Equation 43 rather than Equation 41 to model the suspended sediment data.

### 3. Model Verification

The combined wave and suspended sediment model was verified using the 15-d record of wind speeds (Figure 17) and suspended sediment concentrations (Figure 18). The results of verification runs for two different values of $H_{co}$ are presented in Figure 18. In general, the model does an excellent job of reproducing the observed concentrations, both during the two major storms and the smaller wind events, using the entire $H_{co}$ range that was found to give acceptable model calibrations. A summary of the verification results are given in Table 6. The only systematic errors between the model prediction and the observations occurred during the deposition periods following the two major storms when model predictions decreased more rapidly than the observed data. We can identify three potential causes for these discrepancies within the framework of the suspended sediment model that has been developed. These are horizontal transport, bed cohesiveness, and differential settling. The implications of each are briefly discussed.

The relatively long period of time associated with particle settling after a large storm event suggests that if horizontal transport did effect the suspended sediment concentrations observed at the field site, it most likely occurred during these times. Assuming a mean horizontal advective velocity of 10 cm/s, a horizontal movement of 1 km requires about 3 h. A visual inspection of Figure 18 suggests that a time scale for settling on the order of 10 h might be appropriate. In this case it would be possible for sediment to travel roughly half way across Keszthely Bay before settling. Since both the forcing and the bed sediment properties vary somewhat over this spatial scale-horizontal transport may account for this behavior. Unfortunately, without a synoptic measurement program at several spatial locations, it is not possible to do much more than speculate about the importance of this factor.

As discussed previously, the formulation for $\Phi$ used in this study does not capture a number of the characteristic behaviors that laboratory experiments have shown are associated with the erosion and deposition of cohesive sediment beds, e.g., time-varying values for $\theta$, $H_{co}$, and $n$; deposition as a function of the initial concentration; and different values of $c_{ss}$ during erosion and deposition. Nonetheless, the simple model performed surprisingly well. It is possible that the neglect of some of these processes, and in particular the use of the same expression for $c_{ss}$ during erosion and deposition, caused the overprediction during post-storm deposition periods. However, it is not likely that this is the entire reason, because the model matches the observations quite closely during the initial stages of deposition and only deviates after about 60% or more of the sediment has left suspension.

The final possibility, differential settling, would seem to be the most likely to cause the observed post-storm model discrepancies. Since larger particles settle faster than smaller ones, the average settling velocity is expected to decrease in time during a prolonged period of deposition, which is exactly what appears to occur. While it may be possible to reproduce the observed suspended sediment behavior by explicitly including several different size fractions in the model, given the fact that there is already one free parameter in the model, we feel unjustified in introducing multiple-size fractions into the present model. To prove conclusively that differential settling is the cause of the model discrepancy, it would be necessary to measure the average settling velocity of the suspended particles during and after the storm. This is extremely difficult, if not impossible, to accomplish, since coagulation changes particle sizes in water samples that are saved for any length of time before analysis, and the use of on board-settling devices, such as an Owen tube, was made impossible by the rough weather experienced during the storm.

We note that if the reduced settling rate is taken into account, the model agrees much more closely with the data during the small resuspension events that occurred after the second storm.

## D. CONCLUSIONS

The results of this investigation show that episodic increases in the suspended sediment concentration in Lake Balaton are forced by wind-generated surface waves. It is possible to do a good job of modeling the significant wave height using Equation 48, while an average wave period during substantial wind events could be obtained from Equation 49. The suspended sediment concentration can also be modeled quite convincingly using a model based on wave-induced bottom stress, that has four parameters which can be calibrated. After removing parameter covariances, one free parameter remained. A unique value could not be selected for this parameter based on either the calibration data or a 15-d data set used for model verification. It is often suggested that this type of model behavior indicates that there are too many parameters in the model and that one or more parameter can be eliminated. In the present case, eliminating $H_{co}$ is equivalent to setting it equal to zero. We believe that this is not really eliminating $H_{co}$ but rather assigning it a value that may not have a reasonable physical meaning. Therefore, the model is left with the free parameter. The only systematic model deviation from observed data was toward the end of periods of prolonged deposition and is most likely a result of the constant particle-size distribution that is assumed in the model.

While the sediment-size distribution and properties suggest that a model that includes the processes of sediment erosion and deposition must take into account the cohesive nature of the sediment bed, the success of the present model contradicts this idea. Rather, the fact that its parameter values are constant in time and that it predicts a $c_{ss}$ value that is equal during both net erosion and deposition suggests that the sediments were noncohesive. The reasons for this behavior are not clear. It is possible that our use of a concentration measurement made at the mid-depth as being equivalent to the depth-averaged concentration introduces enough error into the model to obscure the cohesive effects. However, the model's success over the 15-d verification period would tend to argue against this. A second possibility is that while the major portion of the sediment bed is cohesive, there may be a thin layer of "fluff" or sediment particles that are kept from becoming a part of the cohesive bed. A mechanism to prevent cohesion to the bed is the bottom shear exerted by the mean current created by the nearly continuous seiching motion of the lake.

The suspended sediment model developed herein is simple enough that it can easily be integrated into a water quality model. However, we do not expect that parameter values will remain constant if this model is applied in a different part of the lake over a different type of sediment. Therefore, to be used in a water quality model of the entire lake, several separate sets of parameter calibrations must be made. There are also locations in the lake where the assumption that the waves are in local equilibrium with the wind and that horizontal sediment transport is negligible cannot be justified.

At present there are efforts underway to include suspended sediment models into water quality models both at VITUKI for Lake Balaton in Hungary and in the Netherlands for several of the shallow Dutch lakes. This should lead to water quality models that more completely represent the processes determining the water quality of shallow lakes.

## ACKNOWLEDGMENTS

The authors are indebted to many co-workers for their suggestions, ideas, and reviews. Particularly, we wish to thank Dr. Lázslo Somlyódy and his co-workers at VITUKI, Dr. Ference Máté and his co-workers at the Balaton Limnological Research Institute, and Richard Baker for his work on the comparison of phosphorus model formulations. This research was funded by a series of grants from the National Science Foundation.

## REFERENCES

1. **Somlyódy, L. and van Straten, G.,** *Modeling and Managing Shallow Lake Eutrophication with Application to Lake Balaton,* Springer-Verlag, Berlin, 1986.
2. **Luettich, R. A., Jr. and Harleman, D. R. F.,** A comparison of water quality models and load reduction predictions, in *Modeling and Managing Shallow Lake Eutrophication with Application to Lake Balaton,* Somlyódy, L. and van Straten, G., Eds., Springer-Verlag, Berlin, 1986, 323.
3. **Shanahan, P. and Harleman, D. R. F.,** Linked Hydrodynamic and Biogeochemical Models of Water Quality in Shallow Lakes, Report Number 268, Ralph M. Parsons Laboratory for Water Resources and Hydrodynamics, Department of Civil Engineering, Massachusetts Institute of Technology, Cambridge, MA, 1982.
4. **Shanahan, P. and Harleman, D. R. F.,** A linked hydrodynamic and biogeochemical model of eutrophication in Lake Balaton, in *Analysis of Ecological Systems: State-of-the-Art in Ecological Modeling,* Lauenroth, W. K., Skogerboe, G. V., and Flug, M., Eds., Elsevier, New York, 1983, 837.
5. **Shanahan, P. and Harleman, D. R. F.,** Transport in lake water quality modeling, *ASCE J. Environ. Eng.,* 110, 42, 1984.
6. **Shanahan, P., Harleman, D. R. F., and Somlyódy, L.,** Wind-induced water motion, in *Modeling and Managing Shallow Lake Eutrophication with Application to Lake Balaton,* Somlyódy, L. and van Straten, G., Eds., Springer-Verlag, Berlin, 1986, 204.
7. **Shanahan, P. and Harleman, D. R. F.,** Lake eutrophication model: a coupled hydrophysical-ecological model, in *Modeling and Managing Shallow Lake Eutrophication with Application to Lake Balaton,* Somlyódy, L. and van Straten, G., Eds., Springer-Verlag, Berlin, 1986, 256.
8. **Luettich, R. A., Jr.,** Sediment Resuspension in a Shallow Lake, Sc.D. thesis, Massachusetts Institute of Technology, Cambridge, MA, 1987.
9. **Muszkalay, L.,** Characteristic Water Motions in Lake Balaton (in Hungarian), VITUKI, Budapest, Hungary, 1966.
10. **Herodek, S.,** Phytoplankton changes during eutrophication and P and N metabolism, in *Modeling and Managing Shallow Lake Eutrophication with Application to Lake Balaton,* Somlyódy, L. and van Straten, G., Eds., Springer-Verlag, Berlin, 1986, 183.
11. **Somlyódy, L. and Jolánkai, G.,** Nutrient loads, in *Modeling and Managing Shallow Lake Eutrophication with Application to Lake Balaton,* Somlyódy, L. and van Straten, G., Eds., Springer-Verlag, Berlin, 1986, 125.
12. **Leonov, A. V.,** Transformations and Turnover of Phosphorus Compounds in the Lake Balaton Ecosystem, 1976-1978, Working Paper WP-82-27, International Institute for Applied Systems Analysis, Laxenburg, Austria, 1982.
13. **Leonov, A. V. and Vasiliev, O.,** Simulation and analysis of phosphorus transformations and phytoplankton dynamics in relation to the eutrophication problem of Lake Balaton, in *Progress in Ecological Engineering and Management by Mathematical Modeling,* Dubois, D. M., Ed., Editions CEBEDOC, Liege, Belgium, 1981, 627.
14. **van Straten, G.,** Hypothesis testing and parameter uncertainty analysis in simple phytoplankton-P models, in *Modeling and Managing Shallow Lake Eutrophication with Application to Lake Balaton,* Somlyódy, L. and van Straten, G., Eds., Springer-Verlag, Berlin, 1986, 287.
15. **Kutas, T. and Herodek, S.,** A complex model for simulating the Lake Balaton ecosystem, in *Modeling and Managing Shallow Lake Eutrophication with Application to Lake Balaton,* Somlyódy, L. and van Straten, G., Eds., Springer-Verlag, Berlin, 1986, 309.

16. **van Straten, G.,** Lake eutrophication models, in *Modeling and Managing Shallow Lake Eutrophication with Application to Lake Balaton,* Somlyódy, L. and van Straten, G., Eds., Springer-Verlag, Berlin, 1986, 35.

17. **Tóth, L.,** Feeding behaviour of *Daphnia cucullata* SARS in the easily stirred up Lake Balaton as established on the basis of gut content analyses, *Arch. Hydrobiol.,* 101, 531, 1984.

18. **Rigler, F. H.,** Radiobiological analysis of inorganic phosphorus in lake water, *Verh. Int. Verein. Limnol.,* 16, 465, 1966.

19. **Lean, D. R. S.,** Movements of phosphorus between its biologically important forms in lake water, *J. Fisheries Res. Board Can.,* 30, 1525, 1973.

20. **Dobolyi, I. and Herodek, S.,** On the mechansim reducing the phosphate concentration in the water of Lake Balaton, *Int. Rev. Ges. Hydrobiol.,* 65, 339, 1980.

21. **Istvánovics, V., Herodek, S., and Entz, B.,** The effect of sediment dredging on the water quality in Lake Balaton II: The phosphorus and nitrogen cycles (in Hungarian with an English summary), *Hidrol. Közl.,* 63, 125, 1983.

22. **DiToro, D. M. and Matystik, W. F., Jr.,** Mathematical Models of Water Quality in Large Lakes. Part 1: Lake Huron and Saginaw Bay. Report No. EPA-600/3-80-056, U.S. Environmental Protection Agency, Washington, D.C., 1979.

23. **Wang, M.-P. and Harleman, D. R. F.,** Hydrothermal-Biological Coupling of Lake Eutrophication Models, Report Number 270, Ralph M. Parsons Laboratory for Water Resources and Hydrodynamics, Department of Civil Engineering, Massachusetts Institute of Technology, Cambridge, MA, 1982.

24. **Herodek, S., Vörös, L., and Tóth, F.,** The mass production of phytoplankton and the eutrophication of Lake Balaton III. The Balatonszemes basin in 1976–77 and the Siófok basin in 1977 (in Hungarian with an English summary), *Hidrol. Közl.,* 62, 220, 1982.

25. **Bella, D. A. and Grenney, W. J.,** Finite-difference convection errors, *ASCE J. Sanitary Eng. Div.,* 96, 1361, 1970.

26. **Fischer, H. B., List, E. J., Koh, R. C. Y., Imberger, J., and Brooks, N. H.,** Mixing in Inland and Coastal Waters, Academic Press, New York, 1979, 125–130.

27. **Gourlay, A. R. and Mitchell, A. R.,** The equivalence of certain alternating direction and locally one-dimensional difference methods, *SIAM J. Numerical Anal.,* 6, 37, 1969.

28. **Gourlay, A. R. and Mitchell, A. R.,** A classification of split difference methods for hyperbolic equations in several space dimensions, *SIAM J. Numerical Anal.,* 6, 62, 1969.

29. **Gourlay, A. R. and Mitchell, A. R.,** On the structure of certain alternating direction implicit (ADI) and locally one-dimensional (LOD) difference methods, *J. Inst. Math. Appl.,* 9, 80, 1972.

30. **Verboom, G. K.,** The advection-dispersion equation for an an-isotropic medium solved by fractional-step methods, in *Mathematical Models for Environmental Problems,* Brebbia, C. A., Ed., John Wiley, New York, 1976.

31. **Lijklema, L., Gelenscér, Szilágyi, F., and Somlyódy, L.,** Sediment and its interation with water, in *Modeling and Managing Shallow Lake Eutrophication with Application to Lake Balaton,* Somlyódy, L. and van Straten, G., Eds., Springer-Verlag, Berlin, 1986, 156.

32. **Gelenscér, P., Szilágyi, F., Somlyódy, L., and Lijklema, L.,** A Study on the Influence of Sediment in the Phosphorus Cycle of Lake Balaton, Collaborative Paper CP-82-44, International Institute for Applied Systems Analysis, Laxenburg, Austria, 1982.

33. **Máté, F.,** Bottom sediment investigations in Lake Balaton, unpublished report, Balaton Limnological Research Institute, Tihany, Hungary, 1986.

34. **Rouse, H.,** Modern conceptions of the mechanics of fluid turbulence, *Trans. ASCE,* 102, 463, 1937.

35. **Ippen, A. T.,** A new look at sedimentation in turbulent streams, *J. Boston Soc. Civil. Eng.,* 58, 131, 1971.

36. **Partheniades, E.,** Unified view of wash load and bed material load, *ASCE J. Hydraulics Div.,* 103, 1037, 1977.

37. **Mehta, A. J., Parchure, T. M., Dixit, J. G., and Ariathurai, R.,** Resuspension potential of deposited cohesive sediment beds, in *Estuarine Comparisons,* Kennedy, V. S., Ed., Academic Press, New York, 1982, 591.

38. **Parchure, T. M.,** Erosional behavior of deposited cohesive sediments, Ph.D. thesis, University of Florida, Gainesville, FL, 1984.

39. **Mehta, A. J., and Partheniades, E.,** An investigation of the depositional properties of flocculated fine sediments, *ASCE J. Hydraulic Res.,* 13, 361, 1975.

40. **Lee, D., Lick, W., and Kang, S. W.,** The entrainment and deposition of fine-grained sediments in Lake Erie, *J. Great Lakes Res.,* 7, 224, 1981.

41. **Lick, W.,** Entrainment, deposition, and transport of fine-grained sediments in lakes, *Hydrobiologia,* 91, 31, 1982.

42. **Krone, R. B.,** Flume Studies of the Transport of Sediment in Estuarial Shoaling Processes, Final Report, Hydraulic Engineering Laboratory and Sanitary Engineering Research Laboratory, University of California, Berkeley, CA, 1962.

43. **Partheniades, E.,** Results on recent investigations on erosion of deposition of cohesive sediments, in *Sedimentation,* Shen, H. W., Ed., Water Resources Publications, Littleton, CO, 1972, chap. 20.

44. **Krone, R. B.,** A Field Study of Flocculation as a Factor in Estuarial Shoaling Processes, Technical Report No. 19, Committee on Tidal Hydraulics, U.S. Army Corps of Engineers, Vicksburg, MS, 1972.

45. **Lavelle, J. W., Mofjeld, H. O., and Baker, E. T.,** An in situ erosion rate for a fine-grained marine sediment, *J. Geophys. Res.*, 89, 6543, 1984.

46. **Williams, A. J., III,** BASS, an acoustic current meter array for benthic flow-field measurements, *Mar. Geol.*, 66, 345, 1985.

47. **Jonsson, I. J.,** Wave boundary layers and friction factors, in *Proc. Tenth Conference on Coastal Engineering*, Tokyo, Japan, 1966, 127.

48. **Györke, D.,** Formation of suspended sediment concentration in shallow lakes and reservoirs (in Hungarian), Report 7411, 2-323, VITUKI, Budapest, Hungary, 1978.

49. **Somlyódy, L.,** Preliminary study on wind induced interaction between water and sediment for Lake Balaton (Szemes Basin), in *Second Proc. Joint MTA/IIASA Task Force Meeting on Lake Balaton Modeling*, van Straten, G., Herodek, S., Fischer, J., and Kovács, I., Eds., Veszprém, Hungary, 1980, 26.

50. CERC, *Shore Protection Manual*, Vol. 1, U.S. Army Coastal Engineering Research Center, Ft. Belvoir, VA, 1977.

51. **Wu, J.,** Wind-stress coefficients over sea surface from breeze to hurricane, *J. Geophys. Res.*, 87(C12), 9704, 1982.

52. **Luettich, R. A., Jr. and Harleman, D. R. F.,** A comparison between measured wave properties and simple wave hindcasting models in shallow water, *J. Hydraul. Res.*, 28, 299, 1990.

53. **Luettich, R. A., Jr., Harleman, and D. R. F., and Somlyódy, L.,** Dynamic behavior of suspended sediment concentrations in a shallow lake perturbed by episodic wind events, *Limnol. Oceanogr.*, 35, 1046, 1990.

Chapter 4

# LAKE PLESHEYEVO — A CASE STUDY

## A. A. Voinov

## TABLE OF CONTENTS

# I. INTRODUCTION: THE LAKE AND ITS ENVIRONMENT

About 150 km to the north of Moscow there is one of the "pearls" of the Golden Ring of Russia — the town Pereslavl-Zalessky. The town stands on Lake Plesheyevo facing it with its monasteries, convents, and churches. The lake is known as the birthplace of the Russian fleet: on Lake Plesheyevo Tsar Peter the Great built his "amusement" fleet, which was the beginning of the Russian Navy. Besides its historical and recreational value, the lake is the habitat of an endemic species of lake herring, Pereslavl cisco (*Coregonus albula pereslavicus Borisov*), known as the Tsar's herring since as early as 1668, fishing of this species was prohibited by a special government law.

The lake (Figure 1) has an oval form, 9 km in length, 5 km in width, with an area of about 50.8 km$^2$. Its maximum depth is 26 m and the area of its watershed is 375 km$^2$. Nineteen small tributaries enter the lake, the largest of which, the River Trubezh, has a discharge of 2.0 to 4.8 m$^3$/s. There is one outflow, River Vyoksa, which eventually brings the lake waters to the Volga River.

The first complex studies of the lake ecosystem were carried out in the 1920s by R. Rossolimo, who analyzed the hydrological, hydrothermal, and chemical features of the lake. At that same time the planktonic community was studied. More recently, the water balance of the lake was studied by V. Rotmistrov. An overview of the lake budget is presented in Table 1.

For approximately 134 days each year, the lake is covered with ice. Starting from the beginning of June, a thermocline is formed at a depth of 8 to 12 m, with water temperatures of 12 to 15°C above it and 5 to 8°C below. This stratification usually extends to mid-September, maximum temperatures of 17 to 22°C being reached in July.

At present the lake ecosystem is being impacted by pollution from various urban sources, associated with the inflow of the River Trubezh, and also directly from the sewage of the town of Pereslavl. A number of agricultural farms are the other sources of non-point-source pollution to the lake. From 1970 to 1980 the application of fertilizers increased by a factor of 20, the load of herbicides and pesticides being a factor of 6 and 2.5 greater, respectively. For several years, besides nutrients, the Trubezh River carried discharges from a chemical plant containing heavy metals and toxic elements. Luckily in the 1970s these discharges were transferred outside the watershed. However, this cut off 27% of the Trubezh catchment area, diminishing the total inflow to the lake.

Another anthropogenic impact is concerned with the hydrological regime of the lake. A number of wells have been drilled to provide Pereslavl and its industry with fresh water. They take 13,000 m$^3$/d of water from the ground, thus changing the ground water inflow to the lake. Moreover, clean water is pumped directly from the lake metalimnion. Since ground waters contribute significantly to the water balance and affect the thermal regime, these impacts may turn out to be crucial for the ecosystem, even more so because there is still a trend towards increasing withdrawals of water. As a result of these impacts, together with the global rainfall trends after the 1950 to 1963 period of increasing water levels, from 1963 to 1976 a gradual fall in water level was recorded. In order to compensate for the losses, the mouth of the Vyoksa was dammed. This slightly changed the residence time and the water level increased by 17 cm.

Alterations to the lake hydrology, together with the mounting nutrient load, affected the trophic state of the ecosystem. Blooms of blue-green algae were observed and obvious symptoms of eutrophication were reported. The problems of the lake received much press attention: central and local newspapers discussed the future of the ecosystem. It was mentioned that the species content of algae was changing, indicating that the lake was changing from a mesotrophic to a eutrophic state. Anoxic conditions were observed below 10 to 15 m and winter fish kills were indicated for the first time in the lake history. The fishery was badly damaged, the total fish productivity being reduced 3 to 3.5 times.

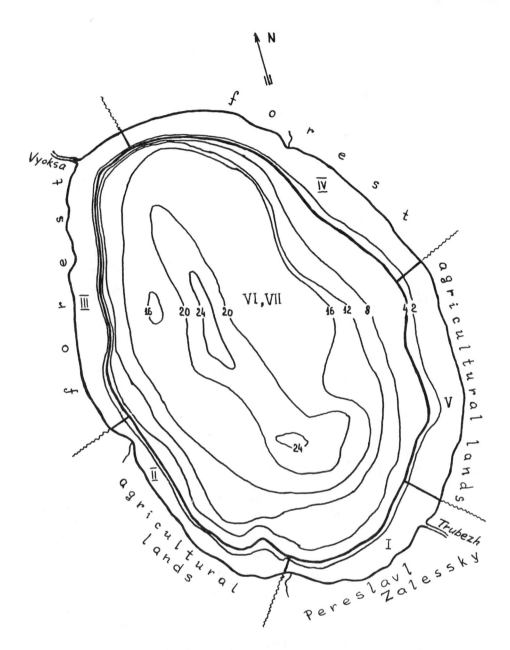

FIGURE 1. Lake Plesheyevo, its watershed, morphometry, and spatially homogeneous segments.

In 1978 a research program was started on the lake by the Institute of Biology of Internal Waterbodies, U.S.S.R. Academy of Sciences. Highly intensive and complex investigations were started in 1982, and at that same time mathematical modeling of the lake ecosystem was included into the research program. On average, monthly observations were made for the lake hydrochemistry and biology. Phytoplankton, production, decomposition, zooplankton, bacteria, macrophytes, etc. were measured. Hydrology and nutrient load were studied. All the results were summarized in an extensive monograph.[10] Unfortunately, the monitoring of various ecosystem components was not initially organized within the framework of a single research project. As a result only the 1983, 1984, and some of the 1985 data turned out to be suitable for the purposes of simulation modeling.

**TABLE 1**
**Lake Plesheyevo Water Balance**

| | | |
|---|---|---|
| Volume | ($10^6$ m$^3$) | 558.96 |
| Inflow | ($10^6$ m$^3$/year) | % |
| Surface runoff | 83 | 44 |
| Groundwaters | 80 | 42 |
| Rainfall | 26 | 14 |
| Total | 189 | |
| Outflow | | |
| Evaporation | 24.5 | 13 |
| Underground | 4.5 | 2 |
| Surface | 160 | 85 |
| Total | 189 | |
| Residence time (years) | | 2.99 |

The simulation model consists of seven segments: a hydrodynamic block simulates the wind-induced currents in order to calculate the material exchange between segments. The ecological block models the dynamics of 21 variables, representing the main biotic component and material cycles in the ecosystem. The model is formulated in terms of the Simulation Modeling System for Aquatic Bodies (SIMSAB), which is a conversational software system designed for simulation modeling of various aquatic ecosystems. In the following section we give a brief overview of SIMSAB and its features. Then the general structure of the model and its variables are discussed. In Section IV the ecological block of the model is described. The links between ecosystem components are formalized and the limiting factors specified. Section V presents the hydrodynamic block of the model and the monitor, which brings together the dynamics of variables in different lake segments, considered as ecologically homogeneous regions. The following sections are concerned with model analysis. The standard procedure of model calibration and verification is described in Section VI, and the model's sensitivity, robustness, and stability are analyzed. Numerous scenarios are then presented and some conclusions about the ecosystem buffer capacity are drawn. A number of variables turn out to be redundant in quasi-stationary conditions, which allows us to develop a simplified version of the model for long-term analysis. In Section VIII a simplified model of the lake — based on the first, more complex — model is outlined and a special procedure is suggested, which combines simplified and detailed models to simulate long-term scenarios with possible structural changes in the ecosystem.

## II. SIMSAB IDEOLOGY

Methods of simulation modeling not only serve the purposes of quantitative processing of information for forecasts and decision making, but also provide a means of interpretation of experiments to formulate and verify hypotheses about operational principles for ecosystems and their separate units. Perception of methods of simulation modeling by mathematically deficient natural sciences — biology, chemistry, ecology — is very much impeded by a lack of an interface, which could, just as in the case of statistics, provide workers ignorant of mathematics and programming with a means to apply mathematical methods of analysis, allowing them to build and modify mathematical models independently.

SIMSAB is an attempt to create such an interface. Being only loosely linked to postulates of a particular school of mathematical modeling, the modeling interface is flexible enough to cover models of different kinds. Those principles that make up the invariant part of the system are listed below and they are open for criticism.

### A. SPATIAL STRUCTURING OF ECOSYSTEMS
It is assumed that the whole water body can be split into a number of ecologically

homogeneous segments (compartments). Within each of the segments, ecosystem components are presented as spatially averaged variables (numbers, concentrations). It should be noted that such a structuring is quite natural for a practical worker, who usually deals with data measured on a fixed network of stations. In this case it is only natural to think that each of the stations (especially if it produces data significantly different from those observed at the adjacent ones) describes an ecologically uniform region and thus represents a certain segment of the model.

## B. TEMPORAL DISCRETENESS OF PROCESSES

It is assumed that a sequence of events can be distinguished on the time axis, each event standing for some processes realized in the ecosystem. These processes are assumed to be realized instantly, and no processes occur in between. Then the functioning of an ecosystem can be described as follows: at a given time, $t$, boundary segments receive a nutrient load; next, for instance, at time $t + dt$, concentrations of various components are altered due to material transfer along trophic chains or in hydrobiochemical cycles (ecological block); then at time $t + 3dt$, the wind-induced currents are formed, determining the water exchange between segments (hydrodynamical block); next at time, say, $t + 7dt$, the whole water body is mixed up by water exchange and diffusive fluxes between segments, correspondingly changing the concentrations of ecological components; then at time $t + 10dt$, again some nutrient load enters the water body, and so on. Note that, generally speaking, the contents of each of the events is arbitrary.

## C. VARIABLES, PARAMETERS, AND FORCING FUNCTIONS

The model is specified by a number of variables, which represent those ecosystem components in whose dynamics we are interested, expressible as a time series. Model parameters are the constants that appear in the formalizations of interactions between variables. This is satisfactory if the parameters are measured, or at least estimated, in experiments. However, it is more commonly the case that most of them are selected during model calibration.

Forcing functions specify the time-variable effects that are independent of the ecosystem. These may be climatic factors, anthropogenic impacts, or, in short, all the functions that affect the ecosystem with no feedback from it. Obviously the set of forcing functions is defined by the boundaries of the modeled object in the representation of the researcher: by varying these boundaries and hence the cause-effect relationships, forcing functions can be turned into variables and vice versa.

## D. DYNAMICS OF ECOSYSTEM VARIABLES

Within a segment, the dynamics of variables is described by a system of ordinary differential equations. In order to formulate them, one should specify their right-hand sides. In SIMSAB-88 this is performed in terms of a special flow language. Generally speaking, the right-hand sides can be formulated in a file as a user-written FORTRAN program. Applying the SIMSAB methods, then the right-hand sides are calculated as sums of flows, specifying the material transfers from one ecosystem component into another. These flows may be described either in terms of SIMSAB, using its library of functions, or as fragments of a FORTRAN program. In any case, SIMSAB results in a number of FORTRAN subroutines that calculate the right-hand sides of systems of differential equations and present various segments of the water body. The user who does not know any FORTRAN is restricted in his model formulations by the SIMSAB *library of functions,* which incidentally includes all the most widely used functions.

The SIMSAB monitor brings together all the subroutines, including those that compute the wind-induced currents and the material exchange between segments, and those that provide an input of data and an explanation of results. The general structure of SIMSAB is presented in Figure 2.

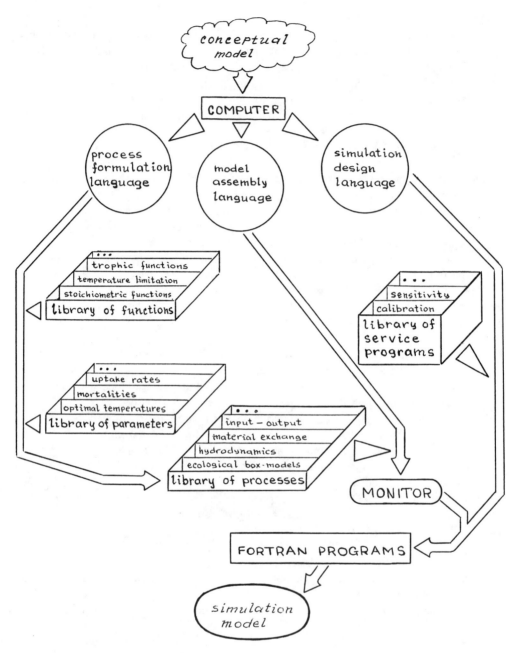

FIGURE 2.  General structure of SIMSAB.

## III. SPATIAL REPRESENTATION AND VARIABLES

The general heterogeneity of biotic and abiotic characteristics of Lake Plesheyevo can be more or less accurately represented within the framework of seven segments (boxes), which will be considered to be ecologically homogeneous. First we split the lake with respect to depth, separating the littoral zone, which is thought to be no more than 4 m deep, from the rest of the lake — the pelagic zone. Next, taking into account the major pollution sources in the watershed as well as the area and species content of the macrophyte belt, we split the littoral zone into five segments (Figure 3). Since the lake is rather deep and there is a stable stratification throughout

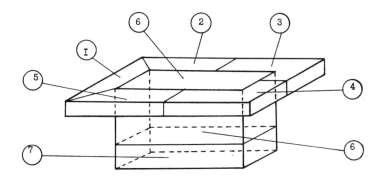

FIGURE 3. Segments of the lake model.

the summer, the pelagic zone is further divided into two segments: (1) the epilimnion plus the metalimnion, above the thermocline at a depth of 10 m, and (2) the hypolimnion, below the thermocline.

Like other natural ecosystems, in Lake Plesheyevo there is a great variety of species of phytoplankton and zooplankton, bacteria, fish, etc. It is surely impossible and senseless to include all the species observed into the model. However, their choice and aggregation remains to be one of the most speculative stages of ecological modeling, being mostly an art, based on intuition and educated guesses, rather than science. Luckily, within the framework of SIMSAB, it is quite easy to make structural changes in the model, to introduce or neglect variables. Therefore we start from a rather detailed model, targeted first of all at the problems of eutrophication and water quality simulation.

The following phase variables were chosen to represent the ecosystem components.

FISH ($R$) represents the 16 species encountered in the lake. At present the fish concentration is about 100 kg/ha, mostly (60 to 70%) small species such as bleak, roach, and perch.[12] In the ratios of these species, zooplankton prevails, alternating with phytoplankton, although a large contribution from detritus is also possible. The proportion of commercially valuable species is negligible. Therefore we may assume that there is no use in detailed modeling of the lake fishery, since in its present state it can be roughly estimated within the framework of a general eutrophication model.

ZOOPLANKTON ($Z$) is an aggregated variable presenting the 128 species of the lake, among which rotifers dominate with respect to numbers, and crustaceans with respect to biomass. In the last 50 years the total size of the zooplankton population has increased three times[18] and fairly high zooplankton concentrations having been observed in winter,[14] which is also evidence of eutrophication.

PHYTOPLANKTON will be modeled by three variables in order to trace the seasonal succession of algal species in the lake. In spring, low temperatures, increasing illumination, and an abundance of nutrients produce a peak of DIATOMS ($A1$), their biomass attaining 7 mg/l. Starting from the late 1970s, blooms of BLUE-GREENS ($A2$) have been observed in June to July. In mid-July the usual summer peak of phytoplankton is due to the PYROPHITE ALGAE ($A3$). These three variables present the 334 species of the Lake Plesheyevo phytoplankton. The occurrence of blue-green blooms, mounting phytoplankton biomasses, and population numbers clearly indicate that the process of eutrophication is occurring in the lake.

Almost all of the shallow zones in the lake are covered by macrophyte vegetation, especially in the mouth of the Trubezh and in the northwest of the lake. The area of the macrophyte belt has been increasing at such a rate that during the last 5 years of the 1970s it trebled. Two variables, one for SUBMERGED MACROPHYTES ($MS$) and the other one for EMERGING MACROPHYTES ($ME$), are included in the model. In comparison with phytoplankton, the cell

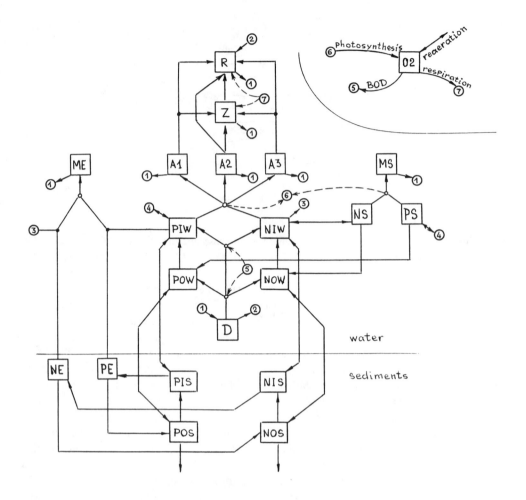

FIGURE 4. Material flows in the littoral segments.

quota of which proved to be negligible in models of natural ecosystems,[3] macrophytes have greater specific times and higher accumulation capacity,[16] and therefore their dependence upon the limiting nutrient is much less. To take this into account, we assume a two-stage uptake process for macrophytes, introducing into the model the NITROGEN and PHOSPHORUS CELL QUOTA (*NS* and *PS*) for submerged macrophytes and the NITROGEN and PHOSPHORUS IN ROOTS (*NE* and *PE*) of the emerging macrophytes.

Dying biota turn into DETRITUS (*D*), which is then recycled due to decomposition, producing INORGANIC NITROGEN (*NIW*) and PHOSPHORUS (*PIW*) IN WATER and the more labile forms of ORGANIC NITROGEN (*NOW*) and PHOSPHORUS (*POW*) IN WATER. These in turn also undergo destruction, slowly turning into *NIW* and *PIW*, on the one hand, and sinking to the bottom, where they are considered as separate variables: ORGANIC NITROGEN (*NOS*) and PHOSPHORUS (*POS*) IN SEDIMENTS, together with INORGANIC NITROGEN (*NIS*) and PHOSPHORUS (*PIS*) IN SEDIMENTS.

The decomposition and water-sediment interactions are regulated by the DISSOLVED OXYGEN ($O_2$) concentration, which is produced mostly as a result of the photosynthesis of producers.

Bacteria, which are an important factor of any aqueous ecosystem, are not among the model state variables, firstly, because there was not enough experimental information to consider the

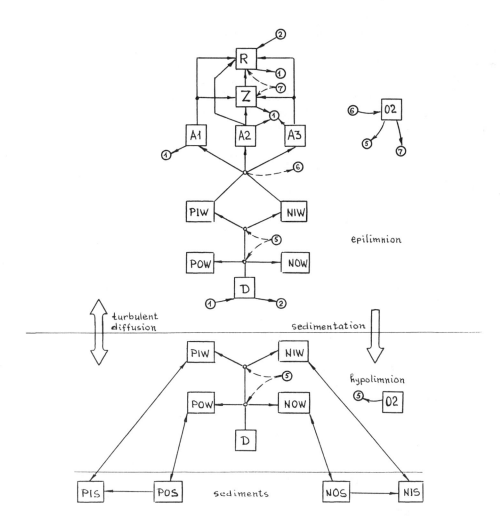

FIGURE 5. Material flows in the pelagic segments.

limiting role of bacteria in various biochemical processes in the lake (nitrification, denitrification, decomposition, etc.) and, secondly, because in most cases the bacteria have high specific rates and can adjust their populations rapidly to changing environmental conditions. Therefore we may assume that they do not limit the processes in the ecosystem, and we associate the bacterial pool of biomass with detritus, which is usually the substrate for them.

Thus the model includes 21 variables. The five littoral segments are presented by the full vector of variables. The epilimnion plus metalimnion segment has no bottom and therefore lacks the variables that describe the sediments and the macrophytes, which means that only 11 variables remain. The biota is neglected in the hypolimnion segment but the sediments are taken into account, and hence we get ten variables in this segment. The flow diagrams for the littoral and pelagic segments are shown in Figures 4 and 5, respectively. Bold lines indicate the material transformations, while the broken lines stand for the "informational" links, showing the regulatory effects of one process or variable upon another.

The forcing functions of the model are the climatic conditions: water temperature ($T$), illumination ($E$), and the wind direction ($ALF$) and velocity ($W$), which are used together with the values of the inflow ($QZI$) and outflow ($QZO$) to model the wind-induced currents and the material exchange between segments. Another group of forcing functions presents the interactions between the water body and the watershed. Since there were no reliable data about toxic inflows, we take into account only the nutrient pollutions coming from the watershed. These may be point pollutions (River Trubezh, primarily) or distributed pollutants. Among them we are to distinguish the nitrogen and phosphorus constituents of organic and inorganic forms. Making use of the variable names listed above, we get the following set of functions: $LNID_i$, $LPID_i$, $LNOD_i$, $LPOD_i$, $LNID_i$, $LPIP_i$, $LNOP_i$, and $LPOP_i$, where $i = 1...5$ is the number of the appropriate littoral segment and $D$ stands for distributed, while $P$ stands for point pollutions.

## IV. DEFINITION OF ECOLOGICAL BOX MODELS

The purpose of this section is to explain some of the SIMSAB notations and functions so that the reader can understand the meta-language employed later (in Tables 3 to 5) to formulate the ecological models for different segments of the lake.

The syntax details of this language can be found elsewhere;[2] here we just outline the main points:

- $Q[X,Y]$ stands for the flow from variable $X$ to $Y$.
- $QIN[X]$ and $QOUT[X]$ stand for the inflow into $X$ from outside the system and outflow from $X$ to within, respectively.
- $Q[*,X]$ and $Q[X,*]$ stand for the sums of all flows entering $X$ and leaving $X$, respectively.
- $ denotes a SIMSAB library function, for instance, $FT2(T,X)$, $RS(X,Y)$, etc.
- #VAR:NAME.EXT is a model parameter: VAR (optional) stands for the variable name, NAME is the parameter name, and EXT (optional) is an extension, which can be another variable name for parameters in binary interactions or any other identifier.
- If a SIMSAB function, returning several values, is used, then these values will be assigned to the flows in curly brackets {...} on the left-hand side.

For the remainder, the basic FORTRAN syntax rules are adopted (C starts the comments, entries start from the seventh position, etc.).

The definition of flows is opened by the specification of the box model name, the names of the variables and forcing functions, and the name of the parameters library, which is a specially structured file storing the information about various parameters obtained in experiments or taken from previous experience of ecosystem modeling. After that all the positive flows of the box model should be specified.

The resulting differential equations of the model that describe the system itself have the form

$$dX_i/dt = \sum_{j=1}^{n}\left(Q[X_j,X_i] - Q[X_i,X_j]\right) + QIN[X_i] - QOUT[X_i] \qquad (1)$$

where $i = 1...n$ and where $n$ is the number of variables.

SIMSAB allows one to build models operating with qualitative, conceptual ideas about ecosystem material transformations. This means that restricting oneself to the functions from the SIMSAB library, one can describe the processes in the ecosystem as a superposition of various elementary processes, which are already mathematically formalized and programed. Table 2 contains the subset of SIMSAB functions used in the model of Lake Plesheyevo. In the following discussion, some examples are presented to show how processes are formalized in SIMSAB.

Let us now consider the process of algal production. It is most common to assume that phytoplankton growth is limited by temperature, light, and one of the nutrients, nitrogen or phosphorus in our case. Let us suppose that the limiting effect of environmental conditions is multiplicative, while the nutrients limit growth according to Liebig's principle, that is, the uptake flow of the nutrient that is at a minimum determines the growth rate, and all the other uptake flows are recalculated with respect to the minimal flow according to the stoichiometric C:N:P ratio.

The temperature in natural ecosystems can attain inhibitory values, especially for diatom species. Therefore it seems more pertinent to use a bell-shaped function $FT2 rather than $FT4 from Table 2. The light limitation function may be $FIINT, with the attenuation calculated by the function $SHAD, taking into account the attenuation by algae, suspended detritus, and submerged macrophytes. The uptake of a nutrient should be specified by a so-called trophic function, which specifies the flow of the nutrient from the environment into the algal organism. Let us assume the traditional S-shaped curve for this process, presented by a SIMSAB function $RS2. In Table 2 we can find a function that will recalculate the flows of nutrient uptake in accordance with Liebig's principle: that will be $LIM1. As a result we get the following formalization for the growth of, say, diatoms

$$\{Q[PIW,A1],Q[NIW,A1]\} = \$FT2(T,A1)$$

$$* \$FIINT(E,A1,\#H,\$SHAD(A1,A2,A3,MS,D))$$

$$* \$LIM1(A1,PIW,NIW,\$RS2(PIW,A1),\$RS2(NIW,A1)) \qquad (2)$$

$$QIN[A1] = \#A1{:}C.C * Q[PIW,A1]/\#A1{:}C.PIW \qquad (3)$$

where $QIN[A1]$ describes the uptake of the nonlimiting nutrients (carbon, etc.) from outside the system, which is necessary, since we measure biota as dry biomass. Let us note that only three parameters are implicitly present in this formulation (stoichiometric coefficients for diatoms #A1:C.C and #A1:C.PIW and the depth of the photic zone #H). All the other parameters will be requested by the system at the stage of translation, providing the user with some hints about the minimal and maximal values of the appropriate parameters registered in previous studies.

For another example, let us give a detailed formalization of the macrophyte processes: first the emerging macrophytes, *ME*. Their life cycle can be schematically presented in the following way:

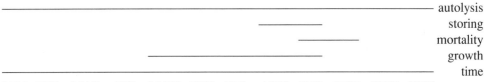

|  |  |  |  |  |  |  |  |  |  |  |  | autolysis |
| | | | | | | | | | | | | storing |
| | | | | | | | | | | | | mortality |
| | | | | | | | | | | | | growth |
| | | | | | | | | | | | | time |

JAN   FEB   MAR   APR   MAY   JUN   JUL   AUG   SEP   OCT   NOV   DEC

In contrast with phytoplankton, growth is considered as a two-stage process. Firstly, available nutrients are accumulated in roots, which take them up from the sediments. This process is limited by temperature and the available nutrient in sediments:

$$\{Q[PIS,PE],Q[NIS,NE]\} = \$FT2(T,ME) * \{\$RS2(PIS,PE),\$RS2(NIS,NE)\} \qquad (4)$$

At the second stage, macrophytes mostly take up the nutrients accumulated in roots, but can also get them directly from water in conditions of oversupply. Besides temperature, this process is limited by light in a special way. While macrophytes are submerged, they are shaded by all the suspended material according to the integral Di Toro function $FIINT. However, when they emerge above the water (this is interpreted as reaching a threshold biomass of #ME:CEXIT), Then the light limitation can be described by the classical Steele's function $FI1. Since there is no special function in SIMSAB for this process, a FORTRAN fragment is used:

**TABLE 2**
**SIMSAB Functions for Lake Plesheyevo Model**

| Name | Legend | Figure |
|------|--------|--------|
| $RS2(X,Y)$ | s-shaped trophic function<br>$= mu*X^n*Y/(k^n + X^n)$,<br>$mu$ - maximal uptake rate (1/d),<br>$k$ - half-saturation coefficient (mg/l),<br>$n$ - slope | |
| $FT2(T,X)$ | bell-shaped temperature function<br>$= \begin{cases} ft0**((1-T/topt)**st), & \text{if } T \leq topt \\ ftm**((T-topt)/(tmax-topt))**st, & \text{otherwise,} \end{cases}$<br>$ft0$ - temp.factor at 0°C,<br>$ftm$ - temp.factor at maximal temp.,<br>$topt$- optimal temperature (°C),<br>$tmax$- maximal temperature (°C),<br>$st$ - slope of the curve | |
| $FT4(T,X)$ | Vant-Hoff temperature function<br>$= 2**((T-to)/10)$,<br>$to$ - reference temperature (°C) | |
| $FTMOR(T,X)$ | temperature function for mortality<br>$= \begin{cases} 1+g1*(tmin-T)^2, & T<tmin, \\ 1, & tmin<T<tmax, \\ 1+g2*(T-tmax)^2, & tmax<T, \end{cases}$<br>$tmin$- minimal temperature (°C),<br>$tmax$- maximal temperature (°C),<br>$g1,g2$- rate coefficients | |
| $FO4(OX,X)$ | anaerobic mortality function<br>$= \begin{cases} 1, & \text{if } OX > oan \\ 1+cm*(oan-OX)^2, & \text{if } OX \leq oan, \end{cases}$<br>$oan$ - threshold to anaerobic conditions (mgO$_2$/l),<br>$cm$ - rate coefficient | |
| $FOSW(OX,X)$ | aerobic/anaerobic function<br>$= 1/[1+exp(-ss*(OX-ana))]$,<br>$ss$ - steepness of curve,<br>$ana$ - aerobic/anaerobic threshold (mgO$_2$/l) | |
| $REA(OX,T,kr)$ | reaeration function<br>$= kr*(os-OX)$, where<br>$os$ - O$_2$ saturation coefficient(mg/l)<br>($os = 14.61996-0.4042*T ++ 0.00842*T^2$<br>$-0.00009*T^3$ (Wang, et al., 1978))<br>$kr$ - reaeration coefficient (1/d) | |
| $SHAD(X1,...,Xk)$ | light attenuation function<br>$= ksw+ks1*X1+...+ksk*Xk$,<br>$ksw$ - extinction coeff. for water,<br>$ksi$ - extinction coeff. for $i^{th}$ component ($i = 1,...,k$) (l/mg) | |

## TABLE 2 (continued)
## SIMSAB Functions for Lake Plesheyevo Model

| Name | Legend | Figure |
|------|--------|--------|
| *$FIINT(E,X,h,ksh)* | integral light limitation func. | |

$FIINT(E,X,h,ksh)$ — integral light limitation func.
$$= 2.71828/ksh*h*$$
$$*[exp(-Eo/eopt)-exp(-E/eopt)],$$
where $Eo = E*exp(-ksh*H)$,
$ksh$ - extinction coefficient($= \$SHAD$),
*eopt*- optimal illumination
   (cal/cm$^2$/d),
$h$ - depth of the segment (m)

$LIM1(Y,X1,...,Xk,\$R1(X1,Y),...,\$Rk(Xk,Y))$

uptake function for Liebig's principle
   of limiting factors
$\{..\} = ci*$
   $*min\{\$R1(X1,Y)/c1,..,\$Rk(Xk,Y)/ck\}$,
$i = 1,..,k$ - number of limiting factors
$Xi$ - limiting factor (nutrient),
$ci$ - stoichiometric coefficient,
$Ri$- trophic function for uptake of
   $i^{th}$ nutrient $Xi$

$ST1(Y,X1,...,Xk,Q)$    stoichiometric decomposition of flow $Q$
from $Y$ to $X1,...,Xk$
$\{..\} = Q*ci/(c1+..+ck)$,
$i = 1,..,k$ - number of recipients,
$ci$ - stoichiometric coefficient

$REL(X1,..,Xk;Y1,..Yn;..;Z)$    function of feeding with electivity
and switching from one group of feeds
$(X)$ to another $(Y)$ etc.
(for details see Svirezhev et al.,
   1984)

The light limitation can be described by the classical Steele's function *$FI1*. Since there is no special function in SIMSAB for this process, a FORTRAN fragment is used:

```
IF (ME .LT. #ME:CEXIT) THEN

   CLIM = $FIINT(E,ME,#H,SH)

ELSE CLIM = $FI1(E,ME,0.0,0.0)

ENDIF
```

Taking into account the temperature limitation we get:

$$CLIM = CLIM * \$FT2(T,ME) \tag{5}$$

The growth is further limited according to Liebig's principle. Therefore the uptake of nutrients from water will be

$$\{Q[PIW,ME],Q[NIW,ME]\} = CLIM$$

$$* \$LIM1(ME,PIW,NIW,\$RS2SW(PIW,ME),\$RS2SW(NIW,ME)) \tag{6}$$

Note that another trophic function from the SIMSAB library was applied here. *$RS2SW* takes into account that the uptake of dissolved nutrients becomes possible only under conditions of oversupply, that is, when the concentrations of *PIW* and *NIW* become higher than a certain level.

The growth due to nutrients accumulated in roots should be proportional to the root biomass, *RB*, not the plant biomass, *ME*. Therefore we first estimate the root biomass, assuming that it can be represented by stoichiometrically proportioned root nitrogen and phosphorus plus the content of nonlimiting nutrients:

$$\{Q1,Q\} = \$LIM1(ME,PE,NE,PE,NE)$$

$$RB = Q1 + Q + Q1 * \# ME{:}C.C \tag{7}$$

The resulting flows will be formulated as

$$\{Q[PE,ME],Q[NE,ME]\} = CLIM * \$LIM1(ME,PE,NE,\$RS2(PE,ME)$$

$$\$RS2(NE,ME))/ME * RB \tag{8}$$

$$QIN[ME] = \#ME{:}C.C * (Q[PIW,ME]+Q[PE,ME])/\#ME{:}C.PIW \tag{9}$$

When temperature becomes unfavorable for vegetation, macrophytes start to die off, turning into detritus. An appropriate function *$FTMOR* can be found in the library. Mortality itself is assumed to be a linear function of the biomass, with the mortality coefficient *#ME:MOR:*

$$Q[ME,D] = \$FTMOR(T,ME) * \#ME{:}MOR * ME \tag{10}$$

In emerging macrophytes, there is a dual relationship between roots and plants. At the end of the season, just before dying off, macrophytes can store some of the nutrients in their roots. Temperature is assumed to be the trigger in this process also, but the temperature function has to be somewhat modified. The distribution of biomass between phosphorus and nitrogen in roots is performed by function *$ST1* in proportion with the stoichiometric coefficients.

$$Q = (\$FTMOR(T,ME,\text{stor}) - 1.0) * \#ME.STOR * ME \tag{11}$$

$$\{Q[ME,PE],Q[ME,NE],QOUT[ME]\} = \$ST1(ME,PE,NE,C,Q) \tag{12}$$

The lower-case comment "stor" in *$FTMOR* indicates that the parameters inserted in the function will be different from the mortality case.

In the process of autolysis, a portion (*#ME:AB*) of the emerging macrophytes biomass is transformed into nutrients in the water according to their stoichiometric content in the biomass. This process is also temperature limited. However, inhibitory temperatures are hardly ever attained, and therefore the *$FT4* function seems to be most suitable in this case:

$$\{Q[ME,PIW],Q[ME,NIW],Q\} = \$FT4(T,ME)$$

$$* \$ST1(ME,PIW,NIW, C,\#ME{:}AB * ME) \tag{13}$$

$$QOUT[ME] = QOUT[ME] + Q \tag{14}$$

And finally we present the processes of mortality and autolysis of roots:

$$\{Q[PE,POS],Q[NE,NOS]\} = \$FT4(T,ME) * \#ME:MORR * \{PE,NE\} \qquad (15)$$

$$\{Q[PE,PIS],Q[NE,NIS]\} = \$FT4(T,ME) * \#ME:AB * \{PE,NE\} \qquad (16)$$

Let us now turn to the submerged macrophytes. For them we have assumed a two-stage growth process with accumulation of nutrients in cells, since the role of root uptake is negligible in this case. The cell uptake is limited by temperature, and only a certain portion (#*PS.MAXCON*, #*NS.MAXCON*) of the *MS* biomass can be used as a reservoir for phosphorus and nitrogen. All this can be formalized as

$$\{Q[PIW,PS],Q[NIW,NS]\} = \$FT2(T,MS) *$$

$$\{(1-PS/MS/\#PS:MAXCON) * \$RS2(PIW,MS)$$

$$(1-NS/MS/\#NS:MAXCON) * \$RS2(NIW,MS)\} \qquad (17)$$

The growth of submerged macrophytes is a Liebig-limited process of utilization of the cell quota with temperature and light limitation similar to the algal case:

$$\{Q[PS,MS],Q[NS,MS]\} = \$FT2(T,MS) *$$

$$\$FIINT(E,MS,\#H,\$SHAD(A1,A2,A3,MS,D))$$

$$* \$LIM1(MS,PS,NS,\$RS2(PS,MS),\$RS2(NS,MS)) \qquad (18)$$

$$QIN[MS] = \#MS:C.C * Q[PS,MS]/\#MS:C.PIW \qquad (19)$$

The flow from *MS* to detritus is composed of mortality, which is also a linear function of the biomass, and the metabolic losses, which are proportional to the total inflow of material into the biomass of submerged macrophytes:

$$Q[MS,D] = \#MS:MOR * MS + \#MS:MB * Q[* ,MS] \qquad (20)$$

Likewise we present mortality and autolysis of cells:

$$\{Q[NS,NIW],Q[PS,PIW]\} = \$FT4(T,MS) * \#MS:AB * \{NS,PS\} \qquad (21)$$

$$\{Q[NS,NOW],Q[PS,POW]\} = \#MS:MOR * \{NS,PS\} \qquad (22)$$

Based on the examples above, one can decipher all the other processes, formalized in Tables 3 to 5, which present the ecosystem models of the littoral, epilimnetic, and hypolimnetic boxes, respectively.

## V. THE MONITOR AND ITS FUNCTIONS

The formulation of flows defines a system of ordinary differential equations that model the ecologically homogeneous segments of the water body. This system has to be solved, information about the initial conditions and the forcing functions being required as input. The material exchange between segments should be calculated and the results somehow presented to the user, etc. All these and other operations are performed by the SIMSAB MONITOR, which actually specifies the sequence of various events modeled by the program.

## TABLE 3
## Formulation of Flows in the Littoral Segments

```
MODEL L
VAR R,Z,A1,A2,A3,MS,ME,NS,PS,NE,PE,D,POW,NOW,
         PIW,NIW,PIS,NIS,POS,NOS,O2;
FORC T,E;
PARAM PLESH.PRM;
C***************************GROWTH*********************************
C   feeding of fish with switching from zooplankton to
C   phytoplankton and then to detritus
        {Q[Z,R],Q[A1,R],Q[A2,R],Q[A3,R],Q[D,R]}=
*                 $FT2(T,R)*$REL(Z;A1,A2,A3;D;R)
C   feeding of zooplankton
        {Q[A1,Z],Q[A2,Z],Q[A3,Z],Q[D,Z]}=
*                 $FT2(T,Z)*{$RS2(A1,Z),$RS2(A2,Z),$RS2(A3,Z),$RS2(D,Z)}
C   uptake of nutrients by diatoms and pyrophites
        SH=$SHAD(A1,A2,A3,MS,D)
        {Q[PIW,A1],Q[NIW,A1]}=$FT2(T,A1)*$FIINT(E,A1,#H,SH)
*           *$LIM1(A1,PIW,NIW,$RS2(PIW,A1),$RS2(NIW,A1))
        QIN[A1]=#A1:C.C*Q[PIW,A1]/#A1:C.PIW
        {Q[PIW,A3],Q[NIW,A3]}=$FT2(T,A3)*$FIINT(E,A3,#H,SH)
*           *$LIM1(A3,PIW,NIW,$RS2(PIW,A3),$RS2(NIW,A3))
        QIN[A3]=#A3:C.C*Q[PIW,A3]/#A3:C.PIW
C   uptake of nutrients by blue-greens with no nitrogen limitation
        {Q[PIW,A2],Q[NIW,A2]}=$FT2(T,A2)*$FIINT(E,A2,#H,SH)*
*                 {$RS2(PIW,A2),$RS2(NIW,A2)}
        Q1=#A2:C.NIW*Q[PIW,A2]/#A2:C.PIW
C   DMIN1 is a standard FORTRAN function to find minimum
        Q[NIW,A2]=DMIN1(Q1,Q[NIW,A2])
        QIN[A2]=#A2:C.C*Q[PIW,A2]/#A2:C.PIW+Q1-Q[NIW,A2]
C   uptake of nutrients from bottom by emerging macrophytes
        {Q[PIS,PE],Q[NIS,NE]}=$FT2(T,ME)*
*                 {$RS2(PIS,PE),$RS2(NIS,NE)}
C   growth of emerging macrophytes
C       due to uptake of nutrients from water
C        (light limitation is calculated differently before and
C         after plants emerge from water )
        IF (ME .LT. #ME:CEXIT) THEN
            CLIM=$FIINT(E,ME,#H,SH,submer)
        ELSE CLIM=$FI1(E,ME,0.D0,0.D0,emerg)
        CLIM=$FT2(T,ME)*CLIM
        {Q[PIW,ME],Q[NIW,ME]}=
*           CLIM*$LIM1(ME,PIW,NIW,$RS2SW(PIW,ME),$RS2SW(NIW,ME))
C   due to nutrients accumulated in roots in proportion to
C       the root biomass RB, not the plant biomass ME
        {Q1,Q}=$LIM1(ME,PE,NE,PE,NE)
        RB=Q1+Q+Q1*#ME:C.C
        {Q[PE,ME],Q[NE,ME]}=CLIM*
*           $LIM1(ME,PE,NE,$RS2(PE,ME),$RS2(NE,ME))/ME*RB
        QIN[ME]=#ME:C.C*(Q[PIW,ME]+Q[PE,ME])/#ME:C.PIW;
C   cell quota of submerged macrophytes
        {Q[PIW,PS],Q[NIW,NS]}=$FT2(T,MS)*
*           {(1-PS/MS/#PS:MAXCON)*$RS2(PIW,MS),
*             (1-NS/MS/#NS:MAXCON)*$RS2(NIW,MS)}
C   growth of submerged macrophytes
        {Q[PS,MS],Q[NS,MS]}=$FT2(T,MS)*$FIINT(E,MS,#H,SH)
*                 *$LIM1(MS,PS,NS,$RS2(PS,MS),$RS2(NS,MS))
        QIN[MS]=#MS:C.C*Q[PS,MS]/#MS:C.PIW
C***************************MORTALITY*****************************
C   mortality and metabolism of zooplankton and fish
        Q[Z,D]=#Z:MOR*Z+#Z:MORS*Z*Z
        {Q[Z,NIW],Q[Z,PIW],QOUT[Z]}=$ST1(Z,NIW,PIW,C,Q[*,Z]*#Z:MB)
```

**TABLE 3 (continued)**
**Formulation of Flows in the Littoral Segments**

```
        Q[R,D]=$FO4(OX,R)*#R:MOR*R+Q[*,R]*#R:MB
C    mortality and metabolism of producers
        Q[A1,D]=#A1:MOR*A1+Q[*,A1]*#A1:MB
        Q[A2,D]=#A2:MOR*A2+Q[*,A2]*#A2:MB
        Q[A3,D]=#A3:MOR*A3+Q[*,A3]*#A3:MB
C    autolysis of phytoplankton
        {Q[A1,PIW],Q[A1,NIW],QOUT[A1]}=$ST1(A1,PIW,NIW,C,#A1:AB*A1)
        {Q[A2,PIW],Q[A2,NIW],QOUT[A2]}=$ST1(A2,PIW,NIW,C,#A2:AB*A2)
        {Q[A3,PIW],Q[A3,NIW],QOUT[A3]}=$ST1(A3,PIW,NIW,C,#A3:AB*A3)
C    storage of nutrients in roots of emerging macrophytes
        Q=($FTMOR(T,ME,stor) – 1.0) * #ME.STOR * ME
        {Q[ME,PE],Q[ME,NE],QOUT[ME]}=$ST1(ME,PE,NE,C,Q)
C    mortality of macrophytes
        Q[ME,D]=$FTMOR(T,ME)*#ME:MOR*ME
        Q[MS,D]=#MS:MOR*MS + #MS:MB*Q[*,MS]
C    mortality and autolysis of cells
        {Q[NS,NIW],Q[PS,PIW]}=$FT4(MS,T)*#MS:AB*{NS,PS}
        {Q[NS,NOW],Q[PS,POW]}=#MS:MOR*{NS,PS}
C    mortality and autolysis of roots
        {Q[PE,POS],Q[NE,NOS]}=#ME:MORRO*{PE,NE}
        {Q[PE,PIS],Q[NE,NIS]}=$FT4(ME,T)*#ME:AB*{PE,NE}
C    autolysis of submerged macrophytes
        {Q[ME,PIW],Q[ME,NIW],Q}=$ST1(ME,PIW,NIW,C,#ME:AB*ME)
        QOUT[ME]=QOUT[ME]+Q
C*****************DESTRUCTION AND WATER-SEDIMENT INTERFACE****************
C    destruction of detritus
        Q1=$FT4(T,D)*$FOSW(O2,D)
        {Q[D,NOW],Q[D,POW],QOUT[D]}=
     *           $ST1(D,NOW,POW,C,Q1*#D:DLABIL*D)
        {Q[D,NIW],Q[D,PIW],Q}=$ST1(D,NIW,PIW,C,Q1*#D:DES*D)
        QOUT[D]=QOUT[D]+Q
        Q[NOW,NIW]=#NOW:DES*Q1*NOW
        Q[POW,PIW]=#POW:DES*Q1*POW
C    sedimentation of phosphorus
        Q[PIW,PIS]=$FOSW(O2,PIW)*#PIW:SED*PIW
C    diffusion and sedimentation of labile organics
        Q[NOW,NOS]=#TURBOD*NOW/#H1+#NOW:SED*NOW
        Q[POW,POS]=#TURBOD*POW/#H1+#POW:SED*POW
        Q[NOS,NOW]=#TURBOD*NOS/#H1/#H
        Q[POS,POW]=#TURBOD*POS/#H1/#H
C    diffusion and burying of nutrients in sediments
        Q[NIW,NIS]=#TURBOD*
     *      (NIW – #NIW:DISSOL*(1–$FOSW(O2,NIW,turbod)))
        Q[PIW,PIS]=#TURBOD*
     *      (PIW – #PIW:DISSOL*(1–$FOSW(O2,PIW,turbod)))
        QOUT[NIS]=#NIS:BURY*NIS
        QOUT[PIS]=#PIS:BURY*PIS
C    burying of sediment organics
        QOUT[NOS]=#NOS:BURY*NOS
        QOUT[POS]=#POS:BURY*POS
C    destruction of bottom detritus
        Q[POS,PIS]=#POS:DEST*$FOSW(O2,POS)*POS
        Q[NOS,NIS]=#NOS:DEST*$FOSW(O2,NOS)*NOS
C****************************OXYGEN****************************
C    photosynthesis and reaeration
        QIN[O2]=#O2:PHOTO*(Q[*,A1]+Q[*,A2]+Q[*,A3]+Q[*,MS])+
     *           $REA(O2,T,#REA)
C    respiration and oxygen consumption for destruction
        QOUT[O2]=#R:RESP*R+#Z:RESP*Z+#O2:DEST*
     *      (Q[POS,PIS]+Q[NOS,NIS]+Q[POW,PIW]+Q[NOW,NIW]+QOUT[D])
END
```

## TABLE 4
## Formulation of Flows in the Epilimnion

```
MODEL E
VAR R,Z,A1,A2,A3,D,POW,NOW,PIW,NIW,O2;
FORC T,E;
PARAM PLESH.PRM;
C································GROWTH································
C    feeding of fish with switching from zoo- to phytoplankto
C    and then to detritus
       {Q[Z,R],Q[A1,R],Q[A2,R],Q[A3,R],Q[D,R]}=
   *              $FT2(T,R)*$REL(Z;A1,A2,A3;D;R)
C    feeding of zooplankton
       {Q[A1,Z],Q[A2,Z],Q[A3,Z],Q[D,Z]}=
   *              $FT2(T,Z)*{$RS2(A1,Z),$RS2(A2,Z),$RS2(A3,Z),$RS2(D,Z)}
C    uptake of nutrients by phytoplankton
       SH=$SHAD(A1,A2,A3,D)
       {Q[PIW,A1],Q[NIW,A1]}=$FT2(T,A1)*$FIINT(E,A1,#H,SH)
   *         *$LIM1(A1,PIW,NIW,$RS2(PIW,A1),$RS2(NIW,A1))
       QIN[A1]=#A1:C.C*Q[PIW,A1]/#A1:C.PIW
       {Q[PIW,A3],Q[NIW,A3]}=$FT2(T,A3)*$FIINT(E,A3,#H,SH)
   *         *$LIM1(A3,PIW,NIW,$RS2(PIW,A3),$RS2(NIW,A3))
       QIN[A3]=#A3:C.C*Q[PIW,A3]/#A3:C.PIW
C    uptake of nutrients by blue-greens with no nitrogen limitation
       {Q[PIW,A2],Q[NIW,A2]}=$FT2(T,A2)*$FIINT(E,A2,#H,SH)*
   *              {$RS2(PIW,A2),$RS2(NIW,A2)}
       Q1=#A2:C.NIW*Q[PIW,A2]/#A2:C.PIW
       Q[NIW,A2]=DMIN1(Q1,Q[NIW,A2])
       QIN[A2]=#A2:C.C*Q[PIW,A2]/#A2:C.PIW+Q1–Q[NIW,A2]
C································MORTALITY································
C    mortality and metabolism of zooplankton and fish
       Q[Z,D]=#Z:MOR*Z+#Z:MORS*Z*Z
       {Q[Z,NIW],Q[Z,PIW],QOUT[Z]}=$ST1(Z,NIW,PIW,C,Q[*,Z]*#Z:MB)
       Q[R,D]=$FO4(OX,R)*#R:MOR*R+Q[*,R]*#R:MB
C    mortality and metabolism of producers
       Q[A1,D]=#A1:MOR*A1+Q[*,A1]*#A1:MB
       Q[A2,D]=#A2:MOR*A2+Q[*,A2]*#A2:MB
       Q[A3,D]=#A3:MOR*A3+Q[*,A3]*#A3:MB
C    autolysis of phytoplankton
       {Q[A1,PIW],Q[A1,NIW],QOUT[A1]}=$ST1(A1,PIW,NIW,C,#A1:AB*A1)
       {Q[A2,PIW],Q[A2,NIW],QOUT[A2]}=$ST1(A2,PIW,NIW,C,#A2:AB*A2)
       {Q[A3,PIW],Q[A3,NIW],QOUT[A3]}=$ST1(A3,PIW,NIW,C,#A3:AB*A3)
C    destruction of detritus
       Q1=$FT4(T,D)*$FOSW(O2,D)
       {Q[D,NOW],Q[D,POW],QOUT[D]}=
   *              $ST1(D,NOW,POW,C,Q1*#D:DLABIL*D)
       {Q[D,NIW],Q[D,PIW],Q}=$ST1(D,NIW,PIW,C,Q1*#D:DES*D)
       QOUT[D]=QOUT[D]+Q
       Q[NOW,NIW]=#NOW:DES*Q1*NOW
       Q[POW,PIW]=#POW:DES*Q1*POW
C································OXYGEN································
C    photosynthesis and reaeration
       QIN[O2]=#O2:PHOTO*(Q[*,A1]+Q[*,A2]+Q[*,A3]+$REA(O2,T,#REA)
C    respiration and oxygen consumption for destruction
       QOUT[O2]=#R:RESP*R+#Z:RESP*Z+
   * #O2:DEST*(Q[D,POW]+Q[D,NOW]+Q[POW,PIW]+Q[NOW,NIW]+QOUT[D])
END
```

## TABLE 5
### Formulation of Flows in the Hypolimnion

```
MODEL H
VAR D,POW,PIW,NOW,NIW,PIS,NIS,POS,NOS,O2;
FORC T,E;
PARAM PLESH.PRM;
C    destruction of detritus
        Q1=$FT4(T,D)*$FOSW(O2,D)
        {Q[D,NOW],Q[D,POW],QOUT[D]}=$ST1(D,NOW,POW,C,Q1*#D:DLABIL*D)
        {Q[D,NIW],Q[D,PIW],Q}=$ST1(D,NIW,PIW,C,Q1*#D:DES*D)
        QOUT[D]=QOUT[D]+Q
        Q[NOW,NIW]=#NOW:DES*Q1*NOW
        Q[POW,PIW]=#POW:DES*Q1*POW
C    sedimentation of phosphorus
        Q[PIW,PIS]=$FOSW(O2,PIW)*#PIW:SED*PIW
C    diffusion and sedimentation of labile organics
        Q[NOW,NOS]=#TURBOD*NOW/#H1+#NOW:SED*NOW
        Q[POW,POS]=#TURBOD*POW/#H1+#POW:SED*POW
        Q[NOS,NOW]=#TURBOD*NOS/#H1/#H
        Q[POS,POW]=#TURBOD*POS/#H1/#H
C    diffusion and burying of nutrients in sediments
        Q[NIW,NIS]=#TURBOD*
    *        (NIW – #NIW:DISSOL*(1–$FOSW(O2,NIW,turbod)))
        Q[PIW,PIS]=#TURBOD*
    *        (PIW – #PIW:DISSOL*(1–$FOSW(O2,PIW,turbod)))
        QOUT[NIS]=#NIS:BURY*NIS
        QOUT[PIS]=#PIS:BURY*PIS
C    burying of sediment organics
        QOUT[NOS]=#NOS:BURY*NOS
        QOUT[POS]=#POS:BURY*POS
C    destruction of bottom detritus
        Q[POS,PIS]=#POS:DEST*$FOSW(O2,POS)*POS
        Q[NOS,NIS]=#NOS:DEST*$FOSW(O2,NOS)*NOS
C******************************OXYGEN******************************
C    oxygen consumption for destruction
        QOUT[O2]=#O2:DEST*
    *        (Q[POS,PIS]+Q[NOS,NIS]+Q[POW,PIW]+Q[NOW,NIW]+QOUT[D])
END
```

The monitor makes use of the *library of procedures*, which contains special FORTRAN subroutines that carry out the necessary operations. Some of the procedures are defined directly in the monitor, but others have to be predefined, just like the flows for the box models. One such procedure, *$HYDRO,* presents the hydrodynamical block of the model.

The hydrodynamical block of the Plesheyevo model uses the SIMSAB procedure, which is a version of the steady-state Ekman-type model of wind-induced currents.[5] $HYDRO needs the data about wind patterns, which were taken from Poddubnyi and Litvinov.[13] Next it asks about the lake morphometry. The grid net of $50 \times 54$ points with a step of 155 m was chosen, which was quite sufficient for a good approximation of the shore line and the depth distribution. During the stagnation period in summer, the depth of the pelagic zone was assumed to be equal to the thermocline depth. In addition, the data about the inflows have to be specified. Weekly averaged inflows from the River Trubezh were used.

As a result, the hydrodynamical block produces patterns of wind-induced currents, such as those shown in Figure 6. The average currents are 0.15 to 0.25 m/s. The upper layer flows along the wind; at a depth of 4 to 8 m the direction of flow is reversed. The roles of the Trubezh inflow and Vyoksa outflow proved to be insignificant.

However, within the framework of a segment model, all the hydrodynamical results are superfluous and may be of interest just by themselves. The total model actually needs information only about the water exchange between segments, which can be represented by the integrals of the unit flows with respect to the cross-sections between any two segments.

The monitor function $MIX makes use of the information about the water discharges among segments stored in a special file. It is all the same for $MIX, whether these discharges were calculated by $HYDRO or by some other procedure, or if they were measured and input into the file.

It is assumed that the material is transferred by lake currents, due to diffusive and convective fluxes. The concentration of components in each segment is altered in proportion to the amount of allochthonous material and in inverse proportion to the segment volume. The contribution of diffusion along the horizontal axis is negligible as compared to the effect of wind-induced currents. Along the vertical axis, diffusion and sedimentation are the dominating processes. As a result, assuming that the upper and lower cross-sections of a segment have the same area, we get the following formula to recalculate the concentration $X_i$ of component $X$ in the $i^{th}$ segment:

$$X_i(t+dt) = X_i(t) * V_i(t)/V_i(t+dt) + S * (X_a - X_i) * dt/H_i$$

$$+ \left(\Sigma_{j \in J_1} Xj * Q_{ij} - X_i * \Sigma_{j \in J_2} Q_{ij}\right) * dt/V_i(t+dt)$$

$$+ 2 * D_{ai} * (X_a - X_i) * dt/H_i/(H_i + H_a)$$

$$+ 2 * D_{ib} * (X_b - X_i) * dt/H_i/(H_i + H_b) \tag{23}$$

Here $V_i$ is the volume of the $i^{th}$ segment (m$^3$), $S$ is the sedimentation rate (m/s), $X_a$ and $X_b$ are the concentrations of component $X$ in the segments above and below the $i^{th}$ one, respectively (g/m$^3$); $H_i, H_a$, and $H_b$ are the depths of the $i^{th}$ segment and of those above and below, respectively (m); $dt$ is the time step (s); $Q_{ij}$ is the discharge between the $i^{th}$ and the $j^{th}$ segments (m$^3$/s); $J_1$ and $J_2$ are the sets of segment numbers for the inflows and outflows of segment $i$, respectively; and $D_{ai}$ and $D_{ib}$ are the diffusion coefficients at the upper and lower boundaries of $i^{th}$ segment, respectively (m$^2$/s).

It should be noted that not all of the model components can be transferred from one segment to another. Macrophytes, for instance, are assumed to be attached to the ground. Sediments are also fixed to a certain segment. Fish can migrate by themselves and are not carried by advective and diffusive fluxes. Therefore only ten of the model variables, A1, A2, A3, Z, D, PIW, NIW, POW, NOW, and O2, are redistributed among segments.

In the function $MIX, material is exchanged not only within the water body, but also between the water body and its environment — the watershed, the bottom, and the atmosphere. In this way the nutrient load to various segments is taken into account.

Other functions of the monitor presented in Table 6 are self-explanatory and do not need any preliminary specifications and information.

## VI. MODEL CALIBRATION AND VALIDATION

The process of modeling is only partially completed by the formation of the formalized representation of the object as a set of equations transformed into a FORTRAN program. In this form, in fact, models can only claim to be "representations of the accumulated ... empirical information about the ecosystem and can solve only tasks of quantitative interpretation of the observations and verification of their mutual consistency."[1] Actually a big system — the water body — is substituted by another big system — the model — the dynamics of which remain to be studied.

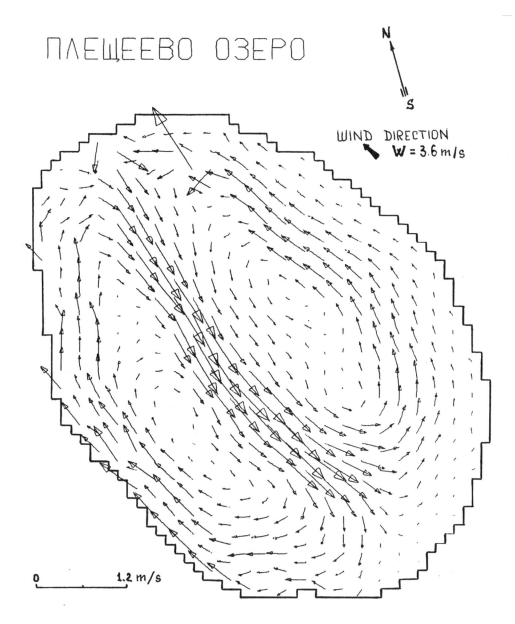

ПЛЕЩЕЕВО ОЗЕРО

FIGURE 6.   Circulation in Lake Plesheyevo.

Recently, increasing attention has been drawn to problems of model analysis,[6,7,15,17,19] including estimates of their sensitivity, adequacy, and accuracy of forecasts to various methods of their calibration. Only a well-analyzed model can serve as a tool of ecosystem research and provide new knowledge about aquatic bodies.

The SIMSAB-88 procedure of model analysis includes a number of modules that can:

- Estimate model sensitivity with respect to parameters and forcing functions;
- Estimate model stability and robustness (that is, define the extreme deviations of parameters and forcing functions that do not result in ecologically absurd model trajectories, i.e., those tending to infinity or becoming negative);
- Calibrate models according to experimental data (adjust the most essential parameters to get the best fit of model trajectories to the experimental time series).

**TABLE 6**
**The MONITOR for Lake Plesheyevo Model**

```
MONITOR
MOD L SEG 1,2,3,4,5;
MOD E SEG 6;
MOD H SEG 7;
SEGMENT 7 MIXING Z,A1,A2,A3,D,POW,NOW,PIW,NIW,O₂;
FLOW 7 MIXING D,POW,NOW,PIW,NIW;
C  Input of the initial data for all the 7 segments
INITIAL
    $READ(INIT1) VARL1
    $READ(INIT2) VARL2
    $READ(INIT3) VARL3
    $READ(INIT4) VARL4
    $READ(INIT5) VARL5
    $READ(INIT6) VARE6
    $READ(INIT7) VARH7
DYNAMIC TEND=365;
C  Every 10 days input of the climatic data from files:
C  illumination, temperature for the epi- and hypolimnion
ST=1,10;
    $READ(ILL)  EL1
    $READ(TEMP1) TL1
    $READ(TEMP2) TH7
C  Once in a month input of nutrient concentrations in inflows:
C  1-Trubezh, 2-distributed from forests, 3-point from forests,
C  4-distributed from agro, 5-point from agro, 6-from atmosphere.
ST=1,30;
    $READ(LOAD) FLOW1,FLOW2,FLOW3,FLOW4,FLOW5,FLOW6;
C  Every 5 days input of the hydrodynamical scenario
ST=1,5;
    $HYDRO1 PLEHYD.RES
C  Every day calculation of ecodynamics for all 7 segments
C  and accumulation of results for the plotter
ST=1,1;
    $FOR_PLOT VARL1,VARL2,VARL3,VARL4,VARL5,VARE6,VARH7
    $ADAMS L,VARL1,FORCL1,20
    $ADAMS L,VARL2,FORCL1,20
    $ADAMS L,VARL3,FORCL1,20
    $ADAMS L,VARL4,FORCL1,20
    $ADAMS L,VARL5,FORCL1,20
    $ADAMS E,VARE6,FORCL1,20
    $ADAMS H,VARH7,FORCH7,20
C  Recalculation of concentrations due to water exchange
    $MIX
TERMINAL
C  Output of results on a plotter
    $PLOT
END

DATA
INIT1: 1,0.01,0.1,0.3,0.5,1.6,0.08,0.005,30,8;
INIT2: 1, 0.1,0.1,0.3,0.5,1.6, 0.08, 0.01,60,5;
INIT3:
INIT4:
INIT5:
INIT6:
INIT7:
```

**TABLE 7**
**Parameters with the Highest Model Sensitivity**

| Parameter | Ecological meaning | Sensitivity |
|---|---|---|
| Class 1 | | |
| Z:MORS | Self-limitation | 1871.158 |
| ME:MORRO | Mortality coefficient | 1404.148 |
| | | |
| Class 2 | | |
| NE:MU.NIS | Maximal uptake rate of | 289.844 |
| R:MOR | Mortality | 245.571 |
| A1:MU.PIW | Maximal uptake rate of | 116.315 |
| TURBOD | Coeff. of turbulent diffusion | 107.893 |
| A2:MU.NIW | Maximal uptake rate of | 93.685 |
| PE:MU.PIS | Maximal uptake rate of | 82.422 |
| ME:MU.NE | Maximal uptake rate of | 75.023 |
| A2:MOR | Mortality | 49.038 |
| MS:STT2 | Steepness of temperature curve | 41.853 |
| ME:STT2 | Steepness of temperature curve | 38.269 |
| Z:MOR | Mortality | 32.007 |
| A3:STT2 | Steepness of temperature curve | 26.522 |
| A1:MOR | Mortality | 20.108 |
| A1:STT2 | Steepness of temperature curve | 16.068 |
| Z:MU.A3 | Maximal uptake rate of A3 | 15.134 |
| ME:MU.NIW | Maximal uptake rate of | 14.24 |
| A3:MU.NIW | Maximal uptake rate of | 13.042 |
| MS:IOPT | Optimal illumination | 12.196 |
| A2:FT0 | Temperature factor at 0°C | 12.129 |
| COMMON:KSHWAT | Extinction coeff. for water | 11.103 |
| A2:STT2 | Steepness of temperature curve | 8.262 |
| Z:K.D | Half-saturation coefficient for D | 7.922 |
| Z:TOPT | Optimal temperature | 5.168 |
| A1:FT0 | Temperature factor at 0°C | 5.021 |
| R:FT0 | Temperature factor at 0°C | 3.881 |
| A2:IOPT | Optimal illumination | 3.419 |
| A3:IOPT | Optimal illumination | 3.231 |
| A3:K.PIW | Half-saturation coefficient for | 3.127 |
| MS:TOPT | Optimal temperature | 3.011 |
| A3:TOPT | Optimal temperature | 2.443 |
| A3:FT0 | Temperature factor at 0°C | 1.462 |
| A1:K.PIW | Half-saturation coefficient for | 1.172 |
| | | |
| Class 3. | All other parameters. | |

While translating the model, SIMSAB displays, for each parameter, the interval of its possible variations registered in the literature or in experiments and stored in the SIMSAB parameters library. The user is then asked to specify the exact value of the parameter that should be used in the initial model runs. The Plesheyevo model needs 214 ecological parameters. Only a small portion of them (23) was directly measured at the lake. Therefore we could mostly rely upon expert estimations and published data for similar ecosystems. Most of the parameters were estimated according to published data.[4,8,11] However, the exact parameter values could be found only at the stage of model calibration.

Before calibrating a model, a sensitivity analysis may be most useful (see also Chapter 5). Analyzing the sensitivity of the Plesheyevo model to variations in parameters, we could divide the set of parameters into three groups with respect to their effect upon the model trajectories (Table 7). The highest sensitivity was observed for the parameters representing the uptake rates and mortalities.

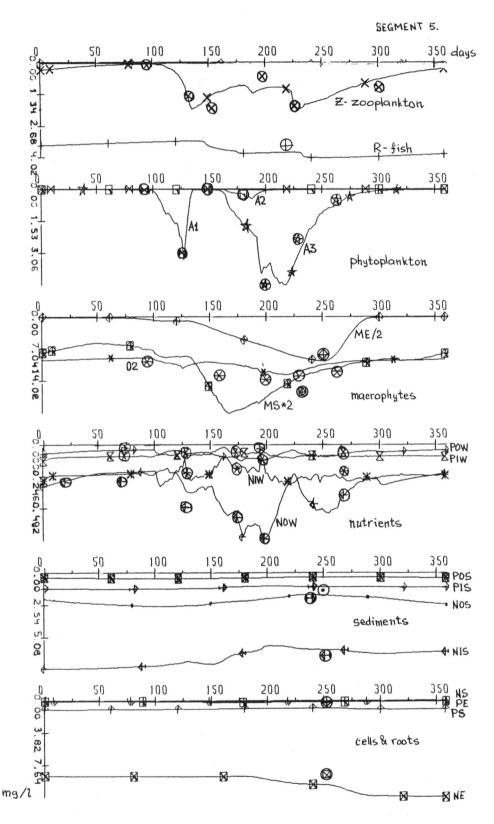

FIGURE 7. Verification of the model based on the data for 1984. (Experimental data are circled.)

There are no efficient methods for calibration of models with so many variables and parameters. The only method that works in such models is the "good old" trial-and-error method based on educated guesswork. The sensitivity data may be of great help, since the number of fitted parameters may be essentially reduced. We may first fix all the parameters except those in class 1, representing the essential parameters. These are varied within the intervals of their possible variations in such a way that the trajectories become closest to the experimental data. In case we fail to improve the fit any further and it still does not satisfy us, we can start to perturb the parameters from class 2. This interactive human-computer method proves to be more effective in approaching a quasi-optimal set of parameters than any of the more rigorous methods of optimization. In this way the Plesheyevo model trajectories were fitted to the time series observed at the lake in 1983. All the parameters were then fixed and calculations were repeated for the forcing functions of 1984. The results obtained are compared with the experimental data in Figure 7. One can see that the fit is rather good, which supports the validity of the model. There may be different quantitative criteria of the fit, which are usually made of various sums of squares. One such criterion was suggested by Theil:[21]

$$h = \frac{\sqrt{\Sigma(X_{obs} - X_{sim})^2/n}}{\sqrt{(\Sigma X_{obs}^2/n)} + \sqrt{(\Sigma X_{sim}^2/n)}} \tag{24}$$

Here $X_{obs}$ is the observed value, $X_{sim}$ is the simulated value, and $n$ is the number of experimental points. This index was employed to compare different sets of parameters at the calibration stage. Obviously, $0 \le h < 1$ and $h = 0$ when there is a perfect fit of the model to the data. In our case the best mean value of $h$ obtained for the second segment was 0.218, varying for different variables from 0.186 to 0.247. For the validation experiments of 1984, we obtained $h = 0.239$, which is acceptable. A similar test with the 1985 data is not as good but is still acceptable (Figure 8). In this case $h$ was equal to 0.265.

Further improvement of the model calls for additional experimental data, which we lack. In this case the analysis of model stability and robustness become even more important for its further application.

## VII. MODEL ANALYSIS: SCENARIO RUNS

Simulation models are mostly designed for predictions, which means that they may be run under very different alterations of the forcing functions, presenting several possible combinations of climatic factors and anthropogenerated impacts. A plausible set of forcing functions is called a scenario. It is intuitively clear that a good model should operate for a sufficiently wide range of scenarios, producing plausible forecasts. Besides forcing functions, the parameters are the other source of uncertainty. Perturbations of parameters, especially of the crucial ones, always result in some variations in model trajectories. Just as in the case of forcing functions, for a good model there should be at least some domain of perturbations in which the model behavior remains ecologically sound, and the larger the domain the better the model. This domain of "safe" perturbations in parameters and forcing functions we call the domain of model stability. In mathematical terms this can be expressed as follows.

Suppose $P = (p_1 \ldots p_k)$ is the vector of parameters, $Y(t) = (y_1(t) \ldots y_m(t))$ is the vector of forcing functions, and $X(t) = (x_1(t) \ldots x_n(t))$ is the vector of model trajectories, $n$ being the number of variables. Then the domain of stability is specified by polyhedra $Q = \{q_{1i} \le \pi \le q_{2i}; i = 1 \ldots k\}$ and $Z = \{z_{1j}(t) \le y_j(t) \le z_{2j}(t); j = 1 \ldots m\}$ such that for any $P \in Q$ and $Y(t) \in Z$ the phase trajectories remain bounded within the positive octant, that is $X(t) \in C = \{0 < c_{1i} \le x_i(t) \le c_{2i} < \infty; i = 1, \ldots, n\}$. In other words the domain of model stability is the domain of the Lagrange stability of the model.[20] The size of this domain characterizes the model robustness.

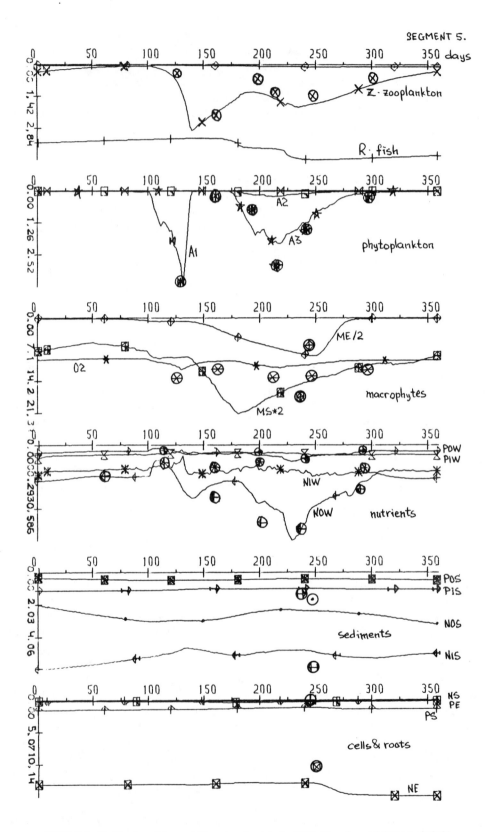

FIGURE 8.   Verification of the model based on the data for 1985. (Experimental data are circled.)

**TABLE 8**
**Parameters with Domain P Smaller than the Interval of Registered Variations**

| Parameter | Registered interval | Value in model | | P interval |
|---|---|---|---|---|
| Z:MORS | (0.0005, 0.05) = | 0.002 | 0.001 | 0.016 |
| R:MOR | (0.0001, 0.005) = | 0.0006 | 0.0001 | 0.001 |
| A1:MOR | (0.0001, 0.05) = | 0.02 | 0.001 | 0.03 |
| NE:MU.NIS | (0.2, 0.9) = | 0.231 | 0.2 | 0.55 |
| ME:MORRO | (0.001, 0.035) = | 0.002 | 0.001 | 0.006 |
| TURBOD | (0.15, 1.7) = | 0.25 | 0.15 | 0.7 |
| A2:MU.NIW | (0.1, 5) = | 0.41 | 0.2 | 0.45 |
| A1:MU.PIW | (0.5, 8.4) = | 0.13 | 0.1 | 0.9 |
| A2:MOR | (0.0001, 0.05) = | 0.0044 | 0.0001 | 0.009 |
| Z:MOR | (0.001, 0.05) = | 0.01 | 0.003 | 0.02 |

Unfortunately again there is no mathematically rigorous procedure to find the domain of model stability for sufficiently complicated models. The trial-and-error method was again applied. First the set $Q$ was analyzed for the Plesheyevo model. Certain information has been obtained at the stage of sensitivity analysis. It turned out that for the class 3 of insignificant parameters the stability domain completely includes the intervals of possible variations. A similar situation occurred in case of some of the neutral parameters (class 2): even the minimal and maximal possible values did not result in impossible dynamics. However, for some of the essential parameters, the stability domain proved to be quite small. These parameters are listed in Table 8. Nevertheless, on the whole the domain $Q$ turned out to be quite large, which means that the model is rather robust with respect to changes in parameter values. Moreover the above analysis allows one to identify those parameters whose values should be measured most precisely and most urgently. On the other hand, if we assume that our model is adequate, we may conclude to which processes and interactions the ecosystem itself is most sensitive. For instance, it is quite obvious from Tables 7 and 8 that the ecosystem has high sensitivity to processes concerned with primary production. This is quite clear intuitively. On the other hand, the high sensitivity to the mortality rate of roots of emerging macrophytes is rather counterintuitive.

Let us now consider the $C$ domain. As mentioned above, the forcing functions in the model are the water temperature ($T$), illumination ($E$), wind direction ($ALF$), and velocity ($W$), the values of the inflow ($QZI$) and outflow ($QZO$), and the functions presenting the nutrient load: $LNID_i$, $LPID_i$, $LNOD_i$, $LPOD_i$, $LNIP_i$, $LPIP_i$, $LNOP_i$, $LPOP_i$, where $i = 1 . . . 5$ is the number of the appropriate littoral segment and $D$ stands for distributed, while $P$ stands for point pollution sources.

It is a very cumbersome, although feasible, task to perform a complete analysis of the domain of possible variations of the forcing functions. As a result of there being too many dimensions in $C$, and because we mostly deal with functions, it is always possible to produce a synergistic effect in one of their infinite number of combinations, such that the model behavior turns out to be unpredictable. However, even some preliminary experiments with domain $C$ may be quite useful.

For instance, in the Plesheyevo model variations of temperatures around the basic scenario of 1984: $T(t) = T_o(t) + T_1$, where $-1° C \le T_1 \le +1° C$, during the vegetation period (day 100 to 240) resulted in the model dynamics shown in Figure 9. Perturbations of trajectories turned out to be linearly dependent upon the perturbations of temperature. Strangely enough, the effect upon the consumers' populations — fish $R$ and zooplankton $Z$ — turned to be even greater than the effect upon the phytoplankton populations. Similar effects were observed when the illumination $E$ was similarly perturbed. Of course, this could be explained somehow by highly stressed trophic chains in Lake Plesheyevo. The trophic pyramid in the lake is paradoxical, which means that the phytoplankton biomass is approximately the same, and occasionally even lower, than the

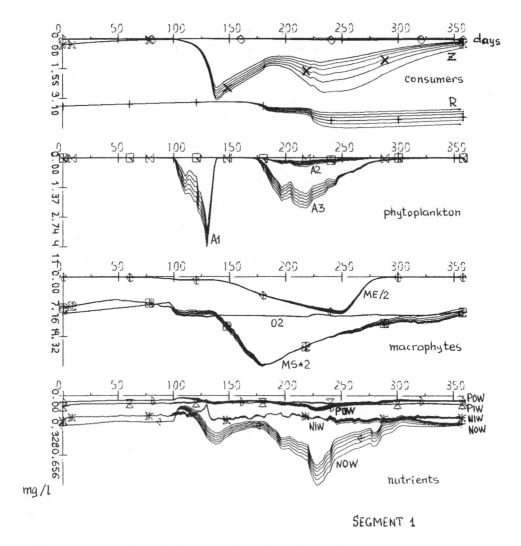

FIGURE 9.   Model trajectories under perturbations of temperature.

zooplankton biomass. As a result, all variations in the phytoplankton population are immediately transferred to the zooplankton dynamics: all perturbations of the resource are indicated by the consumer. A scenario can be invented when the pyramid is reversed: if temperature and illumination are further increased a phytoplankton bloom can be anticipated, with fish and zooplankton inhibited by temperature and oxygen, which may be depleted after the bloom. Dynamics of this kind can be produced by the model (Figure 10), but it is fairly hard to find the exact threshold conditions when such principal reconstructions occur. The threshold or critical behavior of ecosystems in certain regimes is one of their most important features. We will have to return to this topic later.

Still greater perturbations in temperature can destabilize the model. It was found that trajectories tend to infinity if there is a period of optimal temperatures for diatoms longer than 40 days.

Unfortunately we failed to produce the dynamics of Figure 10, by varying the nutrient load instead of temperature. This is a drawback of the model, since in natural conditions the dynamics of Figure 10 can be anticipated without any shifts in temperature, but only due to nutrient enrichment. In general the model sensitivity with respect to perturbations in the nutrient load

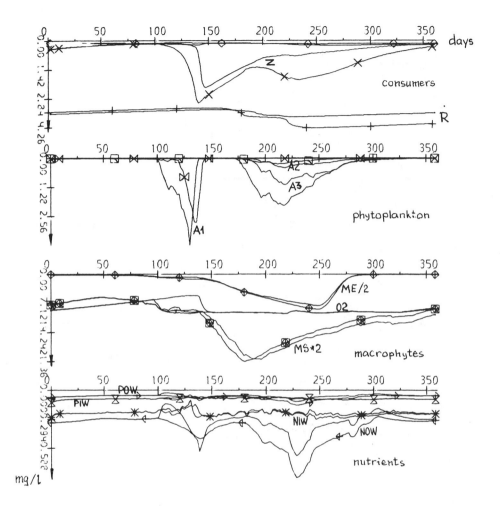

FIGURE 10.   Scenario run. $T=T_o+4$, $E=E_o+550$, for $100 < t < 140$. No wind in this period. (Markers stand on the base run.)

seems to be somewhat low. Increments of 50% in phosphorus loading had no effect upon the model dynamics. A similar increase of the nitrogen load (Figures 11A to D) did alter the model trajectories, especially in segments 1 and 5, which receive most of the nutrient load, with its maximum occurring in spring. This fact can be explained by the existing nitrogen limitation in the lake. The limiting factor was shown to be switched when the nitrogen load was tripled. This may be considered as another critical point in the ecosystem dynamics. The limits of the domain $C$ with respect to the nutrient load functions were surpassed only when the total nutrient load was increased by a factor of eight, which is really too large to be acceptable.

Similar estimates can be made for other forcing functions. Variations of wind parameters produced minor effects upon the model dynamics. Difference between the basic run and the perturbed trajectories was negligible, even when the wind pattern was rotated by 90° or 180°, and when the wind velocity was changed by more than 100%.

Employing various scenario runs we can estimate the borders of domain $C$. From the analysis performed for Lake Plesheyevo, we can conclude that the domain of model stability is sufficiently large to permit studies of very different regimes in ecosystem dynamics. The model is robust enough, maybe even too robust.

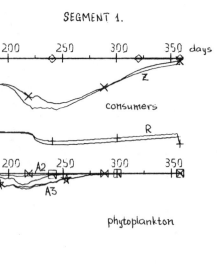

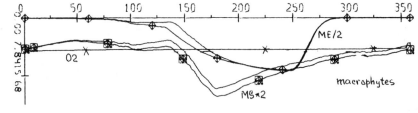

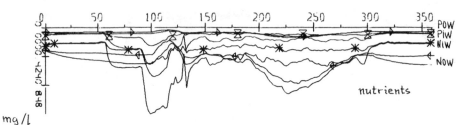

FIGURE 11A-D.   Scenario run: increased nitrogen load from the watershed.

## VIII. SIMPLIFIED MODEL VERSION

In considering the total model dynamics (Figures 7 and 8), it is noticeable that some of the variables are almost constant all over the simulation run, and model sensitivity with respect to processes concerned with these variables is quite low. The model can be suspected to be somewhat redundant. In fact, if we make the concentration of, say, oxygen constant ($O_2 = 9.1$ mg/l), and run the model, the deviations from the basic trajectory will prove to be negligible.

Since SIMSAB allows modifications to be readily introduced into the model, it is simple to estimate the structural sensitivity of models, that is, the sensitivity to variations in model structure: addition or exclusion of variables and/or links, modification of formalizations of interactions. In this context the Plesheyevo model has low sensitivity with respect to a number of variables. Figure 12 presents the results of simulations with 11 fixed littoral variables: $O_2 = 8.3$ mg/l, *NOS* = 2.0 mg/l, *POS* = 0.33 mg/l, *PIS* = 1.11 mg/l, *NIS* = 6.0 mg/l, *PIW* = 0.1 mg/l, *POW* = 0.07 mg/l, *NE* = 7.5 mg/l, *PE* = 0.1 mg/l, *NS* = 0.2 mg/l, *PS* = 1.0 mg/l, and the corresponding pelagic variables. The dynamics of the rest of the variables remains almost the

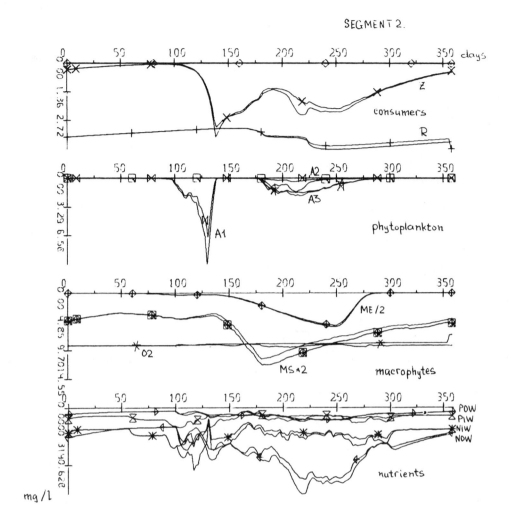

SEGMENT 2.

FIGURE 11B.

same. The simplified model is much easier to analyze. The 3-year scenario in Figure 13 took as much computer time as the 1-year scenarios for the initial model. We may conclude that the initial set of variables was a mistake, and that it may be advisable to construct a simpler model and, in this case, a more accurate model. A simpler model is more acceptable because an accurate model analysis, verification testing, various statistical estimates of forecast reliability, and studies of long-term scenarios of ecosystem development can be performed only for models of low complexity. Only in this case are model results easy to interpret, and typically only such models find applications in decision making.

However, simple models, representing the ecosystem by several aggregated variables, as well as regression models that take no account of the physical and ecological background of processes and mechanisms in a water body, can represent ecosystem dynamics only in the neighborhood of its steady state, overlooking all possible structural modifications in the ecosystem. Any structural rearrangement that was not stipulated by the model results in total inadequacy of the model. Accordingly, the decision making based on simple models is somewhat risky: we can never be sure that certain impacts will not result in structural variations in the ecosystem.

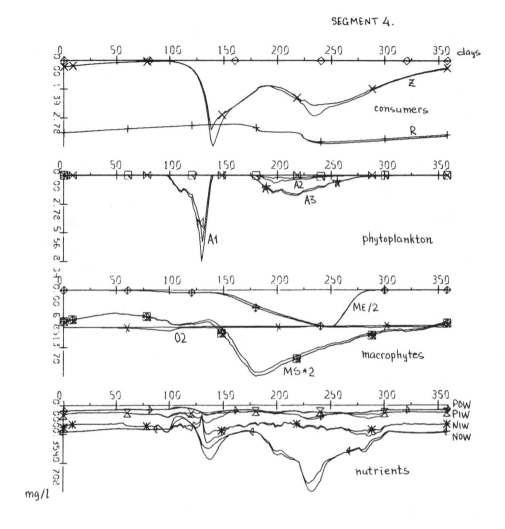

FIGURE 11C.

As a result, we come across a conflict, where, on the one hand, detailed models, which take into account the evolution of the ecosystem, cannot be analyzed and verified in a proper way, and consequently do not seem to be reliable enough. On the other hand, simplified models with totally known behavior turn out to be inadequate in the face of external effects that are strong enough to cause structural modifications of the ecosystem and disturb it from the steady state.

This contradiction can be resolved if we assume the following modeling procedure. First we construct a sufficiently detailed ecosystem model, which is based upon biologically sound representations of processes and takes into account all the variables that are essential for the ecosystem and that are potentially essential according to biological expertise. Actually this model may be regarded as a systemization of all the accumulated data about the ecosystem and is certainly of value by itself. This detailed model is a means to study short-term effects. We can identify the passive variables, that is, the variables that do not affect the model dynamics under certain limited variations in forcing functions. Various critical regimes, when passive variables come into play, can be investigated in order to define the "safe" combinations of forcing functions, which do not perturb the model from its quasi-steady state.

Next we may fix the passive variables to get a simplified model. This model is accurately

SEGMENT 5.

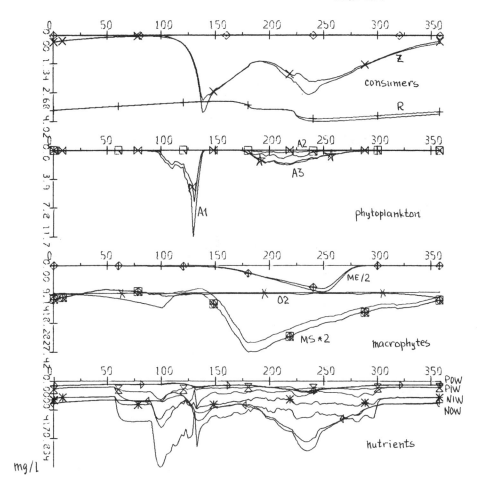

FIGURE 11D.

verified and analyzed. All the necessary scenarios are simulated, always bearing in mind the dangerous external effects, which may cause structural alterations. In case such impacts occur, the detailed model is activated in order to reconsider the model structure and to find out whether the simplified model determined by the passive variables remains the same. In addition, it may be useful to turn to the detailed model intermittently just to check that no new synergetic effects have appeared with time.

In this way, we make a synthesis of the detailed and the simplified models. The simplified model simulates the steady state periods, while the detailed model describes the critical periods with possible structural changes.

This procedure was applied to the Lake Plesheyevo case study. The model described above is a detailed simulator of the water body. Some of the critical regimes have been already identified. One of them is the switching of the limiting factor from nitrogen to phosphorus as a result of intensive nitrogen enrichment. The critical regimes concerned with the sediment variable occur when the nutrient load is diminished and the sediments come into play as a source of the internal load, releasing the accumulated nutrients back into the water column. Another critical regime concerning oxygen dynamics can be obtained if we assume a prolonged under-

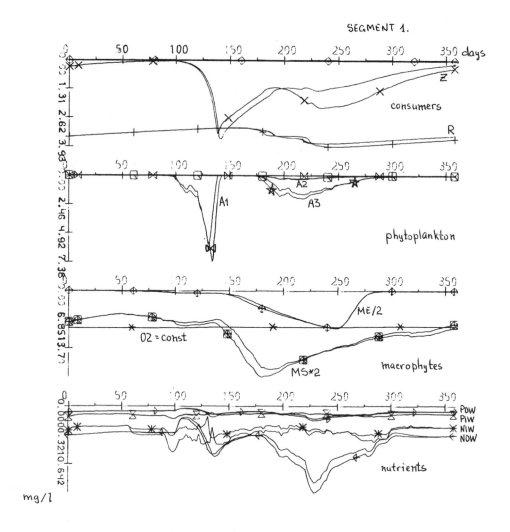

FIGURE 12.    Model run with fixed passive variables compared to the base model run (with markers).

ice period (till the 135th day), impeding reaeration and photosynthesis, followed by lowered temperatures during the vegetation period (days 140 to 250). In this case, dissolved oxygen falls to concentrations that cause fish kills in the lake. We have to realize that the trial-and-error method we employ can give only qualitative estimates of the safe domains of variations in forcing functions. However, the interpretation of phenomenological descriptions of critical events in terms of forcing functions may be quite informative for decision making.

Providing that the variations of forcing functions are safe, some long-term scenarios of ecosystem and watershed development can be analyzed using the model. For example, in Figure 14 another 3-year scenario is presented. Here the existing trends for the nutrient load were assumed, the climatic functions being kept at their average values. At least during these 3 years, there were no dramatic changes predicted in the ecosystem. Therefore, it seems from the model dynamics that after the anthropogenic impacts suffered by the lake in the 1970s, the ecosystem has reached a mesotrophic state, which will be stable for the time being, assuming that there are no more drastic changes in the watershed.

SEGMENT 3.

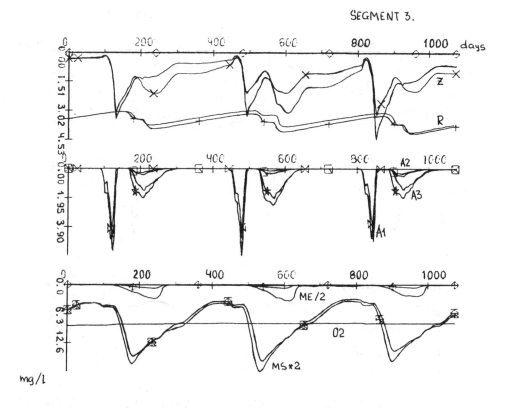

FIGURE 13.   Three-year scenario run: nutrient load decreased by 25% and 50% (with markers).

## IX. CONCLUSIONS

At present, the future development of Lake Plesheyevo ecosystem is not very clear, since accurate scenarios of the watershed development are hard to determine. Aroused public opinion and environmental movements demand that the surroundings of the lake should be preserved. There is a plan to establish a National Historical Park at the site. However, nobody knows what to do with the pre-existing industry of Pereslavl-Zalessky in this case. The legal status of such preserved ecosystems has not yet been completely clarified.

Therefore, now we are at the stage of formulating planning scenarios. Obviously the lake needs permanent surveillance and ecological monitoring. The ecosystem model can serve as a basis for all the monitoring programs at the lake. Along with new experimental findings and accumulation of data, the model should develop in an interactive regime, taking into account new links and variables and deleting the superfluous ones. The SIMSAB framework we made use of allows all kinds of modifications to be readily introduced into the model structure. As it was stated by Jørgensen,[9] "the model is never better than the data on which it is based". It is my opinion that the Plesheyevo model does the best that is now possible with the existing experimental data about the lake. The future development of the case study needs new experiments and observations to be staged.

## ACKNOWLEDGMENTS

My thanks are due to Professor Yu. M. Svirezhev, who gave most helpful comments and advice, and Dr. A. P. Tonkikh, who carried out most of the computer runs. I am very much

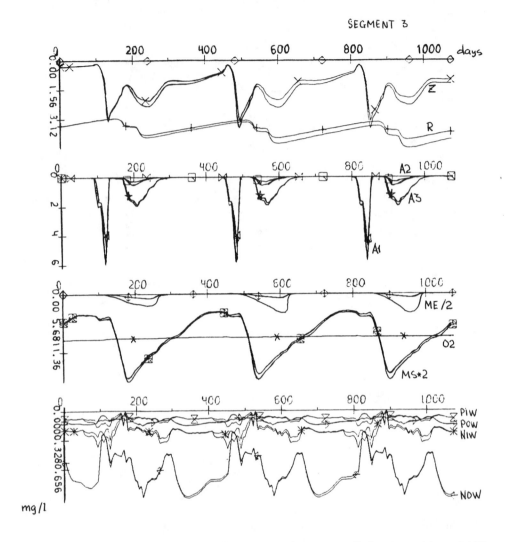

FIGURE 14. Three-year scenario run: ecosystem dynamics with the present trends of nutrient enrichment (~10% annual increase) compared to model run with fixed present nutrient loads (with markers).

obliged to Dr. L. A. Kuchai, Dr. V. L. Sklyarenko, Dr. I. L. Pyrina, Dr. V. A. Ekzertsev and Dr. I. M. Bolonov from the Institute of Biology of Continental Water Bodies for valuable experimental information and stimulative discussions and ideas.

## REFERENCES

1. **Aizatullin, T. A. and Shamardina, I. P.,** Mathematical modeling of ecosystems of continental water bodies (in Russian), in *Biogeocenology Hydrobiology*, Vol. 5, Achievements of Science and Technics, VINITI, USSR Academy of Sciences, Moscow, 1980, 154–228.
2. **Akhremenkov, A. A.,** Modeling complex for aquatic ecosystem simulations (in Russian), Computer Center of USSR Academy of Sciences, Moscow, 1988.
3. **Di Toro, D. M.,** Applicability of cellular equilibrium and Monod theory to phytoplankton growth kinetics, *Ecol. Modeling*, 8, 201–218, 1980.

4. **Di Toro, D. M. and Connolly, J. P.,** *Mathematical Models of Water Quality in Large Lakes. 2. Lake Erie,* U. S. Environmental Protection Agency, Washington D.C., EPA-600/3-80-065, 1980, 230.

5. **Felzenbaum, A. I.,** Theoretical Principles and Calculation Methods for Steady-State Marine Currents (in Russian), Moscow, Academy of Sciences, 1960.

6. **Halfon, E.,** Is there a best model structure? 1. Modeling the fate of a toxic substance in a lake, *Ecol. Modeling,* 20, 135–152, 1983.

7. **Halfon, E.,** Is there a best model structure? 2. Comparing the model structures of different fate models, *Ecol. Modeling,* 20, 153–163, 1983.

8. **Jørgensen, S. E., Friis, M. B., Henriksen, J., Jorgensen, L. A., and Mejer, H. F.,** *Handbook of Environmental Data and Ecological Parameters,* ISEM, Værløse, 1979, 1162.

9. **Jørgensen, S. E.,** State of the art of eutrophication models, in *State of the Art in Ecological Modeling,* Vol. 7, Jørgensen, S.E., Ed., Fair-Print AS, Roskilde,1979, 293–298.

10. Lake Plesheyevo Ecosystem (in Russian), Proceedings of IBVV, Academy of Science USSR, Rybinsk, 1989.

11. **Leonov, A. V.,** *Mathematical Modeling of Phosphorus Compounds in Freshwater Ecosystems (Lake Balaton)* (in Russian), Nauka, Moscow, 1986, 151.

12. **Malinin, L. K. and Linnik, V. D.,** Density and spatial distribution of fish in Lake Plesheyevo (in Russian), in *Functioning of Lake Ecosystems,* Proceedings of IBVV, Academy of Science USSR, Vol. 51(54), Rybinsk, 1983, 125–159.

13. **Poddubnyi, S. A. and Litvinov, A. S.,** On the horizontal circulation of waters in Lake Plesheyevo (in Russian), in *Functioning of Lake Ecosystems,* Proceedings of IBVV, Academy of Science USSR, Vol. 51(54), Rybinsk, 1983, 13–18.

14. **Rivyer, I. K.,** Quantitative and spatial characteristics of winter zooplankton in Lake Plesheyevo (in Russian), in *Functioning of Lake Ecosystems,* Proceedings of IBVV, Academy of Science USSR, Vol. 51(54), Rybinsk, 1983, 62–70.

15. **Shaeffer, D. L.,** A model evaluation methodology applicable to environmental assessment models, *Ecol. Modeling,* 8, 275–295, 1980.

16. **Smirnova, N. N.,** Ecological-physiological features of root systems in aquatic plants (in Russian), *Hydrol. J.,* 16(3), 60–72, 1980.

17. **Somlyody, L.,** *Modeling of a Complex Environmental System: The Lake Balaton Study,* WP-81-108, IIASA, Laxenburg, Austria, 1981, 62.

18. **Stolbunova, V. N.,** Zooplankton as a component of Lake Plesheyevo ecosystem (in Russian), in *Functioning of Lake Ecosystems,* Proceedings of IBVV, Academy of Science USSR, Vol. 51(54), Rybinsk, 1983, 46–70.

19. **Summers, J. K. and McKellar, H. N., Jr.,** A sensitivity analysis of an ecosystem model of estuarine carbon flow, *Ecol. Modeling,* 13, 283–301, 1981.

20. **Svirezhev, Yu. M. and Logofet, D. O.,** *Stability of Biological Communities,* Mir, Moscow, 1982.

21. **Theil, H.,** *Applied Economic Forecasting,* North-Holland, Amsterdam, 1971.

Chapter 5

# MODEL VALIDATION AND SENSITIVITY: CASE STUDIES IN THE GLOBAL CONTEXT

B. Henderson-Sellers and R. I. Davies

## TABLE OF CONTENTS

# I. MATHEMATICAL MODELING

Mathematical modeling is partly a science and partly an art. It encompasses the range of *problem identification*, through *solution*, to *analysis and interpretation* of the results and *model verification/validation*, together with *sensitivity analysis* (Figure 4 of Chapter 1). Problem identification, model building, and interpretation of results have been well discussed in the literature, both in a theoretical framework and through extensive examples and full case studies (see, for example, discussions of eutrophication and thermal stratification models[1-3] and other chapters in this book). Of equal (or even greater) importance, yet less intensively studied, are the areas of validation/verification and sensitivity analysis.

For any numerical model to be considered valid, its realism must be assessed not only for the site for which it was developed (or sites nearby), but also for sites with very different characteristics. Existing models of thermal stratification in lakes have often been designed for a single application/case study and consequently frequently include tuning coefficients to permit simulation of any specific data set. For example, CE-QUAL-R1[4] can only be applied directly to the single lake for which it was originally devised, or to similar lakes following either recalibration[5] or indeed reformulation of the model itself.[6] This model (which is part of the larger water quality model CE-QUAL-R1) has seven tuning (or calibration) coefficients within its structure.[7] To implement the model, it is first necessary to evaluate these seven coefficients by trial and error on a single year's data (a "learning set"). When the seven coefficients have been selected, then a second year's data are used to verify that the model has been relatively well construed for this particular lake. Further simulations may then be possible prognostically, although whether such models can ever predict for conditions beyond their original tuning must remain severely in doubt. The original attempt at transferability of this model was from DeGray Lake in Arkansas (the calibration lake) to Greeson Lake in the same state.[5] More recent experiments have utilized data from Eau Galle reservoir in Wisconsin (and for pumped storage investigations from Carters Lake in Georgia).[6] These authors found that when applying the model to Eau Galle Reservoir, it was necessary to add a subroutine for macrophytes, to extend the description of algae and of diatom photosynthesis, and for the description of inflows it was found that "To better represent these flows, the code was changed ... so that the inflow or outflow was not allowed into those layers representing the borrow pit." Following the recalibration, it is suggested that "all of the major dynamics at Eau Galle were satisfactorily predicted, except for the disappearance during one sampling period of a metalimnetic maxima for nitrite plus nitrate nitrogen."

In this chapter, we discuss the application of a particular simulation model, but not strictly in the case study format adopted in other chapters, i.e., instead of concentrating on a single lake or reservoir, an orthogonal approach will be adopted in which a wide range of lakes will be simulated. These water bodies differ in surface area (from several hundred square kilometers to less than half a square kilometer) and over a range of depths (static characteristics); are situated at different latitudes and altitudes (dynamically controlled by air temperature, radiation fields, extent of winter ice cover, etc); and are of different water types (clear to turbid, as influenced by both the aquatic biota and inorganic suspended sediments). The hypothesis is that, within certain constraints (as for all models), the model can be applied generically across a large database. In terms of decision support systems (DSS), such an investigation can be said to address the second component of a DSS (Figure 1 in Chapter 1), i.e., the database of information required to assist management decisions. Consequently here we investigate the global validity of water quality models. Results are given graphically for a number of lakes and reservoirs across the world. Although much of the comparison with observations is with a relatively short time series (often only a few years, or in some cases, a single year) and is undertaken subjectively, an overall assessment of the model performance can be undertaken successfully. More objective measures of performance are still being developed, and some references are given in the next section towards current work in this area.

# II. MODEL VALIDATION, VERIFICATION, AND SENSITIVITY ANALYSIS

Validation and verification are often used interchangeably. There are, however, two different aspects of model checking to which these two words can usefully be applied: (1) the checking of the model's success in reproducing data associated with the system from which data were abstracted in order to *develop* the model initially (verification); (2) the application of the verified model to a totally different data set, but still within the domain of the model (validation).

Model construction and initial checking often utilizes a data set (often a year) for a specific lake or reservoir. Verification and validation of the model, firstly for other years' data, then for other, totally unconnected, test sites, gives the model developer and model user confidence that the model can be applied successfully — or, conversely, may indicate the limitations of applicability.

In many models, rather informal methods of "parameter tuning" are normally utilized in calibration. More formal statistical methods of time-series analysis are not normally considered, both because of the complexity of the models and the difficulty of obtaining appropriately large data sets, which would allow for statistical identification and estimation. While this is justifiable in the circumstances, it should be realized that the lack of a formal statistical approach and the use of the more arbitrary parameter tuning can limit the credibility of the model.[8] Even when combined with sensitivity analyses and constraints on parameter values, such deterministic tuning is controlled, to some extent, by the modeler's prior prejudice and can be abused. In particular, the development of a model that matches reasonably a rather meager data set is no guarantee that the model is "valid" in wider terms, although it can still possess considerable practical utility.

Model acceptance, therefore, tends to rely on extensive testing, in which simulations and data are compared and evaluated for "correctness". In such comparative assessments, one of the greatest problems is that of subjectivity. It is relatively easy to compare model results of a single dependent variable, $y = f(x)$, by use of, for example, a least squares fit technique; although it is not always clear what (small) value of the test parameter must be achieved before the model is deemed to be "good". Indeed the widespread use of such qualitative terms as good, excellent, and acceptable abound in the water-quality modeling literature,[9] although the translation of such terms to any quantitative statistic would appear to be random. In the case of phosphorus models, the use of the Kirchner-Dillon[10] model has been analyzed[11] in terms of uncertainty analysis to determine model error bounds, a topic also discussed in detail and application elsewhere.[12,13] Other attempts to introduce statistical inferential techniques into water-quality assessment studies have centered on the use of regression analyses and confidence intervals.[14] Further techniques are listed by several authors[9,15] as well as a discussion of various difference measures for a general geophysical problem.[16]

There is, as yet, no agreed objective test for the quality of agreement between simulation and observation of most water quality models. Data are usually obtained as time series, representing space- and time-dependent values of the water quality variable under investigation. As such the data represent a stochastic description of the process, i.e., the data are realizations of random variables. In contrast, many of the water quality models discussed here are deterministic. The objective comparison of essentially random data and deterministically derived simulation values is a current research goal for which the simple application of existing statistical tests for "goodness-of-fit" is inappropriate.[8,9,17,18]

An alternative approach, which explicitly recognizes that models will always be "speculative" to some degree, is the probabilistic simulation model proposed by Hornberger and Spear[19] (see also Young[17]) and used for the modeling of water quality in the Peel Inlet of Western Australia. Here, special Monte Carlo procedures are utilized to carry out a "generalized sensitivity analysis", which can lead either to the generation of hypotheses about the system behavior or can be used to investigate different scenarios in the planning of management activities. Such models are not, however, considered to be "validated" in any sense, since no formal validation against data is attempted, and they cannot, therefore, be used for predictive purposes. Rather, they are simulation tests to extend the thought processes of the modeler within reasonable constraints.

One further problem is that the modeler and the manager responsible for one particular lake may have very different objectives, dependent upon the use to which the water is to be put. Nevertheless, a model simulation assessed as "excellent" should be able to be viewed as such from an objective viewpoint, regardless of the management objective function. It is almost certainly the case that judgement of simulations presented in the literature have an implicit bias towards accepting poor simulation in parameters of less concern. Parallel development of both models and quantitative assessment techniques is currently being undertaken by several authors.

Sensitivity analysis, on the other hand, is a more mechanistic procedure in which the degree to which the model responds to an imposed perturbation is analyzed. If the model exhibits a large response to a small perturbation, then it is likely to be less useful, since it is intolerant of small errors in the perturbing (usually the driving) variables. Secondly, an understanding of the variables and parameters to which the model *is* sensitive allows the field worker to concentrate on minimizing errors in data values collected for those variables, perhaps at the expense of larger errors in observations for the variables to which the model is least sensitive. Such a sensitivity analysis can be undertaken without resort to validation criteria and is a form of testing that should be carried out in parallel to verification and validation tests.

## III. THERMAL STRATIFICATION MODELING

The methodologies described here pertaining to the techniques of mathematical modeling of natural systems (see Chapter 1) are of widespread validity for any thermal stratification model. However, any individual model must be itself subject to the rigor of such an analysis before it can be considered to be useful. In this section, one such stratification model will be outlined, and in subsequent sections preliminary results are analyzed in terms of a large, geographically diverse database of lakes and reservoirs.

As noted in Chapter 1, there are two main approaches to thermal stratification modeling in water bodies: the mixed layer model and the eddy diffusion model. The former is conceptually and mathematically simpler, but the simplifications used may tend to restrict their general applicability, especially for biological modeling. Nevertheless, the models have had wide and successful application in both oceanic and limnological situations. On the other hand, the eddy diffusion approach has until relatively recently suffered from the lack of appropriate analytic descriptions of turbulent mixing. The capacity of this approach for finer resolution and for its Eulerian approach (cf. the Lagrangian approach of the mixed layer models) suggests that (1) it may provide a more useful base for simulating aquatic biological processes and (2) in the context of coupled air–water models, may be more compatible with the numerical schemes used in both the dynamic ocean and dynamic/thermodynamic atmospheric models. However, long-term development of stratification models may be able to benefit from a synergism of these two approaches.

The model selected for study here is the eddy diffusion model EDD1 (Eddy Diffusion Dimension 1).[20] This is a one-dimensional prognostic thermal stratification model that describes the vertical temperature profile in a water body. Although it is only one-dimensional, vertical processes dominate, whilst the code retains the capability to allow advective (horizontal) exchanges of heat and momentum through the water column. This permits simulation of reservoirs used for storm water storage and for pumped storage, as well as the more quiescent water supply reservoirs. It has *no* tuning coefficients, i.e., there are no "arbitrary" parameters for which values have to be selected based upon some optimization against an initial (learning) set of data. All parameter values can be predetermined from site measurements.

The one-dimensional heat transfer equation is given by Equation 10 in Chapter 1 with the value of the eddy diffusion coefficient, $K_H$, given by Equation 9 in Chapter 1, where the neutral value of the eddy diffusion coefficient, $K_{H_0}$, is given by

$$K_{H_0} = \frac{k \, w_{*_s} z}{P_0} \exp\left(-k^* z\right) \tag{1}$$

for a given friction velocity, $w_{*_s}$ and neutral Prandtl number $P_0$ (and where $k$ is the von Kármán constant). The parameter $k^*$ is a nonlinear function of wind speed related to the reciprocal of the Ekman depth, based on Smith[21] and generalized as

$$k^* = 6.6(\sin\theta)^{\frac{1}{2}} U^{-1.84} \tag{2}$$

This representation for $K_{H_o}$ derived from boundary layer concepts,[22] has a subsurface maximum, as observed,[23,24] and that compares well with the the Mellor and Yamada[25] level 5 (MYL5) calculated profile.[26]

The functional form for the non-neutral dependency of the eddy diffusion coefficient is given for stable conditions ($R_i > 0$) as[27]

$$f\left(R_i\right) = \left(1 + 37 R_i^2\right)^{-1} \tag{3}$$

This formula permits more rapid damping of turbulence at high Richardson number than the "classical" form:

$$f\left(R_i\right) = \left(1 + a_3 \, R_i\right)^{-(b_3 + 1)} \tag{4}$$

and is strongly supported by the data of Ueda et al.[28]

In the model EDD1, the surface boundary condition is given by the surface energy exchange (see Chapters 1 and 2), which is calculated at each time step (the size of which is governed by the Courant-Friedrich-Lewy [CFL] numerical stability criterion). Convection is modeled simply in terms of an interactive energy balance.[29] Consequently, the meteorological forcing variables, such as wind speed, air temperature, relative humidity, cloud cover, etc., play an important role. It is important, therefore, to assess how sensitive the model is to potential errors in these measured variable values (for instance, because there is no meteorological observing site near the lake itself). In an initial sensitivity study,[20] it was suggested that wind speed was the most important of the meteorological parameters. In addition, it was found necessary for oligotrophic to mesotrophic lakes to include as accurate a specification of the water turbidity as possible. These preliminary conclusions are further evaluated in the sections below (describing the database itself and then the model results) in the context of simulations from a geographically much wider database of lakes and reservoirs.

## IV. GLOBAL DATABASE

There are two aspects to the global database used in this study: (1) the meteorological data for use in forcing the model and (2) the lake thermal data for validation.

Meteorological data were collected from several sources, including a large number from the U.K. Meteorological Office records. The data required included wind speed (which often had to be converted from non-SI units), solar radiation (often recorded in calories or other hybrid units), relative humidity or vapor pressure, air temperature (often in Fahrenheit and at different heights above the ground), precipitation, cloud amount and/or sunshine hours, and atmospheric pressure (where available).

**TABLE 1**

**Locations of Meteorological Stations Used for Each Lake Cited in this Study**

| Lake | Meteorological station(s) | Latitude | Longitude |
|---|---|---|---|
| Sudbury lakes | Sudbury Airport | 46° 37' N | 80° 48' W |
| | Toronto | 43° 40' N | 79° 23' W |
| Muskoka lakes | Muskoka Airport | 44° 58' N | 79° 18' W |
| | Wiarotor Airport | 44° 45' N | 81° 06' W |
| | Ottawa N.R.C. | 45° 27' N | 75° 37' W |
| DeGray Lake | Little Rock and | 34° 44' N | 92° 14' W |
| | North Little Rock | | |
| South African | Pretoria | 25° 44' S | 28° 11' E |
| lakes | Hartbeespoort Dam | 25° 43' S | 27° 51' E |
| | Buffelspoort Dam | 25° 45' S | 27° 29' E |
| Lake Valencia | Bass Sucre | 10° 14' N | 67° 38' W |
| Windermere | Sellafield | 54° 24' N | 3° 29' W |
| | Hawkshead | 54° 23' N | 3° 00' W |
| | Newton Rigg | 54° 40' N | 2° 47' W |
| | Eskdalemuir | 55° 19' N | 3° 14' W |
| Grafham  Water and | Sutton Bonington | 52° 50' N | 1° 15' W |
| Rutland Water | Wittering | 52° 37' N | 0° 28' W |

For each lake, mean monthly values for each meteorological variable were obtained. For most lakes, site-specific data were not available, so data, sometimes from a site situated at a considerable distance from the lake, had to be used (Table 1). This is not unusual in lake and reservoir model validation studies. Data obtained were sometimes presented as monthly means in the records, and in other instances were averaged from daily data (sometimes as frequent as four times per day). Each set of monthly means were then averaged, for that month, over a number of years (usually 5 to 10) in order to force the model in its hydroclimatic mode. In this mode, the model is most easily run by fitting sinusoidal curves to each meteorological variable (except cloud cover). The appropriateness of this method is illustrated in Figures 1 to 7 for several of the lakes studied here. For example, for the U.K. lakes studied, a sinusoidal variation in solar radiation is perfectly reasonable; whilst for the tropical Lake Valencia, this is less realistic. In the case of cloud cover, which tends to follow a seasonal cycle less well, mean monthly values are used as representative of the midmonth date, with linear interpolation for the remaining days in each month. Further, more detailed, studies using individual data values would require a more extensive meteorological database as well as a limnological database. General accessibility of such data would, of necessity, restrict the study to only a few principal sites. Since the aim of this study is a generic validation of the model component for a DSS, the wider database will be used, retaining consistency in representation across the various annual cycles of the meteorological forcing variables.

Nonmeteorological data pertaining to each lake had to be acquired from other sources. These included location (Table 2), cross-sectional area as a function of depth (Tables 3–12), and extinction coefficient. (No bathymetric data could be obtained for DeGray Lake.) For model use, these data had to be prepared as an algebraic relationship for area as a percentage of the surface area.

With respect to the validation data for the lake temperatures, published sources were used when possible. For other lakes, data have been made available to the authors by personal contacts. In many instances, these form the basis of ongoing collaborative research. Since full reports on these lakes will be made in the literature in conjunction with the originators of the data, only brief overview comments on those lakes will be made here.

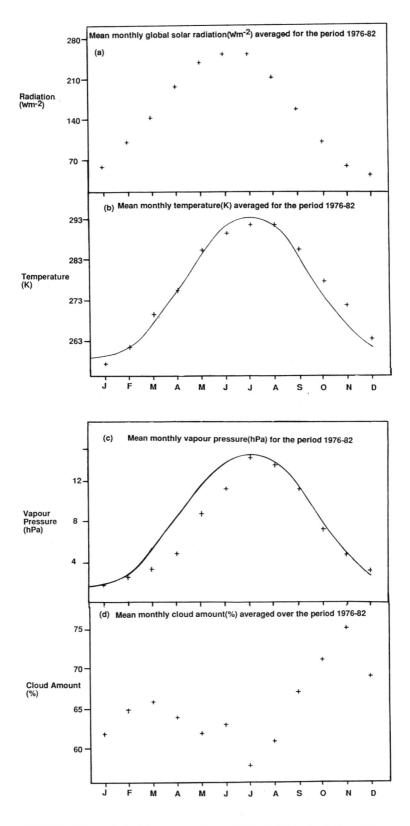

FIGURE 1.  Meteorological data (averaged over 1976 to 1982) for the Sudbury lakes in Canada: (a) radiation data; (b) air temperature data and best fit sine curve; (c) vapor pressure data and best fit sine curve; (d) cloud data (expressed in %) used as monthly averages.

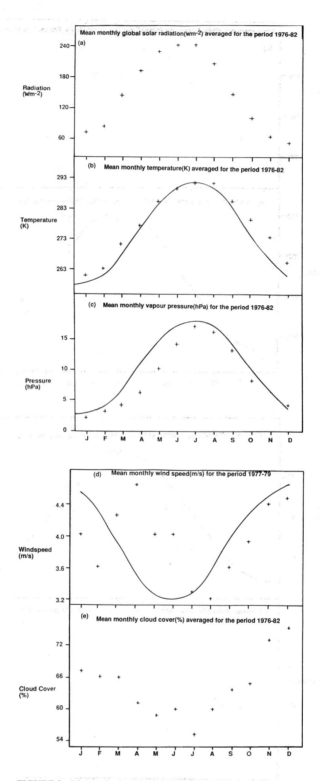

FIGURE 2.   Meteorological data (averaged over 1976 to 1982) for the
lakes in the Muskoka-Haliburton area of Ontario, Canada: (a) radiation
data; (b) air temperature data and best fit sine curve; (c) vapor pressure
data and best fit sine curve; (d) wind speed data and best fit sine curve
(averaged over 1977 to 1979 only); (e) cloud data (expressed in %) used
as monthly averages.

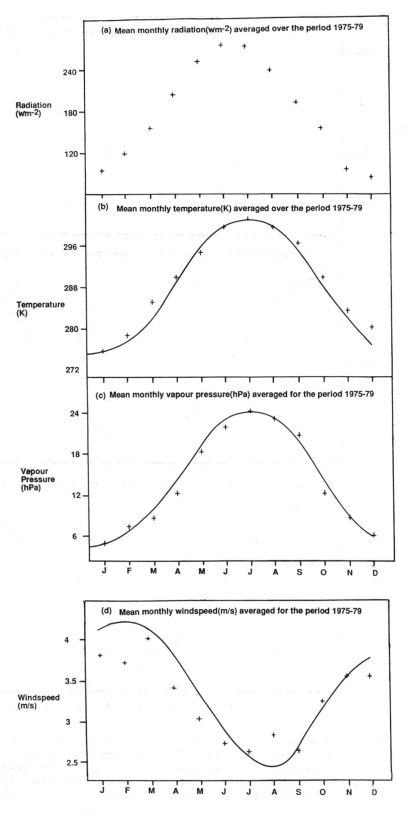

FIGURE 3. Meteorological data (averaged over 1975 to 1979) for DeGray Lake in Arkansas, USA: (a) radiation data; (b) air temperature data and best fit sine curve; (c) vapor pressure data and best fit sine curve; and (d) wind speed data and best fit sine curve.

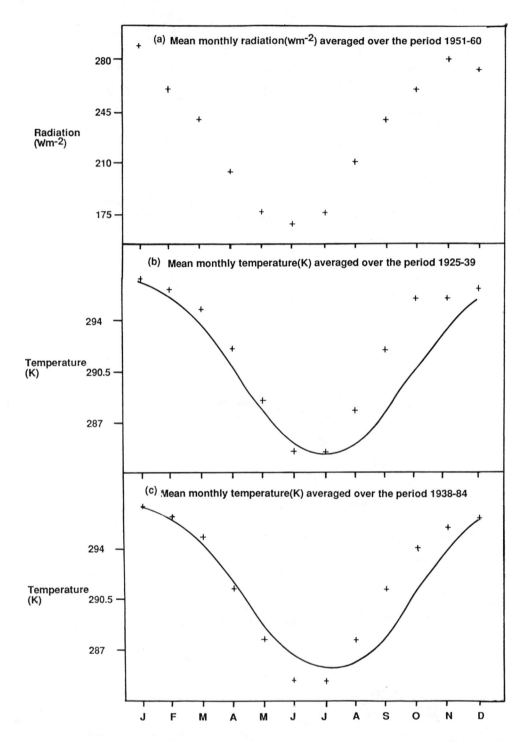

FIGURE 4. Meteorological data for Hartbeespoort and Buffelspoort Dams in South Africa: (a) radiation data (averaged over 1951 to 1960); (b) air temperature data and best fit sine curve (Hartbeespoort Dam, averaged over 1925 to 1939); (c) air temperature data and best fit sine curve (Buffelspoort Dam, averaged over 1938 to 1984); (d) vapor pressure data and best fit sine curve (averaged over 1977 to 1986); (e) wind speed data and best fit sine curve (averaged over 1972 to 1977); (f) cloud data (expressed as %) used as monthly averages (averaged over 1972 to 1977).

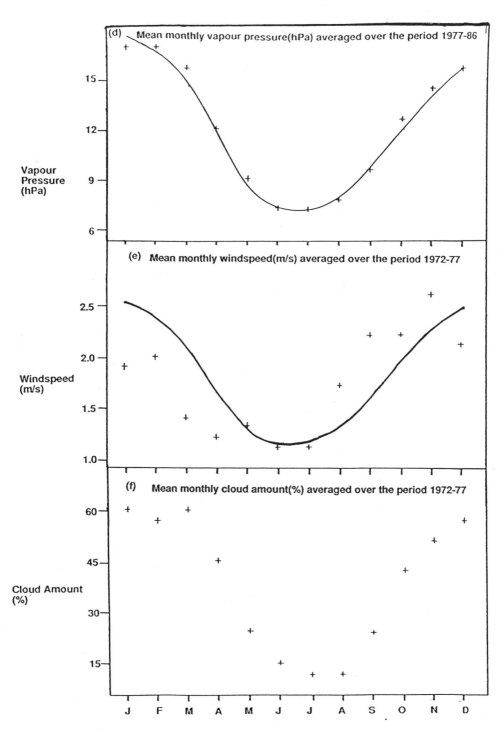

FIGURE 4 (continued).

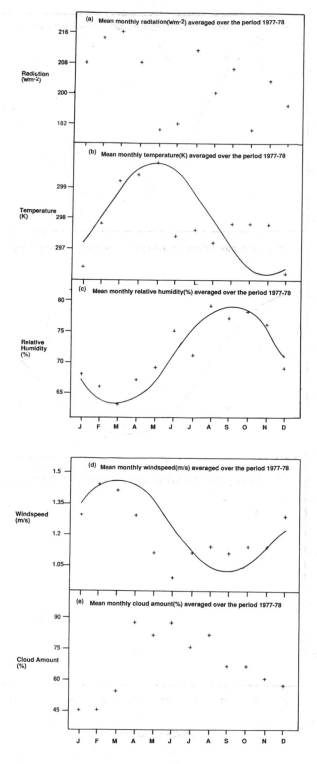

FIGURE 5.  Meteorological data (averaged over 1977 to 1978) for Lake Valencia in Venezuela: (a) radiation data; (b) air temperature data and best fit sine curve; (c) relative humidity data (in %) and best fit sine curve; (d) wind speed data and best fit sine curve; (e) cloud data (expressed in %) used as monthly averages.

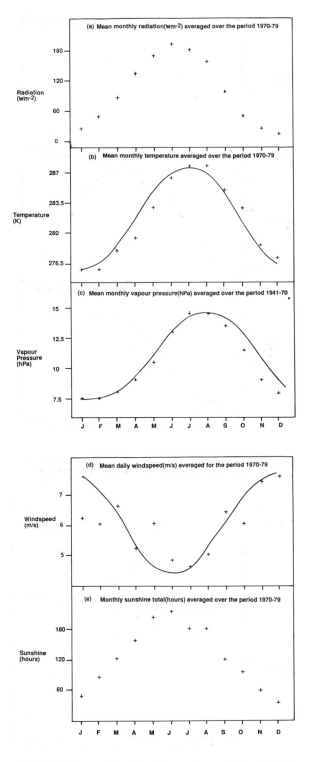

FIGURE 6. Meteorological data (averaged over 1970 to 1979) for Windermere in the U.K.: (a) radiation data; (b) air temperature data and best fit sine curve; (c) vapor pressure data and best fit sine curve (averaged over 1941 to 1970); (d) wind speed data and best fit sine curve; (e) cloud data (expressed by mean daily sunshine totals) used as monthly averages.

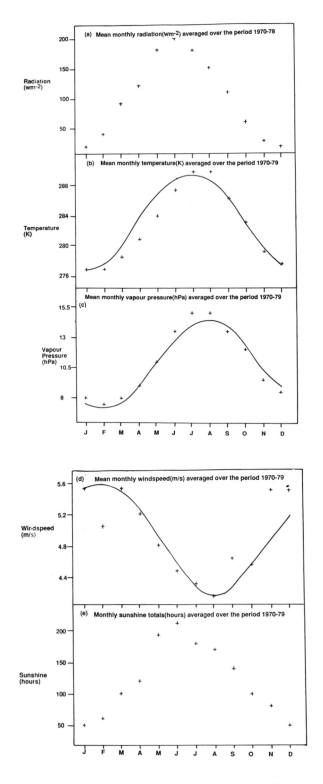

FIGURE 7.  Meteorological data (averaged over 1970 to 1979) for Grafham Water and Rutland Water in the U.K.: (a) radiation data; (b) air temperature data and best fit sine curve; (c) vapor pressure data and best fit sine curve; (d) wind speed data and best fit sine curve; (e) cloud data (expressed by mean daily sunshine totals) used as monthly averages.

## TABLE 2
### Locations of Lakes Used in this Study

| Lake | Latitude | Longitude |
|------|----------|-----------|
| **Muskoka-Haliburton Lakes** | | |
| Buck | 45° 23′ N | 79° 00′ W |
| Solitaire | 45° 22′ N | 79° 00′ W |
| **Sudbury Lakes** | | |
| Clearwater | 46° 22′ N | 81° 03′ W |
| Nelson | 46° 44′ N | 81° 05′ W |
| **U. K. Lakes** | | |
| Windermere | 54° 20′ N | 2° 57′ W |
| Rutland Water | 52° 38′ N | 0° 40′ W |
| Grafham Water | 52° 18′ N | 0° 20′ W |
| **U. S. Lakes** | | |
| DeGray | 34° 15′ N | 93° 15′ W |
| **South African Lakes** | | |
| Hartbeespoort Dam | 25° 46′ S | 27° 50′ E |
| Buffelspoort Dam | 25° 48′ S | 27° 29′ E |
| **South American Lakes** | | |
| Valencia | 10° 13′ N | 67° 30′ W |

## TABLE 3
### Cross-Sectional Profile for the North Basin of Windermere

| Depth (m) | Area (km²) |
|-----------|------------|
| 0 | 8.046 |
| 2 | 6.943 |
| 5 | 6.027 |
| 10 | 5.385 |
| 20 | 4.343 |
| 30 | 3.288 |
| 35 | 2.873 |
| 40 | 2.454 |
| 45 | 1.665 |
| 50 | 1.022 |
| 55 | 0.607 |
| 60 | 0.147 |
| 64 | 0 |

Perhaps the most difficult data to obtain are values for the extinction coefficient. For relatively pristine lakes, a value for clear water of $\eta = 0.1$ to 0.2 is reasonable to assume in the absence of measurements. For the Muskoka lakes, a seasonal variation between a winter value of 0.2 and a summer value of 0.4 was utilized. For the more turbid lakes, a larger value of $\eta$ was more appropriate, as reported in part in the literature.[30,31]

## TABLE 4
### Cross-Sectional Profile for Nelson Lake, Sudbury Region, Canada

| Depth (m) | Area (km²) |
|-----------|------------|
| 0 | 3.090 |
| 2 | 2.530 |
| 4 | 2.210 |
| 6 | 1.970 |
| 8 | 1.710 |
| 10 | 1.480 |
| 12 | 1.230 |
| 14 | 1.030 |
| 16 | 0.837 |
| 18 | 0.660 |
| 20 | 0.522 |
| 22 | 0.442 |
| 24 | 0.327 |
| 26 | 0.262 |
| 28 | 0.228 |
| 30 | 0.195 |
| 32 | 0.175 |
| 34 | 0.138 |
| 36 | 0.113 |
| 38 | 0.0992 |
| 40 | 0.0829 |
| 42 | 0.0708 |
| 44 | 0.0592 |
| 46 | 0.0422 |
| 48 | 0.0258 |
| 50 | 0.0129 |
| 51 | 0.0 |

From Sudbury Environmental Study, Study of Lakes and Watersheds near Sudbury Ontario, SES 009/82, Supplement, Ontario Ministry of the Environment, 1982. With permission.

## TABLE 5
### Cross-Sectional Profile for Clearwater Lake, Sudbury Region, Canada

| Depth (m) | Area (km²) |
|-----------|------------|
| 0 | 0.765 |
| 2 | 0.615 |
| 4 | 0.515 |
| 6 | 0.444 |
| 8 | 0.369 |
| 10 | 0.303 |
| 12 | 0.258 |
| 14 | 0.208 |
| 16 | 0.0794 |
| 18 | 0.0456 |
| 20 | 0.00599 |
| 21 | 0.00120 |
| 21.5 | 0 |

From Sudbury Environmental Study, Study of Lakes and Watersheds near Sudbury Ontario, SES 009/82, Supplement, Ontario Ministry of the Environment, 1982. With permission.

## TABLE 6
### Cross-Sectional Profile for Buck Lake, Muskoka–Haliburton Area of Ontario, Canada

| Depth (m) | Area (km²) |
|---|---|
| 0 | 0.403 |
| 2 | 0.349 |
| 4 | 0.310 |
| 6 | 0.268 |
| 8 | 0.228 |
| 10 | 0.195 |
| 12 | 0.168 |
| 14 | 0.145 |
| 16 | 0.111 |
| 18 | 0.0495 |
| 20 | 0.0603 |
| 22 | 0.0379 |
| 24 | 0.0224 |
| 26 | 0.0155 |
| 28 | 0.00864 |
| 30 | 0 |

From Nicolls, A., et al., Morphometry of the Muskoka–Haliburton Study Lakes, Data Report DR 83/3, Ontario Ministry of the Environment, Dorset, Ontario, 1983. With permission.

## TABLE 7
### Cross-Sectional Profile for Lake Valencia, Venezuela

| Depth (m) | Area (km²) |
|---|---|
| 0 | 350 |
| 5 | 297.5 |
| 10 | 257.3 |
| 15 | 220.5 |
| 20 | 178.5 |
| 25 | 134.8 |
| 30 | 94.5 |
| 35 | 45.5 |
| 40 | 0 |

For temperature data obtained from the literature and other grey literature sources (e.g., Reid et al.[32] for the Muskoka lakes), the values were obtained as isopleth patterns. For visual comparison, as discussed here, this is adequate. For objective analysis, as presaged in Henderson-Sellers,[18] the isopleth patterns were digitized. An objective assessment of the lake simulations presented here forms the topic of an ongoing study.

It should also be stressed that the aim of the study presented here is to illustrate the *general* and *widely applicable* nature of this illustrative stratification model. The comparisons made are to be judged at a high level of abstraction (viz., with respect to depth of thermocline, maximum surface temperatures), since the model is being forced by climatological, rather than meteorological, data so that the model can be used for decision support for managerial planning. In this context, "typical", order-of-magnitude answers for stratification depth and duration, maximum temperatures, hypolimnetic temperatures, etc. are most important with respect to (1)

**TABLE 8**
**Cross-Sectional Profile for Hartbeespoort Dam, South Africa**

| Depth (m) | Area (km²) |
|---|---|
| 0 | 20.344 |
| 0.32 | 19.854 |
| 1.32 | 18.241 |
| 2.32 | 16.585 |
| 3.32 | 15.292 |
| 4.32 | 14.147 |
| 5.32 | 12.940 |
| 6.32 | 11.711 |
| 7.32 | 10.514 |
| 8.32 | 9.341 |
| 9.32 | 8.628 |
| 10.32 | 8.021 |
| 11.32 | 7.484 |
| 12.32 | 6.783 |
| 13.32 | 6.050 |
| 14.32 | 5.307 |
| 15.32 | 4.554 |
| 16.32 | 4.011 |
| 17.32 | 3.415 |
| 18.32 | 3.040 |
| 19.32 | 2.548 |
| 20.32 | 2.044 |
| 21.32 | 1.633 |
| 22.32 | 1.218 |
| 23.32 | 1.055 |
| 24.32 | 1.015 |
| 25.32 | 0.834 |
| 26.32 | 0.707 |
| 27.32 | 0.472 |
| 28.32 | 0.304 |
| 29.32 | 0.179 |
| 30.32 | 0.091 |
| 31.32 | 0 |

From Hartbeespoort Dam, Capacity Determination, Report A2R01, Planning Division, Department of Water Affairs, Pretoria, South Africa, 1979.

their direct influence on dissolved oxygen concentrations, algal growth rates, etc. and (2) appropriate depth of abstraction with respect to these biochemical parameters as well as temperature. In general, the observations presented here are literature values for individual years and, as such, have a relatively high degree of specificity and spatio-temporal variance, which is not possible to represent in the hydroclimatic simulations. Future work will aim to (1) average large data sets and (2) force the model with meteorological data from specific years in order to produce more compatible comparisons.

# V. MODEL VALIDATION

The model EDD1 has been used successfully for a number of lakes worldwide, using local meteorological values to which a sine curve has been fitted. Figure 8 shows observations and simulation results for a warm monomictic/dimictic natural lake at 54°N: Windermere (North basin). Results are shown as hydroclimatic averages, i.e., the climatic response rather than an individual year's behavior. The observations are taken from Lund et al.[33] for the year 1947. This

**TABLE 9**
**Cross-Sectional Profile for Buffelspoort Dam, South Africa**

| Depth (m) | Area (km²) |
|---|---|
| 0 | 1.357 |
| 0.37 | 1.315 |
| 1.37 | 1.206 |
| 2.37 | 1.101 |
| 3.37 | 0.970 |
| 4.37 | 0.877 |
| 5.37 | 0.794 |
| 6.37 | 0.721 |
| 7.37 | 0.626 |
| 8.37 | 0.553 |
| 9.37 | 0.493 |
| 10.37 | 0.424 |
| 11.37 | 0.344 |
| 12.37 | 0.271 |
| 13.37 | 0.211 |
| 14.37 | 0.167 |
| 15.37 | 0.128 |
| 16.37 | 0.095 |
| 17.37 | 0.073 |
| 18.37 | 0.053 |
| 19.37 | 0.034 |
| 20.37 | 0.020 |
| 21.37 | 0.012 |
| 22.37 | 0.002 |
| 23.14 | 0 |

From Buffelspoort Dam, Capacity Determination, Report A2R05, Planning Division, Department of Water Affairs, Pretoria, South Africa, 1979.

**TABLE 10**
**Cross-Sectional Profile for Rutland Water, Midlands, U.K.**

| Depth (m) | Area (km²) |
|---|---|
| 0 | 12.602 |
| 1.52 | 11.254 |
| 3.05 | 10.414 |
| 4.57 | 9.095 |
| 6.10 | 8.002 |
| 7.62 | 6.707 |
| 9.14 | 6.106 |
| 10.67 | 5.363 |
| 12.19 | 4.578 |
| 13.72 | 3.828 |
| 15.24 | 2.970 |
| 16.76 | 2.096 |
| 18.29 | 1.890 |
| 19.81 | 1.528 |
| 21.34 | 1.260 |
| 22.86 | 0.912 |
| 24.38 | 0.624 |
| 25.91 | 0.424 |
| 27.43 | 0.246 |
| 28.96 | 0.0975 |
| 30.48 | 0.0295 |
| 32.00 | 0 |

### TABLE 11
### Cross-Sectional Profile for Grafham Water, Midlands, U.K.

| Depth (m) | Area (km²) |
|:---:|:---:|
| 0 | 6.282 |
| 1 | 6.018 |
| 2 | 5.761 |
| 3 | 5.497 |
| 4 | 5.214 |
| 5 | 4.931 |
| 6 | 4.630 |
| 7 | 4.222 |
| 8 | 3.788 |
| 9 | 3.323 |
| 10 | 2.871 |
| 11 | 2.425 |
| 12 | 1.954 |
| 13 | 1.432 |
| 14 | 1.037 |
| 15 | 0.741 |
| 16 | 0.484 |
| 17 | 0.276 |
| 17.8 | 0.163 |

### TABLE 12
### Cross-Sectional Profile for Solitaire Lake, Muskoka-Haliburton Region of Ontario, Canada

| Depth (m) | Area (km²) |
|:---:|:---:|
| 0 | 0.109 |
| 2 | 0.0899 |
| 4 | 0.0797 |
| 6 | 0.0651 |
| 8 | 0.0539 |
| 10 | 0.0371 |
| 12 | 0.0246 |
| 14 | 0.0152 |
| 16 | 0.0104 |
| 18 | 0.00720 |
| 20 | 0.00448 |
| 22 | 0.00208 |
| 24 | 0.000240 |
| 25 | 0.0 |

From Nicolls, A., et al., Morphometry of the Muskoka-Haliburton Study Lakes, Data Report DR 83/3, Ontario Ministry of the Environment, Dorset, 1983. With permission.

lake rarely experiences ice conditions and is of a large surface area. It is seen that the basic hydroclimate features (metalimnion at 10 to 20 m, winter temperatures and timing of the two overturns) are well simulated. The areas of least precision are the ~ 4K cooler summer surface temperatures and the cooler summer hypolimnetic temperatures. Simulation results for the South basin of Windermere are given in Henderson-Sellers,[22] showing a good agreement with

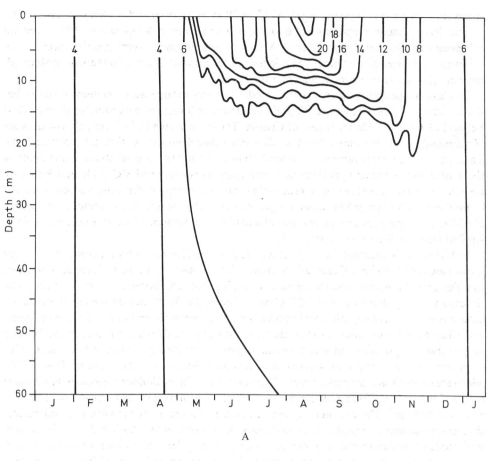

A

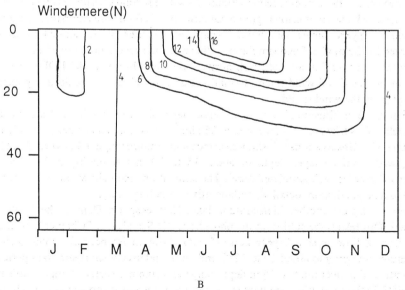

B

FIGURE 8. (A) Observed hydroweather (for 1947) and (B) simulated hydroclimate for the north basin of Windermere in the U.K. (Isopleths are in °C.) (Figure 8A is from Lund, J. W. G., et al., *Phil. Trans. R. Soc.*, 246, 255–290, 1963. With permission.)

observations for a metalimnetic depth of 15 to 20 m, maximum surface temperature ≥17°C, summer hypolimnetic temperature of 6 to 8°C, winter minimum temperature ~3°C, occurring in February, and isothermal conditions (~8°C) recovered in late November/early December. In this instance the area of least precision is in the fall period when the 8°C isotherm is mislocated, probably giving rise to errors of the order of ~0.5K.

To illustrate the appropriateness of this model, not only to large warm monomictic lakes, but also to dimictic lakes with winter icing, Figure 9 depicts 5 years of data from the deep (~50m) Nelson Lake, in the Sudbury region of Canada. The surface area is 3.09 km$^2$, almost an order of magnitude larger than the other Canadian lakes described below. Generally Nelson Lake maximum surface temperatures are around 20 to 22°C and the normal stratification depth is about 10 m, with summer hypolimnetic temperatures between 4 and 6°C. Although these data are not averaged, based on this description we can compare the observations with the hydroclimatically averaged simulation. Agreement is high with maximum surface temperatures of >20°C and hypolimnetic temperatures of 4 to 6°C. The duration of ice-free conditions, from April to October, is also well simulated.

In Figure 10 is depicted DeGray Lake, situated in Arkansas, with a considerably larger surface area of 68.80 km$^2$. Observations from 1979, shown in this figure, illustrate a lake that stratifies strongly, with summer temperatures ranging from 6°C (bottom) to 30°C (surface). The simulation is only slightly cooler (~27°C) than observed in 1979, and the overall stratification pattern is good, with a slightly cool hypolimnion. As noted at the end of the last section, many of the features of the stratification cycle that need to be simulated for use in management decision support (both for planning and operational management) are clearly evident in the model results.

Figures 11 and 12 show the simulation of two fully dimictic lakes in Canada (45 to 46°N), which are of much smaller surface area: Clearwater Lake, in the Sudbury region, with a surface area of 0.765 km$^2$, and Buck Lake, in the Muskoka-Haliburton region of Ontario, with a surface area of 0.403 km$^2$. The excellent agreement here in Figure 11 is between a hydroclimatic simulation and data averaged over several years. All features of the stratification cycle are very well modeled, although the hypolimnion warms slightly less than observed in summer. A similarly excellent agreement is seen in Figure 12 with a second Canadian lake, this time Buck Lake in the Muskoka-Haliburton region of Ontario. In this figure are compared isotherms from 1977 and the hydroclimate simulation. Again, all features of the annual cycle are well represented by the model. Such simulations illustrate the success not only of simulating lakes with winter ice cover, but also of small horizontal extent. In the model EDD1, the effect on mixing of the restricted wind fetch of small lakes is modeled using a decaying exponential modification to the wind speed.

In Figure 13 is illustrated a very different lake: the tropical Lake Valencia at 10°N (Venezuela). This lake has a surface area of 350 km$^2$ and is situated at a height of 420 m above sea level (ASL). Observations[34,35] show a surface temperature range of between about 25.6 and 28.5°C, with a mixed layer depth of about 15 to 20 m terminating in November (both characteristics being well simulated here). The simulation result shown here in Figure 13 is a prelude to a planned, more detailed, collaborative modeling study.

Figure 14 depicts another high-altitude lake, Hartbeespoort Dam in the South African Transvaal. This lake is the subject of the phosphorus DSS analysis of Chapter 8. Isotherms are shown from a little over a 2-year period[30] and show an isothermal winter period, with temperatures dropping to about 10 to 12°C and with a relatively short stratified period with a loose stratification over the 5 to 15 m depth range and summer surface temperatures reaching about 26°C. In the simulation, the winter temperatures are a little too warm at depth (probably because of a poor simulation of the high altitude and clear skies in the winter period), with a reasonable simulation of thermocline depth and annual range in surface temperatures.

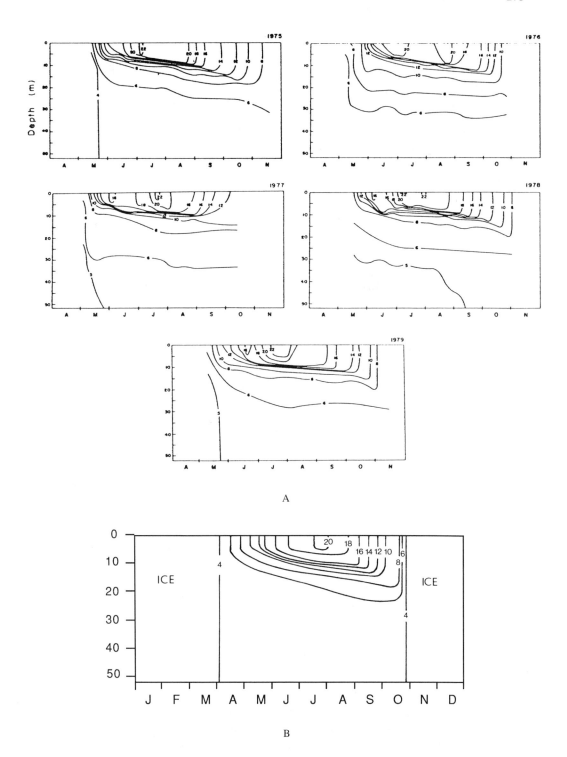

FIGURE 9. Nelson Lake: (A) Observations from 1975 to 1979 and (B) simulated hydroclimate. (Isopleths are in °C.) (Figure 9A is from Sudbury Environmental Study, Studies of Lakes and Watersheds near Sudbury Ontario. Final Limnological Report, SES 009/82, Supplement, Ontario Ministry of the Environment, 1982. With permission.)

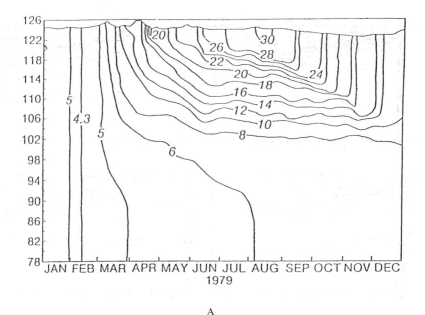

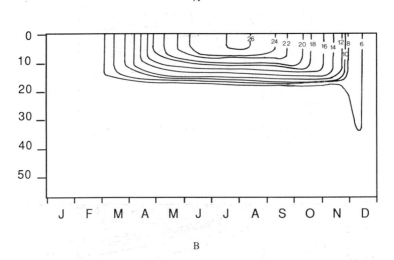

FIGURE 10. (A) Observed hydroweather (for 1979) and (B) simulated hydroclimate for DeGray Lake in Arkansas, U.S. (Isopleths are in °C.) (Figure 10A is from Johnson, L. S., and Ford, D. E., Verification of a One-Dimensional Reservoir Thermal Model, presented at ASCE 1981 Convention and Exposition, St. Louis, MO, October 1981. With permission.)

In Figure 15 is shown a second large South African reservoir, Buffelspoort Dam. The simulation is contrasted with data for the years 1973 to 1975 from Walmsley and Toerien,[31] which show a large variation in water level, slightly masking a stratification depth of around 8 to 12 m with a similar temperature range to that of Hartbeespoort Dam. These data have been roughly averaged in Figure 15B, thus providing a more readily interpretable comparison with the simulation of Figure 15C, which shows a reasonable range for summer temperatures; although in this case the stratification is too sharply defined in the simulation, and, once again, as noted above, the winter temperatures are a little too warm.

Figures 16 and 17 show two further U.K. reservoirs, both filled at least partially by a pumped inflow. Figure 16 shows Rutland Water, constructed in the mid-1970s for water supply, which

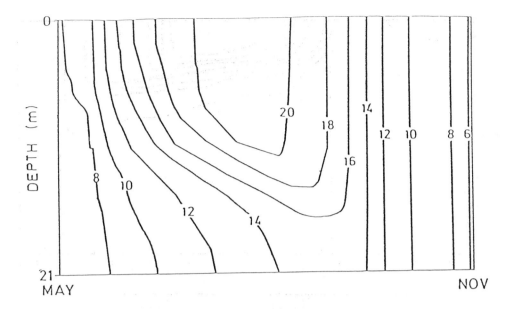

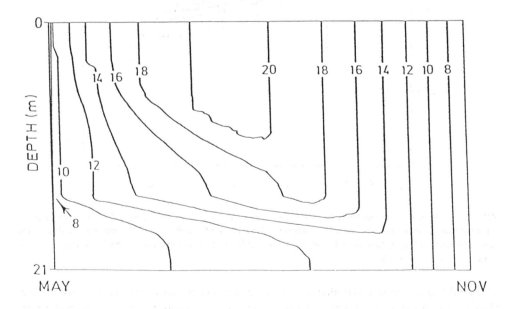

FIGURE 11. (top) Observations averaged from 5 years' data[32] and (bottom) simulated hydroclimate for Clearwater Lake, Ontario. (Isopleths are in °C.) (Figure 11 [bottom] is from Henderson-Sellers, B. and Davies, A. M., *Ann. Rev. Numerical Fluid Mechan. Heat Transf.*, 2, 86–156, 1989. With permission from Hemisphere Publishing.)

is filled by cross-country pipeline. It is of roughly the same size as the natural lake Windermere (which is itself the largest water body in England) and is in a locality where stratification and algal blooms are highly likely. When the lake was built, artificial destratification devices were installed and used in subsequent years (except one). Consequently the database of the *natural* stratification cycle is virtually nil, since it is not clear how a single year's data inbetween 2 years

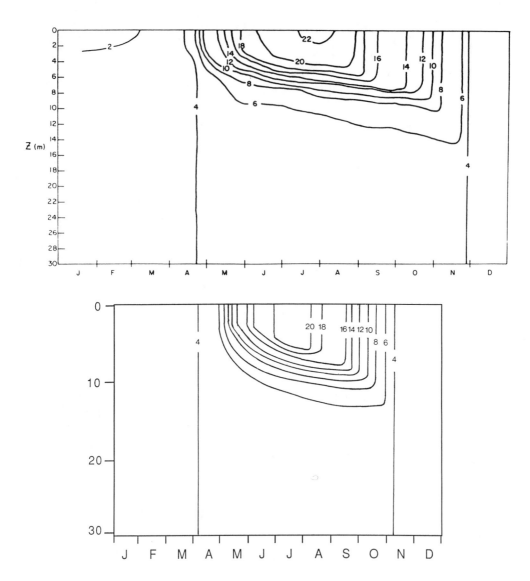

FIGURE 12.  Buck Lake in the Muskoka–Haliburton area of Ontario: (top) observations for 1977 and (bottom) simulated hydroclimate. (Isopleths are in °C.) (Figure 12 [top] is from Reid, R. A., et al., Temperature Profiles in the Muskoka–Haliburton Study Lakes [1976–1982], Data Report DR 83/4, Ontario Ministry of the Environment, Dorset, Ontario, 1983. With permission.)

of destratification can be reconciled with the long-term averaged hydroclimate discussed here. Nevertheless, the stratification pattern depicted in Figure 16 illustrates a stratification cycle not atypical of the location. In contrast, Grafham Water (Figure 17), in the same climatological area of England, is much shallower and shows little signs of stratification, with typical maximum temperatures of 16°C. The data for 1977 show a similar quasi-isothermal temperature structure, but with marginally warmer summer temperatures.

Recently EDD1 has been applied to Barombi Mbo, the largest crater lake in Cameroon (4.15 km²). This lake has been recently closely monitored in conjunction with the studies of the catastrophic degassings of nearby Lakes Nyos and Monoun.[36,37] The simulated hydrothermal regime indicates surface temperatures in the region of 26.5 to 28.5°C, with a thermocline at a depth of approximately 20 m. (The full details of this collaborative modeling project will be presented at a later date.)

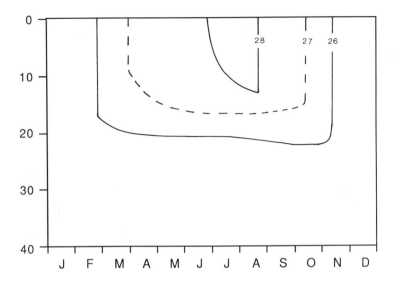

FIGURE 13.    Lake Valencia, Venezuela: simulated hydroclimate. (Isopleths are in °C.)

## VI. SENSITIVITY ANALYSIS

In an extended sensitivity analysis of a "typical" lake,[20] it was found that for realistic perturbations to the meteorological forcing parameters, the thermal structure of a lake is most sensitive to the wind speed, only more so at lower wind speeds. For changes in solar radiation, however, the lake recovered quickly, suggesting that the inclusion of a full surface energy budget permits the accurate simulation of the damping feedbacks present in the total lake system. The results of these extensive sensitivity tests are summarized in Table 13.

It is not particularly instructive to repeat the detailed results of those experiments here; rather an experiment will be conducted to determine whether the consequences of such sensitivity tests can be used to improve the simulations of lakes. One particular example will be used illustratively.

With the basic forcing parameters and lake turbidity characteristics assumed to be generically valid as a first approximation, Solitaire Lake, another lake in the Muskoka-Haliburton area of Ontario, is found to be too warm at depth. This suggests that the stratification as simulated is too weak. Consequently the general value of the water turbidity ($\eta = 0.2$) is replaced in a series of experiments with $\eta = 0.1, 0.3, 0.4, 1.0$ and a seasonally varying value. The results, illustrated in part in Figure 18, suggest that increasing the value of $\eta$ rapidly prevents the hypolimnion from warming. However, any such larger constant value of $\eta$ tends to make the predicted stratification too strong, and it is the simulation with a seasonally varying $\eta$ that appears to provide the most worthwhile improvement, whilst endorsing the necessity of data acquisition for the temporal variability of the extinction coefficient (only incorporated in Figure 18 to a first approximation).

## VII. CONCLUSIONS AND RECOMMENDATIONS

These studies illustrate the wide range of lake types and locations that can be modeled by a single systems model, if it is constructed in such a way that site-dependent tuning is unnecessary, and emphasize the parameter values that need most accurate measurement (i.e., those to which the system is most sensitive). For lake models, it is found that an assessment of the water turbidity and its seasonal variability can have a profound effect on the success (or otherwise) of simulations. This is especially true for lakes in the oligotrophic to mesotrophic range with

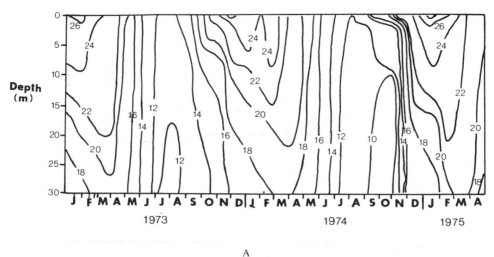

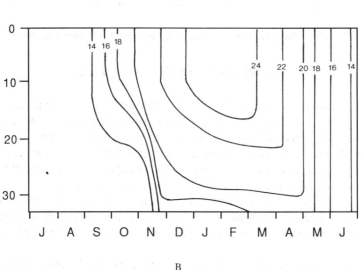

FIGURE 14. (A) Observed hydroweather (for 1973–1975) and (B) simulated hydroclimate for Hartbeespoort Dam in South Africa. (Isopleths are in °C.) (Figure 14A is from Scott, W. E., et al., *J. Limnol. Soc. Sth. Afr.*, 5, 43–58, 1977. With permission.)

relatively clear waters, say η less than about 0.6. Secondly, although not demonstrated here (cf. Henderson-Sellers[20]), errors in measurement at high wind speeds are less likely to impact the lake simulation than errors at low wind speed. Such conclusions lead to the recommendation to observationalists to ensure that data on such parameters as extinction coefficient and wind speed, in addition to radiation levels, air temperature, relative humidity, etc. are collected on a regular basis as part of any limnological study.

## ACKNOWLEDGMENTS

We wish to thank Bill Lewis for data from Lake Valencia; Peter Dillon, Norman Yan, and Barbara Locke for data for the Muskoka and Sudbury lakes; Mike Pearson for data on Grafham Water and Rutland Water; and George Kling for data on the Cameroon lakes.

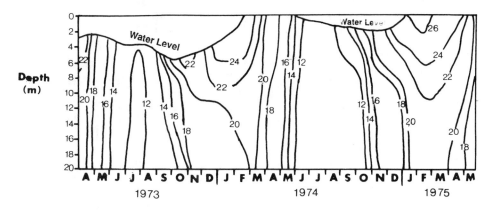

A

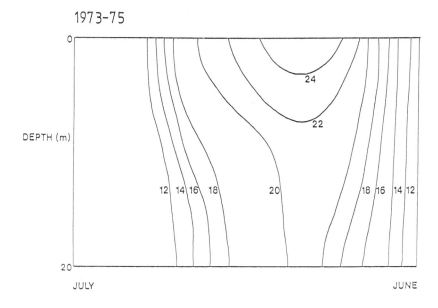

B

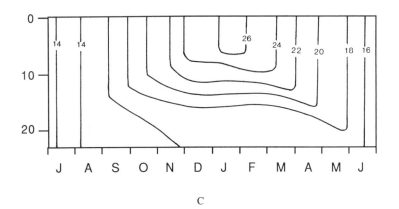

C

FIGURE 15. (A) Observed hydroweather (for 1973–1975 which are averaged in [B]) and (C) simulated hydroclimate for Buffelspoort Dam in South Africa. (Isopleths are in °C.) (Figure 15A is from Walmsley, R. D. and Toerien, D. F., *J. Limnol. Soc. Sth. Afr.*, 5, 51–58, 1979. With permission.)

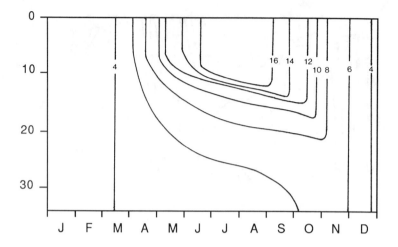

FIGURE 16. Hydroclimatic simulation of Rutland Water in the U.K. (Isopleths are in °C.)

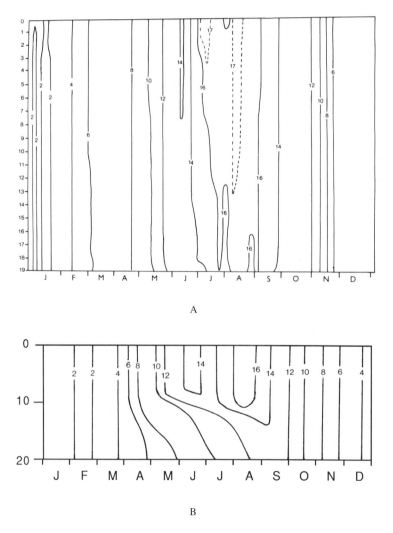

FIGURE 17. (A) Observed hydroweather (for 1979, unpublished data) and (B) simulated hydroclimate for Grafham Water in the U.K. (Isopleths are in °C.)

**TABLE 13**
**The Maximum Impact on Surface Temperatures for Various**
**Perturbation Experiments**

| Variable (perturbation) | Max $\Delta T$ (K) | Comments |
|---|---|---|
| *Wind speed* | | |
| Doubled | −3 | Errors at low wind speed |
| Halved | +5 | more important |
| *Aerodynamic drag coefficient* | | |
| (Various formulations) | | No significant effect |
| *Air temperature* | | |
| Increased by 3K | +2 | Everywhere differences >1.5K |
| Decreased by 3K | −2.5 | Range of 0 to −2.5 K |
| *Cloud cover* | | |
| No cloud | +6 | Large changes |
| Total cloud | −6 | everywhere |
| *Atmospheric emissivity* | | |
| Constant = 0.75 | −2 | |
| Constant = 0.95 | 1.3–1.8 everywhere | |
| *Extinction coefficient* | | |
| η = 0.3 within date range 160–290 plus | | Max. impact at thermocline |
| η = 0.5 within date range 180–250 | ~1 | level of −5 to −7 K |
| η = 0.8 date range 180–260 | ~1 | Max. impact at thermocline level of −5 to −7 K |
| η = 0.4 for T > 288 K plus | | Max. impact at thermocline |
| η = 0.8 for T > 293 K | ~1 | level of −5 to −7 K |
| η = 0.5 all year | −3 to +2 | Max. impact at thermocline level of −5 to −7 K |
| η = 0.8 all year | −2 to +2 | Max. impact at thermocline level of −10 K |
| *Surface albedo* | | |
| 0.06–0.08 ±50% | < 1 | No significant effect |
| *Bathymetry* | | Significant differences especially at depth |
| *Eddy diffusion coefficient* | | |
| Multiplied by 10 | −1 | More significant for |
| Divided by 10 | +3 | reduced eddy diffusion values |

*Note*: For each indicated variable and perturbation, the maximum value of $\Delta T$ (difference between perturbation and control experiment) is given.

From Henderson-Sellers, B., *Appl. Math. Modelling*, 12, 31–43, 1988. With permission.

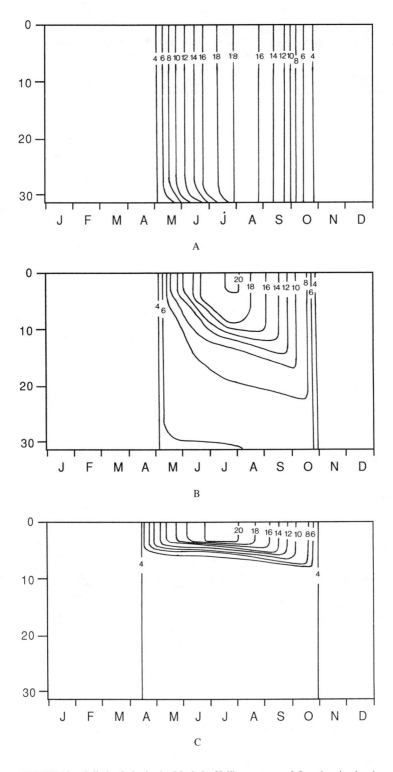

FIGURE 18.   Solitaire Lake in the Muskoka-Haliburton area of Ontario: simulated hydroclimate for (A) $\eta = 0.1$; (B) $\eta = 0.2$; (C) $\eta = 1.0$; (D) a seasonally varying value for $\eta$, and (E) observations for 1978. (Isopleths are in °C.) (From Reid, R. A., et al., Temperature profiles in the Muskoka–Haliburton Study Lakes (1976–1982), Data Report DR 83/4, Ontario Ministry of the Environment, Dorset, Ontario, 1983.With permission. )

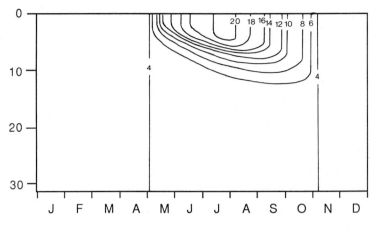

FIGURE 18D.

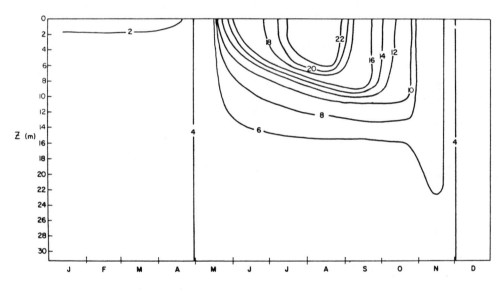

FIGURE 18E.

# REFERENCES

1. **Reckhow, K. H. and Chapra, S. C.,** *Engineering Approaches for Lake Management*, 2 vols, Ann Arbor Sci., Ann Arbor, MI, 1983.
2. **Henderson-Sellers, B.,** *Engineering Limnology*, Pitman, London, 1984.
3. **Henderson-Sellers, B. and Markland, H. R.,** *Decaying Lakes. The Origins and Control of Cultural Eutrophication*, John Wiley, Chichester, England, 1987.
4. **Environmental Laboratory,** CE-QUAL-R1, A numerical one-dimensional model of reservoir water quality, User's Manual, Instruction Report E-82-1, U.S. Army Corps of Engineers, Waterways Experiment Station, CE, Vicksburg, MS, 1982.
5. **Johnson, L. S. and Ford, D. E.,** Verification of a One-Dimensional Reservoir Thermal Model, presented at ASCE 1981 Convention and Exposition, St.Louis, MO, October 1981.

6. **Wlosinski, J. H. and Dortch, M. S.,** Development and evaluation of a model (CE-QUAL-R1) of reservoir water quality, in *Lake and Reservoir Management: Practical Applications*, Proc. 4th Annual Conference and International Symposium (October 16–19, 1984), McAfee, NJ, North American Lake Management Society, 1985, 186–190.

7. **Waide, J.,** personal communication, 1982.

8. **Beck, M. B. and van Straten, G., Eds.,** *Uncertainty and Forecasting of Water Quality*, Springer-Verlag, Berlin, 1983.

9. **Reckhow, K. H. and Chapra, S. C.,** Confirmation of water quality models, *Ecol. Modelling*, 20, 113–133, 1983.

10. **Kirchner, W. B. and Dillon, P. J.,** An empirical method of estimating the retention of phosphorus in lakes, *Water Resour. Res.*, 11, 182–183, 1975.

11. **Reckhow, K. H. and Chapra, S. C.,** A note on error analysis for a phosphorus retention model, *Water Resour. Res.*, 15, 1643–1646, 1979.

12. **Reckhow, K. H.,** A method for the reduction of lake model prediction error, *Water Res.*, 17, 911–916, 1983.

13. **Huson, L. W.,** Definition and properties of a coefficient of sensitivity for mathematical models, *Ecol. Modelling*, 21, 149–159, 1984.

14. **Loftis, J. C., Ward, R. C. and Smillie, G. M.,** Statistical models for water quality regulation, *J. Water Poll. Contr. Fed.*, 55, 1098–1104, 1983.

15. **Wlosinski, J. H.,** *Evaluation Techniques for CE-QUAL-R1: A One-Dimensional Reservoir Water Quality Model*, Miscellaneous Paper E-84-1, U.S. Army Waterways Experiment Station, Vicksburg, MS, 1984

16. **Willmott, C. J., Ackleson, S. G., Davis, R. E., Feddema, J. J., Klink, K. M., Legates, D. R., O'Donnell, J., and Rowe, C. M.,** Statistics for the evaluation and comparison of models, *J. Geophys. Res.*, 90, 8995–9005, 1985.

17. **Young, P. C.,** The validity and credibility of models for badly defined systems, in *Uncertainty and Forecasting of Water Quality*, Beck, M. B. and van Straten, G., Eds., Springer-Verlag, Berlin, 1983, chap. 2.

18. **Henderson-Sellers, B.,** Methodologies for the statistical validation of one-dimensional thermal stratification models for water bodies, *Focus on Modeling Marine Systems*, Vol. 2, 373–385, 1990.

19. **Hornberger, G. M. and Spear, R. C.,** An approach to the analysis of behaviour and sensitivity in environmental systems, in *Uncertainty and Forecasting of Water Quality*, Beck, M. B., and van Straten, G., Eds., Springer-Verlag, Berlin, 1983, chap. 3.

20. **Henderson-Sellers, B.,** Sensitivity of thermal stratification models to changing boundary conditions, *Appl. Math. Modelling*, 12, 31–43, 1988.

21. **Smith, I. R.,** Hydraulic conditions in isothermal lakes, *Freshwater Biology*, 9, 119–145, 1979.

22. **Henderson-Sellers, B.,** New formulation of eddy diffusion thermocline models, *Appl. Math. Modelling*, 9, 441–446, 1985.

23. **James, I. D.,** A model of the annual cycle of temperature in a frontal region of the Celtic Sea, *Estuarine and Coastal Mar. Sci.*, 5, 339–353, 1977.

24. **Filatov, N. N., Rjanzhin, S. V., and Zaycev, L. V.,** Investigation of turbulence and Langmuir circulation in Lake Ladoga, *J. Great Lakes Res.*, 17, 1–6, 1981.

25. **Mellor, G. L. and Yamada, T.,** A hierarchy of turbulent closure models for planetary boundary layers, *J. Atmos. Sci.*, 31, 1791–1806, 1974.

26. **Warn-Varnas, A. C., Dawson, G. M., and Martin, P. J.,** Forecast and studies of the oceanic mixed layer during the Mile experiment, *Geophys. Astrophys. Fluid Dynam.*, 17, 63–85, 1981.

27. **Henderson-Sellers, B.,** 1982, A simple formula for vertical eddy diffusion coefficients under conditions of nonneutral stability, *J. Geophys. Res.*, 87, 5860–5864, 1982.

28. **Ueda, H., Mitsumoto, S., and Komori, S.,** Buoyancy effects on the turbulent transport processes in the lower atmosphere, *Q. J. R. Meteor. Soc.*, 107, 561–578, 1981.

29. **Ryan, P. J. and Harleman. D. R. F.,** Prediction of the Annual Cycle of Temperature Changes in a Stratified Lake or Reservoir: Mathematical Model and User's Manual, MIT Tech. Report no. 137, MIT, Cambridge, MA, 1971.

30. **Scott, W. E., Seaman, W. T., Connell, A. D., Kohlmeyer, S. I., and Toerien, D. F.,** The limnology of some South African impoundments. I. The physico-chemical limnology of Hartbeespoort Dam, *J. Limnol. Soc. Sth. Afr.*, 3, 43–58, 1977.

31. **Walmsley, R. D. and Toerien, D. F.,** A preliminary limnological study of Buffelspoort Dam and its catchment, *J. Limnol. Soc. Sth. Afr.*, 5, 51–58, 1979.

32. **Reid, R. A., Locke, B., and Girard, R.,** Temperature Profiles in the Muskoka–Haliburton Study Lakes (1976–1982), Data Report DR 83/4, Ontario Ministry of the Environment, Dorset, Ontario, 1983.

33. **Lund, J. W. G., Mackereth, F. J. H., and Mortimer, C. H.,** Changes in depth and time of certain chemical and physical conditions and of the standing crop of *Asterionella formosa*, Hass. in the North Basin of Windermere in 1947, *Phil. Trans. R. Soc.*, B, 246, 255–290, 1963.

34. **Lewis, W. M., Jr.,** Temperature, heat and mixing in Lake Valencia, Venezuela, *Limnol. Oceanogr.*, 28, 273–286, 1983.

35. **Lewis, W. M., Jr.,** A five-year record of temperature, mixing, and stability for a tropical lake (Lake Valencia, Venezuela), *Arch. Hydrobiol.,* 99(3), 340–346, 1984.

36. **Kling, G. W., Clark, M. A., Compton, H. R., Devine, J. D., Evans, W. C., Humphrey, A. M., Koenigsberg, E. J., Lockwood, J. P., Tuttle, M. L., and Wagner, G. N.,** The 1986 Lake Nyos gas disaster in Cameroon, West Africa, *Science,* 236, 169–175, 1987.

37. **Kling, G. W.,** Seasonal mixing and catastrophic degassing in tropical lakes, Cameroon, West Africa, *Science,* 237, 1022–1024, 1987.

38. **Nicolls, A., Reid, R., and Girard, R.,** Morphometry of the Muskoka–Haliburton Study Lakes, Data Report DR 83/3, Ontario Ministry of the Environment, Dorset, Ontario, 1983.

39. **Henderson-Sellers, B. and Davies, A. M.,** Thermal stratification modeling for oceans and lakes, *Ann. Rev. Numerical Fluid Mechan. Heat Transf.,* 2, 86–156, 1989.

40. Hartbeespoort Dam Capacity Determination, Report A2R01, Planning Division, Department of Water Affairs, Pretoria, South Africa, 1979.

41. Buffelspoort Dam Capacity Determination, Report A2R05, Planning Division, Department of Water Affairs, Pretoria, South Africa, 1979.

42. **Sudbury Environmental Study,** Studies of Lakes and Watersheds Near Sudbury Ontario. Final Limnological Report, SES 009/82, Supplement, Ontario Ministry of the Environment, 1982.

Chapter 6

# DYNAMIC SIMULATION OF TURBIDITY AND ITS CORRECTION IN LAKE CHICOT, ARKANSAS

Heinz G. Stefan, S. Dhamotharan, Frank R. Schiebe, A. Y. Fu, and John J. Cardoni

## TABLE OF CONTENTS

# I. INTRODUCTION

Lake Chicot is an oxbow lake that was created more than 600 years ago by the meandering of the Mississippi River. It is located in Chicot County in southeastern Arkansas adjacent to the present Mississippi River (Figure 1). As the largest natural lake in Arkansas, it earned an early reputation for its good fishing and recreational value.

A flood in 1927 partially created a natural earth dam, dividing the lake into a large lower lake and a considerably smaller upper lake (Figure 2). This dam was completed by the State of Arkansas. Development of a levee system forced the enlargement of the lake's watershed to its present 350 mi$^2$ (900 km$^2$).

Initially this alteration affected only the volume flow through the lake, drastically reducing the water residence time. Because the watershed was located in one of the most agriculturally productive regions in the world, the land, predominantly comprised of clay and fine silts, quickly became more intensively farmed. The use of agricultural chemicals increased, large amounts of sediment were produced, and the lake began to become severely impacted by this activity.

In the early 1960s the U.S. Congress enacted legislation authorizing the U.S. Corps of Engineers to begin planning a method of restoring the lake. Plans were made to construct three

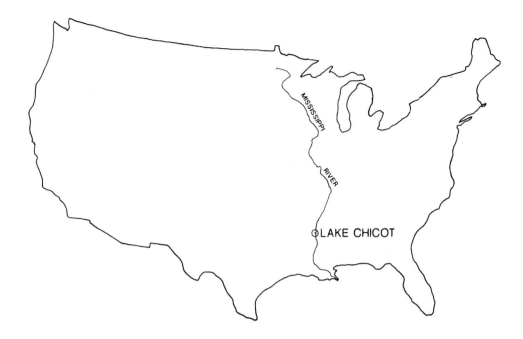

FIGURE 1.    Location of Lake Chicot, Arkansas, U.S.

structures: a dam to prevent poor quality water from entering the lake, a combination gravity flow-pump facility to divert the poor quality water through the levee into the Mississippi River, and a dam on the outflow to regulate lake levels and discharge. The structures were placed into operation during March 1985 and are operational according to a plan to improve the water quality in the lake. These structures and plans have resulted in a nearly complete restoration of the lake.

The U.S. Corps of Engineers Vicksburg District, as the lead agency, considered several alternative schemes to restore the recreational potential of the lake. The Agricultural Research Service, U.S. Department of Agriculture (USDA), was subcontracted to do field investigations. A cooperative effort by the University of Minnestoa, St. Anthony Falls Hydraulic Laboratory, USDA Sedimentation Laboratory in Oxford, MS, and the U.S. Corps of Engineers in Vicksburg, MS, led to the development of a mathematical model with a daily time scale for the simulation of (1) the water temperature, which is the dominant factor that controls the hydromechanics of Lake Chicot; (2) the suspended sediment concentration in the lake, which is the primary cause of the observed objectionable lake turbidity; (3) the trap efficiency of the reservoir; and (4) primary productivity of the lake. The model was useful for the selection of lake management alternatives.

The Lake Chicot model, called RESQUAL II,[40] is an extension of the Minnesota Lake Temperature Model developed by Stefan and Ford[37] and based on a method using energy principles. The model, originally validated for northern dimictic lakes, is found to apply equally well to a monomictic lake such as Lake Chicot, with minor modifications. For suspended sediment concentration, the one-dimensional, unsteady, convective-diffusive transport equation was formulated in a finite difference form by a fully implicit hybrid scheme. The scheme had no restrictions with regard to Peclet number. The numerical scheme results were verified against exact analytical solutions for a special case.

The model was validated for the existing temperature and turbidity conditions with limnological data collected in the lake beginning in July 1976. Water budget analysis showed the possibility of considerable seepage flow into and out of the lake. Seepage flow from the Mississippi River

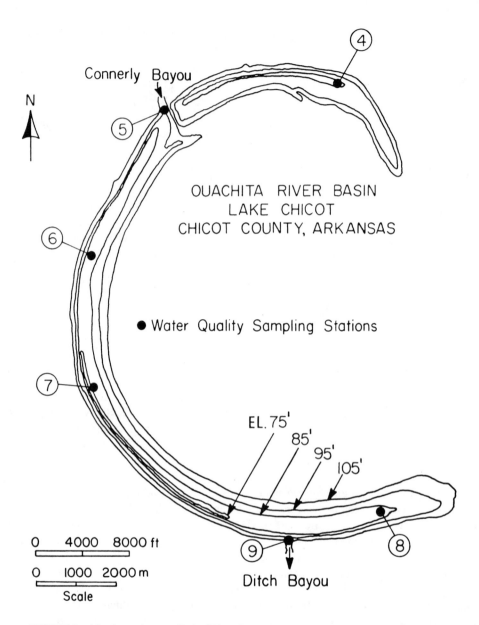

FIGURE 2.    Morphometric map of Lake Chicot. (Bottom contour elevations in feet above mean sea level [MSL], 1 ft = 0.305 m).

to Lake Chicot and vice versa was included in the model. Using the meteorological, hydrological, suspended sediment data, and biological data collected during the water years 1976 to 1979, the water temperature, the suspended sediment concentration, and the reservoir trap efficiency were predicted for several proposed management alternatives, primarily several lake levels and restrictions on inflow. The model formulation and its application will be described below.

## II. LAKE CHICOT CHARACTERISTICS

Lake Chicot consists of a lower lake and an upper lake and has approximately 900 km$^2$ of watershed. The upper Lake Chicot has no principal inflow or outflow, and is divided from the lower lake by an earthen dam. The lower lake, which will be herein referred to as Lake Chicot, is about 18 km in length, with an average width of 0.8 km. Connerly Bayou drains into Lake Chicot, and the outflow is through Ditch Bayou over an outflow weir. The lake morphometry and the sampling stations are shown in Figure 2 (stations 1, 2 and 3 were in the watershed). The lake has a total volume of about $51.9 \times 10^6$ m$^3$, with a corresponding surface area of $13.6 \times 10^6$ m$^2$ at 30.5 m (105 ft) elevation above mean sea level (MSL). Its average depth is 3.8 m.

Water quality data, including temperature, conductivity, and total suspended solids were collected on a biweekly basis beginning in July 1976. Intensive hourly data also was taken for a 10 day period continuously during the summer of 1977.

The inflow of fine suspended sediment from Connerly Bayou into Lake Chicot (Figures 3 and 4) caused a profound change in the appearance, ecology, and recreational use of the lake. To reduce the turbidity and to stabilize the lake level, the U.S. Army Corps of Engineers constructed a new lake outlet structure and a 6500 cfs pumping station to divert inflow from Connerly Bayou to the Mississippi River.[32,34]

The preconstruction water quality of the lake is documented in several publications.[3,4,8,12,19] The material and nutrient budgets of the lake were studied by Swain[35] who also examined correlations between hydrological and water quality parameters in the lake. Since July 1976, the U.S. Department of Agriculture's Sedimentation Laboratory at Oxford, Mississippi, and since January 1980, the U.S.D.A. Water Quality Laboratory in Durant, Oklahoma, monitored the lake's hydrological, chemical, and biological regimes. These studies produced extensive baseline data as part of the preconstruction assessments and were continued in order to document lake changes as a result of inflow diversion and stage management.

To assist in the interpretation of the data, and ultimately in the selection of lake operational management alternatives, a process-oriented, dynamic model simulating several water quality parameters, including transparency of the lake on a daily time scale was developed. Stages of the model development were described at three symposia,[14,15,18] in a paper,[39] and in a final report.[40] The model formulation and its application therefore will only be summarized here. The model is of interest not only for the solution of the Lake Chicot problem, but has potential application in the design and operation of shallow detention basins and shallow reservoirs intended for the entrapment of suspended sediments.

To formulate an appropriate dynamic model that would describe the dependence of transparency (turbidity) of Lake Chicot on inflows as well as in-lake processes, some field measurements and observations had to be reviewed. The most important of these was perhaps the relationship between suspended solids (inorganics) and Secchi depth as a measure of lake clarity (Figure 5). Because of the very small size of the particles, it did not take a large amount of suspended solids to make the water very turbid.

To predict lake clarity, it was necessary to understand particle distribution, i.e., settling, resuspension, and turbulent mixing in the lake. The mixing dynamics of shallow reservoirs or lakes show a strong dependence on air-water interaction through the effects of surface heat exchange and wind.

Temperature cycles in Lake Chicot are quite variable. The relatively shallow (mean depth = 3.8 m), elongated basin, and the surrounding delta are important factors in determining the annual temperature behavior of the lake. Weak thermal stratification occurs and persists until strong winds prevail and mixing occurs. Lake Chicot experiences numerous overturn periods during the year because of the morphometry, surrounding terrain, and climate. Its geographical location is in the region where most lakes are classified as warm monomictic. Figure 6 illustrates the temperature dynamics by presenting the isotherms derived from hourly water temperature measurements during a 10-d period. The rapid variations in surface mixed layer depths are noteworthy, as are the intermittent stratification and complete mixing events. (Station 7, at which the data were obtained, is shown in Figure 2.)

Examples of density profiles in the lake are given in Figure 7. Water temperature, suspended solids, and dissolved solids were taken into consideration when the densities were computed.

FIGURE 3.    LANDSAT image of Lake Chicot west of the Mississippi River on April 14, 1979, band 7. High turbidity (light color) in lower lake and less turbidity in upper lake (dark color).

FIGURE 4.     Aerial view of the dike separating the upper and lower basins of Lake Chicot. The lower and more turbid lake is on the right.

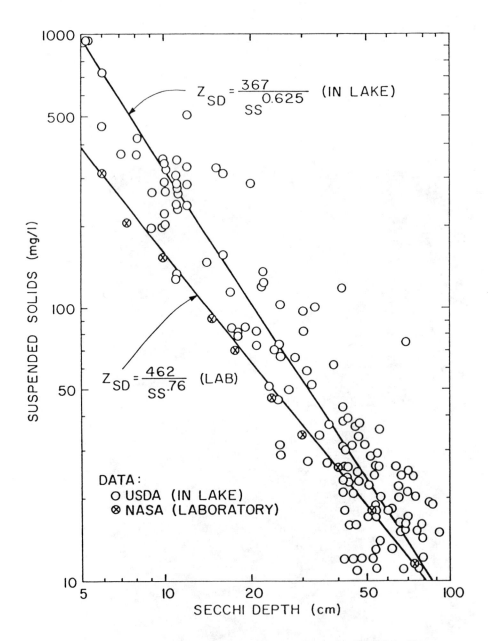

FIGURE 5.    Secchi depth versus suspended sediment concentration (SS) in Lake Chicot.

It is evident that water temperature controls the density stratification of the lake much more than do suspended or dissolved materials. In spring, the inflow carries a high suspended load, as shown by the turbidigraph in Figure 8, but the incoming water from the shallow Connerly Bayou warms and cools faster than the lake water. When the bayou water is warmer than the lake water, the inflowing water can spread over the lake surface; when the bayou water is colder than the lake water, the inflow sinks as turbidity current into the reservoir. One inflow condition can change into another rapidly, as the weather changes. As a result of these mechanisms, the lake has its highest and nearly uniform turbidity at the end of spring. After the onset of seasonal stratification in May, suspended concentrations decrease continuously. The relatively clear water entering the lake in the fall sinks to the bottom. In summary, the inflow can alternate

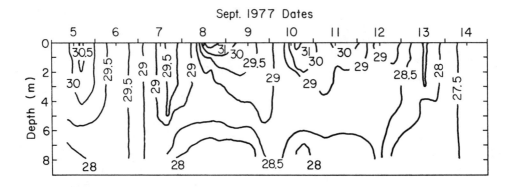

FIGURE 6.    Measured hourly temperature structure at station 7 (Figure 2) in Lake Chicot (isotherms in °C).

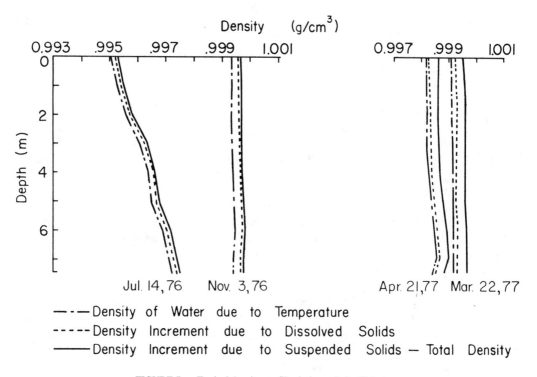

FIGURE 7.    Typical density profiles in lower Lake Chicot.

seasonally as well as daily from overflow to interflow to underflow, as described by Wunderlich and Elder.[46]

To determine whether the model had to be one, two, or three dimensional, temperature profiles with measured isotherms were plotted in a longitudinal section through the lake. Examples are given in Figures 9A and 9B for stratified and well-mixed conditions, respectively. It was concluded that the stratification was uniform enough throughout the lake to justify the use of a one-dimensional model.

## III. MODEL CONCEPT

Model RESQUAL II simulates lake stage, surface mixed-layer depth, water temperature (T), suspended solids (SS), phytoplankton (chl*a*), available dissolved orthophosphorus ($P_a$), nona-

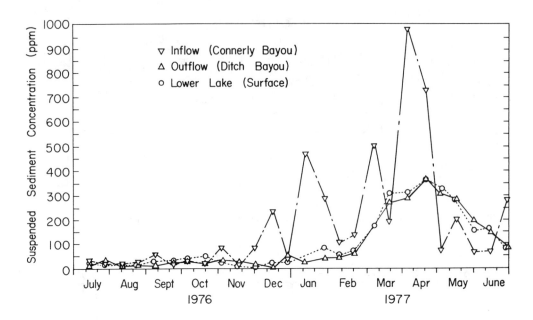

FIGURE 8.    Suspended sediment content of Lake Chicot waters.

vailable particulate phosphorus $(P_n)$, light attenuation coefficient $(k)$, and Secchi depth $(z_{SD})$ in a shallow stratified lake or reservoir. The primary objective is to simulate present and future transparency of the water. An earlier version of the model simulated only T and SS, and was developed by Dhamotharan et al.[14]

The turbidity of Lake Chicot is the result of erosion and runoff from the watershed. It is for this reason that inflow diversion was considered by the Corps of Engineers as a means to reduce turbidity in Lake Chicot. To simulate the response of the lake to any reduction in sediment loading, all significant in-lake processes had to be considered. These include stratification turbulent diffusion, settling, resuspension, and growth kinetics in the case of phytoplankton (Figure 10). The mathematical model description of these processes used the general flow chart shown in Figure 11 for RESQUAL II. Additional elements in RESQUAL II are the groundwater model and the nutrient and phytoplankton subroutines. To account for the temperature and density stratification, a multilayer model (Figure 12) had to be formulated. To account for inflow and outflow, layers were chosen to be of variable thicknesses.

A one-dimensional unsteady water temperature stratification and mixing model (MLTM) developed at the University of Minnnnesota[22,37] for a daily time scale was used as a starting point for the RESQUAL model development. Additions and changes were made in the MLTM model to account for inflows, outflows, and the effects of suspended sediment on the heat transfer processes. Subsequently, an unsteady, mass-transport submodel for suspended sediment was formulated. Results from the temperature stratification dynamics model were used as input to the suspended sediment model. Results from the suspended sediment distribution simulation were necessary, in turn, to specify attenuation and reflection of radiation in the heat transfer relationships in the dynamic temperature model. Subsequently, submodels for a density current inflow, light attenuation, phytoplankton, phosphorus, and Secchi depth were added.

The RESQUAL II model gives the water quality changes as a function of depth and on a day-by-day basis in response to real weather, also specified on a day-by-day basis. It is therefore possible to use the model

1.    To simulate past conditions by hindcasting with measured weather conditions,

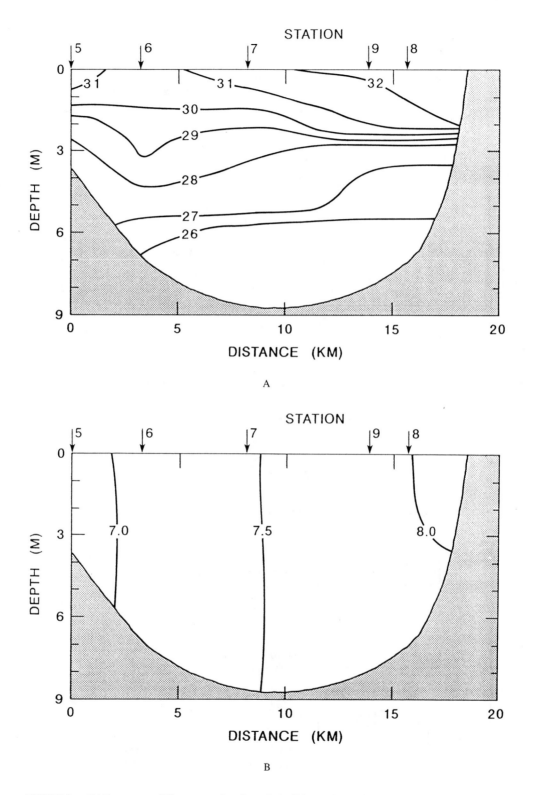

FIGURE 9. (A) Temperature (°C) structure along lower Lake Chicot, July 14, 1976; (B) temperature (°C) structure along lower Lake Chicot, December 2, 1976.

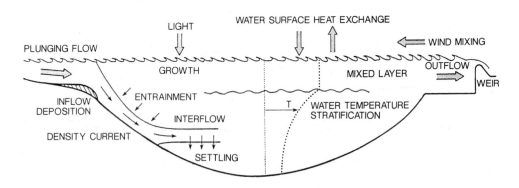

FIGURE 10.   Main processes affecting lake water quality.

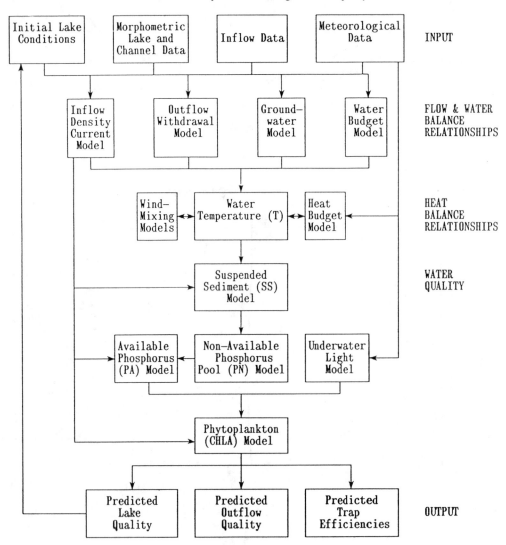

FIGURE 11.   Elements of the model RESQUAL II.

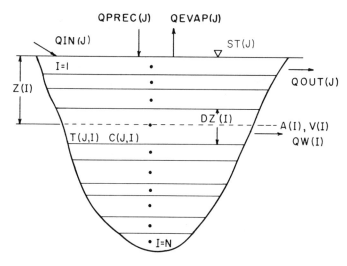

FIGURE 12.  Schematic of lake subdivision into layers I = 1 to I = N and flow rates at time step *J*.

2.    To simulate conditions that would have existed had, e.g., an inflow diversion been implemented, and

3.    To forecast conditions on a real-time basis for different diversion strategies.

## IV. ADVECTIVE FLOW MODEL AND CONSERVATION OF VOLUME AND MASS

### A. INFLOW/OUTFLOW MECHANICS IN A STRATIFIED LAKE OR RESERVOIR

The lake or reservoir is considered to be composed of horizontal layers of variable thickness, and density (Figure 12). Each layer has a mean horizontal area and volume that is determined by the reservoir's morphometry.

The density of each layer of water in the lake or reservoir is determined by its temperature, suspended sediment content, and dissolved solids content. The inflowing water will seek a layer with a density equal to its own. It will augment the volume of that layer, and consequently all the layers above it will be displaced upward. As a layer rises, its horizontal area becomes larger and its thickness consequently diminishes.

Outflow is simulated by withdrawing water from the layers in front of the outlet — in the case of Lake Chicot, a weir.

Numerical computations can be kept simple by considering the reservoir as a stack of discrete volumes to which additions or subtractions are made at each time step.

As the inflow moves into the reservoir and towards its isopycnic layer, it entrains water from each layer through which it passes (Figure 13). The amount of entrainment is a complex function of the flow rate, the density gradients, and other factors.[31] If the inflow is into the surface mixed layer, entrainment can be ignored. Entrainment from deeper layers is specified in a density current subroutine. The characteristics of the density current, i.e., its temperature and suspended and dissolved solid content, are changed by dilution as the current passes from one layer to the next until it reaches its isopycnic layer. The temperature and suspended and dissolved solid content of the isopycnic layer are recalculated, including the thermal energy and the mass of suspended and dissolved solids added by the density current.

### B. WATER BUDGET

The water budget for the reservoir includes surface inflow, outflow, precipitation, evapora-

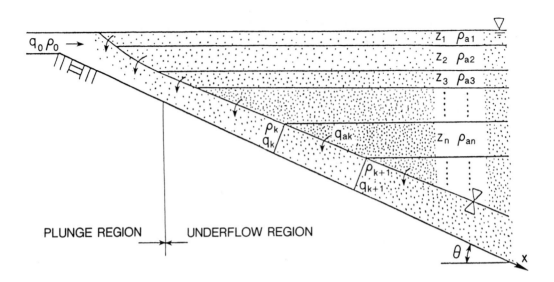

FIGURE 13.   Schematic of density current analysis.

tion, and seepage to and from the surface layer. Evaporative losses are derived directly from the evaporative heat transfer term in the heat budget equation. Groundwater seepage into or out of the lake, and mostly from or to the nearby Mississippi River, is calculated.

Water balance equations for each layer and between time steps $j$ and $j + 1$, typically 1 day apart, are

$$V(i, j+1) = V(i, j) - \left[ Q_e(i) + Q_w(i) + Q_s(i) \right] \Delta t \tag{1}$$

where $V$ = volume of layer, $Q_e$ = flow entrained by density current from layer $i$, $Q_w$ = withdrawal rate, $Q_s$ = seepage rate, and $\Delta t$ = time step. For the surface layer ($i = 1$) the relationship is expanded to include $Q_{ev}(1)$ = evaporative water loss rate and $Q_p(1)$ = volumetric rate of water added by precipitation.

$$V(1, j+1) = V(1, j) - \left[ Q_e(1) + Q_w(1) + Q_s(1) + Q_{ev}(1) - Q_p(1) \right] \Delta t \tag{2}$$

For the isopycnic layer, the mass balance equations are

$$V(ip, j+1) = V(ip, j) - \left[ Q_w(ip) + Q_s(ip) - Q_c \right] \Delta t \tag{3}$$

where $Q_c$ = density current flow rate when it meets layer $ip$.

$$Q_c = Q_i + \sum_{i=1}^{ip-1} Q_e(i) \tag{4}$$

For the layers below the isopycnic layer ($ip < i < N$),

$$V(i, j+1) = V(i, j) - \left[ Q_w(i) + Q_s(i) \right] \Delta t \tag{5}$$

In Lake Chicot, seepage flow and withdrawal affect only the first layer.

## C. LAYER THICKNESS

Layer thicknesses are determined starting with the lowermost layer. A volume-vs.-elevation curve derived from the reservoir morphology is used. The thickness of each layer is

$$\Delta z(i) = \frac{V(i)}{A(i)} \tag{6}$$

where $A(i)$ is the horizontal area taken at the center of the layer $i$. $A(i)$ is a function of the elevation of the center of the layer above the reservoir bottom and therefore is dependent on $\Delta z(i)$. For this reason, an iteration scheme, described in more detail by Dhamotharan,[16] is used to derive the best estimates of $\Delta z(i)$. To safeguard against the accumulation of round-off errors, the sum of all layer thicknesses $\Delta z(i)$ is computed at the end of each time step and compared to the total reservoir depth derived from a hydrological water budget equation. To equalize the two values a correction factor is applied uniformly to all $\Delta z(i)$'s in each timestep.

Initial layer thicknesses are specified by the model user. After the initial time step, layer thicknesses will keep changing. To avoid the development of anomalies, maximum and minimum layer thicknesses are specified. Selection of a maximum layer thickness is guided by total reservoir depth and affordable computational time. A value of 75 cm, one-twelfth of the maximum reservoir depth, was chosen. Layers exceeding the specified maximum value are divded into two or more layers (subroutine SPLIT). The minimum layer thickness chosen was 15 cm, which relates to the maximum possible withdrawal and to the total reservoir depth. Depletion of more than one layer must not occur in any one time step. If the thickness of any one layer falls below the minimum value, it is added to the layer below it (subroutine MERGE). Layers are renumbered and the pertinent morphometric values assigned as layers are generated or eliminated.

### D. OUTFLOW FROM STRATIFIED LAKE

Outflow from Lake Chicot is through Ditch Bayou. A damaged rubble mound dam was replaced by a concrete weir in 1979. The following rating curves were used to calculate the volumetric outflow rate:

$$Q_w = 17.12(S - 101.42)^{2.42} \qquad 10/1/76 \text{ to } 7/15/79$$

$$Q_w = 146.67(S - 101.42)^{1.72} \qquad \text{after } 7/15/79$$

where $Q_w$ = withdrawal flow rate in Ditch Bayou (cfs) and $S$ = lake stage (ft).

The total depth from which the withdrawal flow is taking place has to be calculated. The withdrawal from each individual layer within the withdrawal layer is apportioned according to individual layer thickness:

$$Q_w(i) = Q_w \frac{\Delta z(i)}{z_w} \tag{7}$$

where $\Delta z(i)$ = thickness of an individual layer, and $z_w$ = total withdrawal layer thickness.

### E. INFLOW DENSITY CURRENT

If the density of the inflowing water is higher than that of the first layer, the inflow plunges and continues as a density current (Figure 13). The submodel of the inflow density current follows the analysis by Akiyama and Stefan.[1]

In the analysis, it is assumed that inflow rate per unit width ($q_o$) inflow density ($\rho_o$), layer thicknesses ($z$), ambient density ($\rho_a$), and channel slope ($S$) are known, and that internal backwater effects in the case of flow over a mild slope channel are absent.

The analysis considers first the plunging region (plunging depth), then the dilution of the density current (underflow) as it progresses from layer to layer. Details of this analysis can be found in two reports.[2,40]

## 1. Plunging Region

Following the detailed analysis of Akiyama and Stefan,[2] the plunging depth is evaluated as a function of channel slope ($S$); inflow rate per unit width of inflow channed ($q_{in}$); total friction factor ($f_t$), including bed friction and interfacial friction; and buoyancy ($\varepsilon_{in}$). Water depth at the plunge point $h_p$ and initial dilution at the plunge point $\gamma_{in}$ are found.

## 2. Underflow Region

In the underflow region beyond the plunging region, the dilution of the flow by entrainment is calculated from the continuity equation:

$$Q_{i+1} = Q_i + Q_{ei} = Q_i\left(1 + \gamma_i\right) \tag{8}$$

The entrainment ratio $\gamma_i$ for the underflow is evaluated separately, layer by layer.[1]

## F. GROUNDWATER INFLOW AND OUTFLOW

The volume of groundwater going into or out of a lake must be evaluated on a case-by-case basis and incorporated in the model appropriately. The ground water contribution to the Lake Chicot water budget is found to be dependent on the interaction between the lake and the Mississippi River stages. The groundwater is added or taken out at the surface mixed layer because studies by Winter[45] have shown that seepage connections between a lake and an aquifer are usually most effective near the surface, where contact areas and permeabilities are the largest.

The groundwater flow rate for Lake Chicot is calculated from a regression equation.[42] It combines expressions for two-dimensional flow in an unconfined aquifer and flow in a confined aquifer. Groundwater flow out of the lake is taken from the surface mixed layer. The volume of the mixed layer is adjusted in each time step accordingly. Groundwater flow into the lake is routed to the isopycnic layer in the lake without dilution. Goundwater temperature is set equal to the annual average air temperature. Groundwater contains no sediments or nutrients in the simulation.

## G. PRECIPITATION

Water additions by precipitation are calculated and added to the surface layer.

## H. DILUTION OF WATER QUALITY CONSTITUENTS ASSOCIATED WITH ADVECTIVE TRANSPORT

Associated with the advective transfer of water into and out of an individual layer is the transfer of heat, and suspended and dissolved materials. Water temperatures $T$ and concentration $C$ of individual layers are affected only if the advective flow is into the layer. For the isopycnic layer, $T$ and $C$ are therefore recalculated from the conservation equations:

$$T(ip, j+1) = \frac{T(ip, j)\, V(ip) + T_c(i)\, Q_c(i)\, \Delta t}{V(ip) + Q_c(i)\, \Delta t} \tag{9}$$

$$C(ip, j+1) = \frac{C(ip, j)\, V(ip) + C_c(i)\, Q_c(i)\, \Delta t}{V(ip) + Q_c(i)\, \Delta t} \tag{10}$$

where $V$ = volume of a layer, $Q_c$ = density current flow rate, and $\Delta t$ = time step of computation = 1 d, and the concentrations ($C$) are those of suspended solids (SS), chlorophyll-*a* (chl*a*), available (dissolved ortho) phosphorus ($P_a$), or nonavailable phosphorus ($P_n$). Subscript $c$ refers to the inflow density current as it arrives at layer ($i$).

Interlayer density current flow and the temperature and concentrations of suspended and dissolved solids are calculated successively for each layer by considering the dilution due to entrainment at each step down; they are also used to compute water densities both in the lake and in the density current.

$$T_i = \frac{T_{i-1} \, Q_{i-1} + T_{i,j} \, Qe_i}{Q_{i-1} + Qe_i} \tag{11}$$

$$C_i = \frac{C_{i-1} \, Q_{i-1} + C_{i,j} \, Qe_i}{Q_{i-1} + Qe_i} \tag{12}$$

where $T_i$ = temperature of density current after passing layer $i$; $Q_{i-1}$ = flow rate of density current before reaching layer $i$; and $Qe_i$ = entrainment rate from layer $i$; and $C_i$ = concentration of suspended solids, dissolved solids, and available and nonavailable phosphorus, and chlorophyll-*a*. Surface layer dilution by advection requires an expanded equation,

$$T(1, j+1) = \frac{T(1, j) \, V(1) + \left[ Q_s T_s + Q_p T_p + Q_{in} T_{in} \right] \Delta t}{V(1) + \left( Q_s + Q_p + Q_{in} \right) \Delta t} \tag{13}$$

for water temperature and a similar one for concentrations. Subscripts *s, p,* and *in* refer to groundwater inflow (seepage), precipitation and surface inflow, respectively.

# V. WATER TEMPERATURE STRATIFICATION AND SURFACE ENERGY TRANSFER MODEL

## A. CONCEPT

As a consequence of the predominant influence of water temperature on the midsummer density stratification and vertical mixing in Lake Chicot, the dynamic, one-dimensional temperature prediction model developed by Stefan and Ford[37] and Ford and Stefan[22,23] was used as a starting point. The model uses a system of energy equations, including wind energy input, in addition to various forms of heat energy. It is a particularly suitable model for a shallow lake such as Lake Chicot since it can simulate vertical mixing dynamics using weather input at a time scale of a day or even shorter. Mixing is determined by a stability criterion that compares the total kinetic energy available for mixing with the incremental potential energy of the temperature profile. Thus, mixing is intermittent and occurs only when sufficient wind energy is available. The typical result of the simulation is a daily water temperature profile. The integral energy method emphasizes the net results of wind mixing and heat exchange between the lake and the atmosphere. Meteorological and morphometric data are the only required input data.

The suspended particles causing the objectionable turbidity in Lake Chicot are tiny, flat clay particles of about 1 μm average size. They increase the reflectivity and reemergence from the water body of incoming radiation at the water surface and also the attenuation of radiation penetrating the water column. For use in the temperature model, the dependence of albedo and diffuse radiation attenuation coefficient on suspended sediment concentration had to be established.

The model considers the following:

- Radiation heat transfer at the water surface and absorption in the water column.
- Heat losses from the water surface by back radiation, evaporation, and convection; and the surface mixed layer depth formed by natural convection.
- The surface mixed layer depth produced by wind mixing and natural convection during cooling.
- The heat transfer below the surface mixed layer by turbulent diffusion.

## B. HEAT TRANSFER EQUATION

The one-dimensional transient diffusion equation for heat in a water column (Equation 10 in Chapter 1) is

$$\frac{\partial T}{\partial t} = \frac{\partial}{\partial z}\left(K_z\,\frac{\partial T}{\partial z}\right) + \left(\frac{S}{\rho c V}\right) \tag{14}$$

where $T$ = water temperature, $t$ = time, $K_z$ = vertical exchange coefficient, $z$ = depth, $S$ = solar radiation absorbed at depth $z$, $\rho c$ = specific heat per unit volume, and $V$ = volume of layer. Energy absorbed in the topmost layer is

$$S(1) = (1-r)\beta\,\phi_s + \phi_{an} - \phi_{br} - \phi_e - \phi_c \tag{15}$$

where $\phi_s$ = solar radiation received; $\beta$ = near-surface absorption coefficient, after Dake and Harleman[13] = 0.4; $r$ = reflectivity; $\phi_{an}$ = net atmospheric radiation; $\phi_{br}$ = back radiation; $\phi_e$ = evaporative heat flux; and $\phi_c$ = convective heat flux.

The remaining radiation, $(1-\beta)(1-r)\,\phi_{sn}$, is attenuated exponentially with depth. The amount $S$ absorbed at depth $z$ is therefore

$$S(i) = k(1-\beta)(1-r)\phi_s\,e^{-kz} \tag{16}$$

where $k$ = attenuation coefficient (m$^{-1}$).

In the surface mixed layer, water temperatures are first calculated layer by layer:

$$T(j,i) = T(j-1,i) + \frac{S}{\rho\,cV(i)} \tag{17}$$

Then the mixed-layer depth due to natural convection is determined from

$$\phi_L = \sum_{i=1}^{N_m}\left[T(j-1,i) - T\left(j,N_m\right)\right]V(i)\,\rho\,c \tag{18}$$

where $N_m$ = number of layers forming the surface mixed layer and $\phi_L$ = total surface heat loss $= \phi_e + \phi_{br} + \phi_c$. The values of $\phi_{an}$, $\phi_e$, $\phi_{br}$, and $\phi_c$ are calculated from empirical relationships.

## C. AIR-WATER ENERGY EXCHANGE
### 1. Solar Radiation

Solar radiation $\phi_s$ is a measured total daily quantity. Values were from the Stoneville weather station. Both the reflectivity $r$ and the attenuation coefficient $\eta$ are functions of the radiation wavelength, angle of incidence of suspended sediment, and color of the water. The dependence on suspended sediment concentration was determined by field and laboratory measurements, and is shown in Figures 14 and 15. Wavelength was dependent only on natural radiation and was assumed to be independent of season. Angle of incidence variations with season were expressed as a function of seasonal radiation intensity. The following relationships were developed to fit Lake Chicot data.[38] Only measurements of incident and upwelling radiation from the water surface were available; therefore, reflectance and albedo had to be assumed as equivalent.

$$r = 0.087 - 6.76 * 10^{-5}\,\text{RAD} + 0.11\left[1 - \exp(-0.01\,\text{SS})\right] \tag{19}$$

where $r$ = reflectance = $\phi_r/\phi_{si}$, $\phi_r$ = reflected solar radiation, RAD = total daily incident solar radiation in cal/cm$^2$/d, and SS = suspended sediment concentration in mg/l. The first two terms

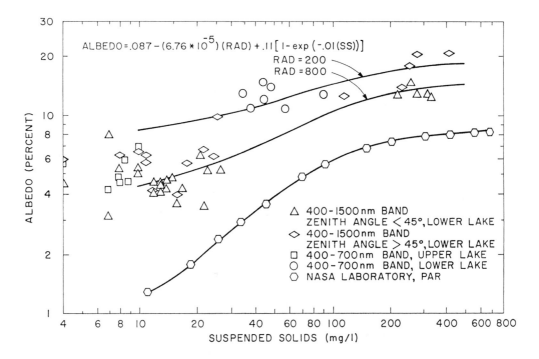

FIGURE 14. Albedo of Lake Chicot (400- to 1500-nm band).

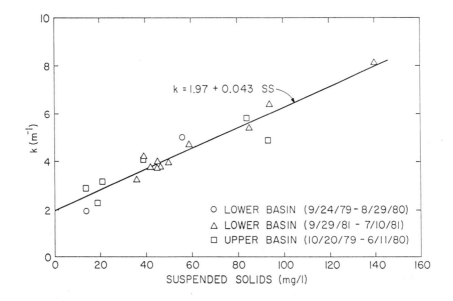

FIGURE 15. Suspended solids versus attenuation *k* for photosynthetically active radiation measured with a spherical collector.

are for clear water, and the third term is from Lake Chicot data[38] and accounts for sediment effects. Reflectance had to be adjusted for seasonal variation of the angle of incidence. This is done by the second term, which was first introduced by Dingman.[17] Figure 14 shows the field data from Lake Chicot, together with fitted equations for two levels of radiation.

Attenuation of radiation was calculated using the attenuation coefficient

$$k = 1.97 + 0.043 \, \text{SS} + 0.025 \, \text{chl}a \tag{20}$$

where SS = suspended sediment concentration, (mg/l) and chl$a$ = chlorophyll-$a$ concentration (µg/l). The above equation was actually derived for photosynthetically active light from Lake Chicot data,[38] but can also be applied to the entire solar spectrum, since much of the long-wave component has been removed in the surface layer.

### 2. Atmospheric Radiation

Net atmospheric (long-wave) radiation was expressed in the usual way, as the fourth power of the absolute air temperature.

### 3. Back Radiation

Back radiation is the atmospheric long-wave radiation emitted by the water surface. The emissivity of water = 0.97.

### 4. Evaporative Transfer

Evaporative heat transfer from the water surface was expressed by the relation

$$\phi_e = \rho L \left( Wftn \right)_z \left( e_{sw} - e_{az} \right) \tag{21}$$

where $e_{az}$ = vapor pressure of the air at height $z$, $e_{sw}$ = saturated vapor pressure at water surface temperature, $Wftn_z$ = a wind function using wind velocity at height $z$, $L$ = latent heat of vaporization for water, and $\rho$ = density of water.

### 5. Convective Heat Transfer

The convective heat transfer from an air-water interface, when evaluated according to Bowen,[6] may be expressed as

$$\phi_c = 0.61 \frac{P_a}{1000} \rho L \, Wftn_z \left( T_s - T_{az} \right) \tag{22}$$

where $T_{az}$ is air temperature at a height $z$ above the water surface, $P_a$ is in mb, and $Wftn_z$ is the same as that for evaporative heat transfer.

A number of empirical wind function formulas $(Wftn)_z$ have been developed for various conditions. A formula used by many investigators[28] for natural water bodies is

$$\left( Wftn \right)_z = a + b \, W_z \tag{23}$$

where $W_z$ = wind velocity at elevation $z$ above the water surface and $a,b$ = empirical constants.

The above heat transfer scheme ignores seasonal heat storage in the lake bed.

### D. WIND MIXING

The deepening of the surface mixed layer by wind shear was considered by a stability criterion.[22,37]

$$\frac{\rho_a u_*^3 \, A(1) \Delta t}{V(m) \, \Delta \rho g \left( z_m - z_g \right)} = \sigma \tag{24}$$

where $\rho_a$ = air density, $u_*$ = wind shear velocity, $V(m)$ = volume of the surface mixed layer, $\Delta \rho(m)$ $- \rho(m+1)$, $z_m$ = depth of the surface mixed layer, and $z_g$ = center of gravity of the surface mixed layer.

$$z_g = \frac{\sum\limits_{i=1}^{m} z(i)\, A(i)\, \Delta z(i)}{\sum\limits_{i=1}^{m} A(i)\, \Delta z(i)} \tag{25}$$

The surface mixed layer depth is attained where $\sigma = 1$.

The effect of wind on vertical diffusivities in the surface mixed layer and below the surface mixed layer was described by an equation of the form:

$$K_z = aW^b \tag{26}$$

where $K_z$ = vertical diffusivity ($m^2$/d), $a,b$ = coefficients, and $W$ = wind velocity (mph). This empirical equation was proposed by Filatov et al.[21] For shallow Lake Ladoga, coefficient $b$ varied from 1.2 to 1.4. The average value, $b = 1.3$, was chosen for Lake Chicot.

Coefficient $a$ was estimated by using the seasonal mean values, $K_z = 400$ $m^2$/d, for the mixed layer and $K_z \approx 1$ $m^2$/d for all layers below, as previously applied.[18] For an average annual wind velocity, $W = 7.73$ mph, $a = 28$ for the mixed layer and $a = 0.1$ for the hypolimnion. Thus

$$K_z = 28\, W^{1.3} \qquad \text{in the mixed layer} \tag{27}$$

$$K_z = 0.1\, W^{1.3} \qquad \text{below the mixed layer} \tag{28}$$

## VI. SUSPENDED SEDIMENT MODEL

In stratified lakes and reservoirs of moderate size, including Lake Chicot, advection in the horizontal direction is rapid, relative to vertical mixing. This was verified in Lake Chicot by measurements of longitudinal temperature gradients,[16] and hence only vertical gradients in suspended sediment concentration $C$ were simulated. Biweekly suspended sediment measurements in the lake also showed that one dimensionality was an acceptable assumption for Lake Chicot. A relationship among suspended sediment concentration profiles, vertical mixing intensity, rate of deposition, and time is

$$A\frac{\partial(C)}{\partial t} + \frac{\partial(WAC)}{\partial z} - WC\frac{\partial A}{\partial z} - \frac{\partial C}{\partial z}\left(AK_z\,\frac{\partial C}{\partial z}\right) = 0 \tag{29}$$

where $C$ = suspended sediment concentration, $W$ = fall velocity of suspended sediment in quiescent water, $A$ = cross-sectional area, and $K_z$ = vertical turbulent diffusivity. The first term in this equation represents the change in sediment content with time, the second term is the rate of transfer by settling from one layer to another, the third term is the rate of deposition on the lake bed, and the fourth term is the vertical turbulent mixing rate. The particle fall velocity was determined after Gibbs et al.[25] For Lake Chicot, a mean particle size was determined as $r_s \approx 1$ μm by Schiebe[33] and confirmed by model calibration. The sediment transport equation accounts for deposition on the shelf. The equation is solved over the entire depth. Solution of the equation requires two boundary conditions, which are

1.    No suspended sediment transfer at the water surface, i.e.,

$$K_z\frac{\partial C}{\partial z} - WC = 0 \qquad \text{at } z = 0 \tag{30}$$

2. No resuspension at the bottom, i.e.,

$$K_z \frac{\partial C}{\partial z} = 0 \qquad \text{at } z = h \qquad (31)$$

Condition (1) is usually well satisfied, while condition (2) requires field verification. In Lake Chicot no resuspension was observed after storms. A uniform concentration distribution $C = C_o$ is specified as the initial condition at $t = 0$ (October 1, 1976, after the fall overturn).

Equation 29 describes a balance between advective transport by settling and diffusive transport by vertical mixing. In a stratified reservoir, vertical exchange coefficients are strongly dependent on depth and wind on the surface. Daily variations in $K_z$ were computed from Equations 27 and 28.

The model also computes a suspended sediment budget and determines the amount of sediment deposited in the reservoir. The apparent trap efficiency is defined as

$$\text{ATE} = \frac{\Sigma \text{ Sediment Inflow} - \Sigma \text{ Sediment Outflow}}{\Sigma \text{ Sediment Inflow}} \qquad (32)$$

The trap efficiency ATE is meaningful only when computed over long time periods. It does not take into account the change in the storage of suspended sediment in the lake.

The RTE real trap efficiency is the apparent trap efficiency ATE minus the change in storage in the lake. It is calculated from

$$\text{RTE} = \text{ATE} - \frac{\text{Change in Storage}}{\Sigma \text{ Sediment Inflow}} \qquad (33)$$

Change in storage is the amount of suspended sediment in the lake at the beginning of the time interval minus the amount at the end.

## VII. PHYTOPLANKTON MODEL

### A. CONCEPTS

Lake Chicot is divided into an upper and lower basin. The lower basin receives turbid inflow from Connerly Bayou, while the upper basin receives primarily local overland runoff. As a result, the two basins represent distinctly different systems in terms of suspended solids concentration and primary productivity. The lower basin is highly turbid due to inorganic suspended solids, and biological productivity in this basin in substantially lower than in the upper basin.[12] The upper basin is high in primary productivity, and much of the turbidity there is due to organic material. Observed seasonal variations in chlorophyll-$a$ and suspended solids[12] are shown in Figure 16 and indicate that primary productivity in the lower basin is limited by the amount of available light, while in the upper basin both available light and nutrients most likely limit productivity. In the lower and upper basins, Secchi depths rarely are greater than 0.50 and 0.70 m, respectively.

Surges in phytoplankton populations occur whenever inorganic sediment concentrations and turbidity have diminished (Figure 16). If flow diversion effectively reduced inorganic sediment concentrations and turbidity in the future, phytoplankton will grow more substantially. It is for this reason that phytoplankton was modeled.

Phytoplankton concentrations were described by a suspended sediment equation, except that fall velocities are smaller than for clay, and terms for biological growth and loss kinetics had to be added. At present, the growth of nuisance algae in Lake Chicot is predominantly controlled by the light available for photosynthesis. Losses of phytoplankton are due to settling and

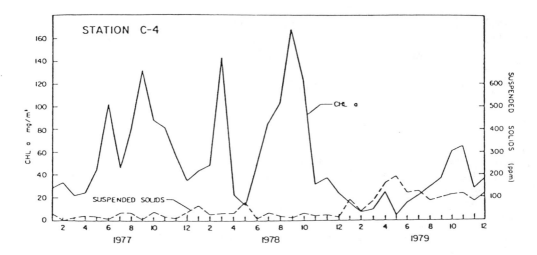

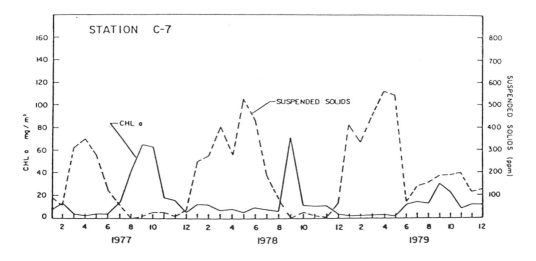

FIGURE 16. Monthly (28-d) surface chlorophyll-*a* (mg/m$^3$) and suspended sediment values for the north (station 4) and south (station 7) basins of Lake Chicot, 1977-1979.[12]

respiration. Grazing by zooplankton was not included separately, because the observed phytoplankton species[3,12,19] are not desirable food sources.

A relationship between productivity rate, light intensity, and temperature was developed from available field measurements by Cardoni and Stefan.[9] Nutrient limitation was not considered in this first analysis. A light limited situation occurred frequently in lower Lake Chicot.[19] An extension of this analysis, including a nutrient limitation (phosphorus), was also given by Cardoni et al.[10]

## B. BASIC EQUATION

The goal was to predict the concentration of phytoplankton that will be present in the future. The parameter used to indicate algal abundance is chlorophyll-*a* concentration [chl*a*]. The model is for use on a daily time scale. The most important input paramenters to the chlorophyll-*a* model are above water surface light intensity, suspended solids concentration, and water temperature, each on a daily time basis. The basic dynamic equation is

$$\frac{\partial[\text{chl}a]}{\partial t} + \frac{1}{A}\frac{\partial\left(AW_c[\text{chl}a]\right)}{\partial z} - \frac{W_c[\text{chl}a]}{A}\frac{\partial A}{\partial z} - \frac{1}{A}\frac{\partial}{\partial z}\left(A\,K_z\,\frac{\partial[\text{chl}a]}{\partial z}\right)$$
$$+ K_2 T[\text{chl}a] - P'[\text{chl}a] = 0 \tag{34}$$

where [chl*a*] = chlorophyll-*a* concentration, mg/m$^3$; $A$ = area at center of layer, m$^2$; $W_c$ = chl*a* fall velocity in quiescent water, m/d; $K_z$ = vertical turbulent diffusivity, m$^3$/d; $K_2$ = respiration/mortality loss coefficient, d$^{-1}$ °C$^{-1}$; $T$ = water temperature, °C; and $P'$ = productivity rate, d$^{-1}$. Equation 34 is of the same form as Equation 29 for the concentration of suspended solids, but includes additional source and sink terms. Also, the fall velocity $W_c$ is different for algal particulates than for the suspended sediment particulates (mostly clay).

## C. PRIMARY PRODUCTIVITY

The productivity of Lake Chicot was measured by Bacon,[3] using the [14]C uptake method, and reported in part by the U.S.D.A.[43] Bacon's data was converted to a specific growth rate by dividing measured rate of [14]C uptake by the concentration of chlorophyll-*a* present in the lake,[33] thus obtaining productivity in units of milligrams carbon/(mg chl*a* h). Unfortunately carbon uptake and chlorophyll-*a* were never measured in the same water samples.

A relationship between productivity rate, light intensity, and temperature was developed by Cardoni and Stefan[9] from all available field measurements.

All light intensity data used in the productivity rate analysis were from the National Weather Service Station in Stoneville, MS, and attenuated with depth in the lake using relationships developed by Stefan et al.[38]

All available productivity data are listed by Cardoni and Stefan.[9] The data display the expected characteristic relationship between productivity and light: (1) an approximately linear increase in growth rate with light intensity at low values of light and (2) a plateau of maximum growth rate at high light intensity. Photoinhibition, i.e., the decrease of growth rate of excessive light exposure of the plant, was not observed. This is not unexpected, since light intensity in Lake Chicot is usually quite low due to rapid attenuation with depth.

The shape of the $P(I)$ curve can be described by a variety of mathematical, and mostly empirical, formulations. Some include the effect of photoinhibition. Comparisons of some of the equations to sets of measured data have given inconclusive results.[20,26] All empirical equations for the $P(I)$ curve require the use of coefficients, usually $F_{\text{max}}$ and the initial slope of the curve.

A Michaelis-Menten type equation was selected.

$$P = P_{\text{max}}\left(\frac{I}{K_I + I}\right) \tag{35}$$

where $P$ = productivity rate at light intensity, $P_{\text{max}}$ = minimum productivity at optimum light intensity, $I$ = light intensity, and $K_I$ = half saturation coefficient. This equation is simple and is sufficient to describe the basic shape of the $P(I)$ curves calculated from Lake Chicot field measurements.

In the $P(I)$ growth rate expression (Equation 35), an optimum growth rate $P_{\text{max}}$ is needed. Light intensity is considered to be the limiting factor in primary productivity; the maximum growth rate is found at optimum light conditions. The maximum growth rate also varies with temperature. An expression relating growth rate to temperature is derived from the Arrhenius equation. Data and fitted equations are given in Figure 17.

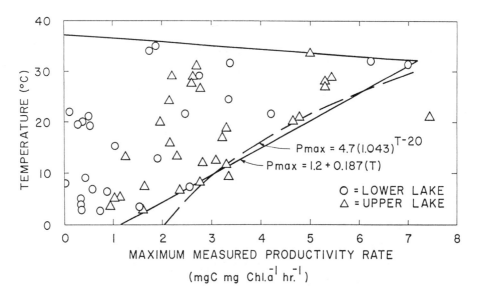

FIGURE 17.    Maximum productivity rate versus temperature in Lake Chicot.

The parameter $K_I$ was estimated directly from the productivity-vs.-light-intensity curves. $K_I$ was taken as the light intensity at one half the maximum productivity rate measured. No relationship between $K_I$ and temperature was found from the Lake Chicot data. A constant value $K_I = 100 \ \mu Em^{-2}s^{-1}$ was the best estimate. Jørgensen et al.[23] quote Gargas as using $K_I = 400$ kcal m$^{-2}$d$^{-1}$, which is equivalent to 45 $\mu Em^{-2}s^{-1}$ using the conversion of Combs.[11]

The relationships retained for simulation of light- and temperature-controlled primary productivity rates in Lake Chicot were

$$P = (1.2 + 0.187T) \frac{I}{(100 + I)} \qquad \text{for } 0 < T < 32°C \tag{36}$$

$$P = (52.9 - 1.43T) \frac{I}{(100 + I)} \qquad \text{for } 32°C < T < 37°C \tag{37}$$

$$P' = \frac{24P}{\phi} \tag{38}$$

where $P$ = primary productivity rate (mg C (mg chl$a$)$^{-1}$h$^{-1}$), $P'$ = primary productivity rate (d$^{-1}$), $I$ = light intensity ($\mu Em^{-2}s^{-1}$), $T$ = water temperature (°C) and $\phi$ = day length correction.

## D. UNDERWATER LIGHT PENETRATION

To apply the $P(I)$ relationships to the prediction of daily photosynthesis in a stratified lake, it is necessary to describe the variation of underwater irradiance as a function of depth and time over the course of a day. A model for underwater irradiance in Lake Chicot was developed by Stefan et al.[38] The input to the model is terrestrial (above-water) total daily solar radiation measurement, as available, e.g., from the Stoneville, MS weather station. Using empirical equations for albedo (Equation 19) and attenuation (Equation 20), and a conversion from energy units to quantum units, a composite relationship is derived for photosynthetically active radiation (PAR) under water.

Radiation available with depth $z$ is calculated as

$$I = I_s (1 - \text{albedo}) e^{-k(z)z} \qquad (39)$$

The variation of irradiance over the length of the daylight is described by a cosine function:

$$I(t) = I_s \left( \frac{\pi}{2} \right) \cos \left( \frac{\pi t}{t_d} \right) \qquad (40)$$

where $I(t)$ = PAR intensity at time $t$ ($\mu E/m^2/s$). Total daily solar radiation measurements are converted to average PAR values by

$$I_s = \frac{27.25 \, \phi_s}{t_d} \qquad (41)$$

where $I_s$ = average photosynthetically active radiation over daylight period, above surface ($\mu Em^{-2}s^{-1}$); $\phi_s$ = measured total daily radiation above water surface (cal cm$^{-2}$ d$^{-1}$); and $t$ = time of day starting with $t = 0$ at solar noon (hours). The length of daylight $t_d$ (h), at latitude 35°N (U.S. Naval Observatory, 1977) is

$$t_d = 12.16 + 2.36 \cos \left[ \left( \frac{2\pi}{365} \right) (172 - D) \right] \qquad (42)$$

where $D$ = Julian day of year (January 1: $D = 1$). For the numerical computation, the daylight period is divided into eight subperiods. Productivity is calculated for each period and averaged over a day. This procedure is repeated for each layer (depth $z$). The details of the computation are given by Cardoni and Stefan.[9]

### E. LOSS RATE AND SETTLING RATE

Loss rate represents the decrease in phytoplankton mass due to normal endogenous respiration and other factors causing phytoplankton mortality (e.g., zooplankton grazing, herbicides, etc.). Loss rate is taken to be a combination of all processes that cause a decrease in phytoplankton mass, except settling. Zooplankton grazing has not been considered independently, since there are insufficient data to make this distinction.

The relationship proposed by O'Connor et al.[30] is for the endogenous respiration rate. Since endogenous respiration represents a significant portion of the total loss rate, a temperature dependence of the form used by O'Connor is used in the model:

$$\text{loss rate} = K_2 T(^\circ C) \qquad (43)$$

The coefficient $K_2$ was determined by calibration with Lake Chicot chlorophyll-*a* measurements to be on the order of 0.005°C$^{-1}$d$^{-1}$.

No direct measurements of phytoplankton settling velocities from Lake Chicot were available. Due to the highly variable nature of this parameter and the difficulty associated with measuring it accurately, the model was calibrated by varying the settling rate within the range of values reported in the literature. By comparison of measured in-lake chlorophyll-*a* concentrations with those predicted by the simulation model, a settling rate on the order of 0.04 m/d was determined.

## VIII. NUTRIENT MODEL

Phytoplankton growth in Lake Chicot was most often controlled by available light. Nutrient concentrations in the lake are generally high, and primary productivity was not significantly

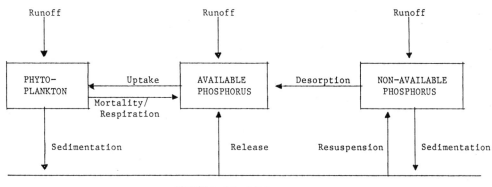

FIGURE 18.   Phosphorus submodel - RESQUAL II.

inhibited by lack of nutrients. The Lake Chicot restoration project was designed to reduce the amount of water and suspended sediment entering the lake. The input of nutrients was therefore also significntly reduced. A nutrient limitation in the growth model is desirable for future conditions when lower nutrient levels may restrict phytoplankton growth.

Phosphorus and nitrogen are the most likely limiting nutrients. Both were considered, but only phosphorus was represented in the model. Phosphorus was chosen because it is more often the limiting nutrient for algal growth in fresh waters. The U.S. Environmental Protection Agency (EPA)[19] suggested that phosphorus may be more significant in controlling growth in Lake Chicot. Baker[5] reviewed more recent Lake Chicot nutrient data and came to the conclusion that nitrogen may also be an important nutrient controlling phytoplankton growth.

The framework and general approach to the modeling of phosphorus and nitrogen cycles and interactions with phytoplankton growth in Lake Chicot were given by Cardoni et al.[10] and Baker,[54] respectively. The cycles of both elements are quite complex, and their dynamic modeling requires substantial numbers of rate coefficients and field and laboratory data that are presently not available. It is also for this reason that only a phosphorus model of a relatively simple form was incorporated into RESQUAL II.[40] The underlying phosphorus flow chart is shown in Figure 18.

## IX. SECCHI DEPTH MODEL

Secchi depth is a comprehensible measurement of water transparency. Lay people can easily relate to the meaning of Secchi depth.

Secchi depths in upper and lower Lake Chicot were analyzed and related to attenuation coefficients in an extensive study.[38,40] Secchi depths in the lower lake were related to total suspended solids (Figure 4), and Secchi depths in the upper lake were related to chlorophyll-*a* (Figure 19). Equations 44 and 45 describe the data in Figure 4 and 19, respectively.

$$z_{SD} = \frac{3.67}{(SS)^{0.625}} \quad (\text{lower lake}) \tag{44}$$

$$z_{SD} = \frac{1.37}{\text{chl}a^{0.258}} \quad (\text{upper lake}) \tag{45}$$

where $z_{SD}$ is the Secchi depth in meters, (SS) is the suspended solids concentration in ppm, and chl*a* is the chlorophyll-*a* concentration in ppb.

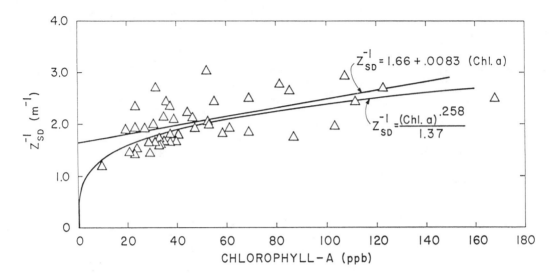

FIGURE 19.   Inverse of Secchi depth ($z_{SD}$) versus chlorophyll-*a* concentration, upper Lake Chicot (SS concentration < 45 ppm).

Development of a predictive model of Secchi depth required incorporation of the cumulative effects of inorganic suspended solids and phytoplankton on transparency. Two alternative methods for predicting cumulative Secchi depth in Lake Chicot were developed. One method involved linear relationships between the inverse of Secchi depth, and suspended solids and chlorophyll-*a* concentrations. The effects of suspended solids and chlorophyll-*a* were isolated by considering the lower and upper basins of Lake Chicot during periods when inorganic suspended solids (SS) and phytoplankton, respectively, dominated turbidity. Figure 19 shows the measured relationships between $1/z_{SD}$ and chlorophyll-*a*. The data from the upper lake were obtained when suspended solids concentrations were less than approximately 45 mg/l. The equations of best fit to the measured sets were

$$\frac{1}{z_{SD}} = 2.16 + 0.0265\,(SS) \tag{46}$$

$$\frac{1}{z_{SD}} = 1.66 + 0.0083\,(\text{chl}a) \tag{47}$$

where SS is in parts per million and chl*a* is in parts per billion. The coefficient for the effect of chl*a*, 0.0083, was lower than presented by Brezonik[7] ($\approx$.03) and Shapiro[36] (0.0146). Brezonik's data is from 55 Florida lakes and Shapiro's data is from Minnesota lakes.

The second approach to formulating a Secchi depth model used a combination of the logarithmic and the linear relationships for Secchi depth presented. Combining Equations 45 and 46 gave a composite relationship for Secchi depth:

$$\frac{1}{z_{SD}} = \frac{(\text{chl}a)^{0.258}}{1.37} + 0.0265\,(SS) \tag{48}$$

To prevent the Secchi depth from going to infinity at zero chl*a* and SS concentrations, a minimum chl*a* concentration of 3 ppb was imposed. This sets the maximum possible Secchi depth at 1.03 m, which is slightly deeper than the maximum Secchi depth measurement from the available Lake Chicot data. Equation 48 is presented graphically in Figure 20.

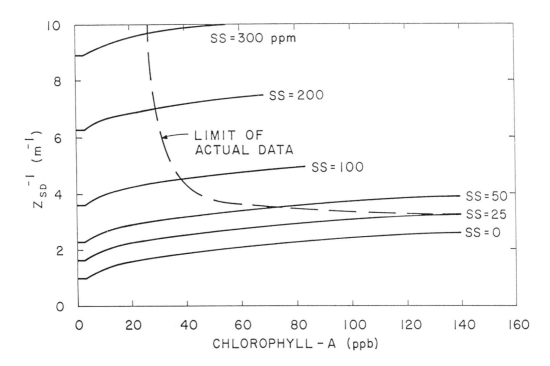

FIGURE 20. Inverse of Secchi depth versus chlorophyll-*a* and suspended sediment concentration composite relationship for Lake Chicot.

# X. COMPUTER PROGRAM RESQUAL II

### A. ORGANIZATION

The submodels identified in Figure 11 and described in the preceding sections were incorporated in computer program RESQUAL II. A detailed description of the program was prepared by Fu.[24]

There are a total of 53 subroutines in RESQUAL II. A complete alphabetical listing and brief description of each routine was assembled by Fu.[24]

Subroutines to facilitate input, computing options, and output are START = control input and computing options and PRNOUT = control output.

The main program calls the submodels in sequence. By not solving all equations simultaneously, the program is simplified and the computing time is considerably reduced.

The uncoupling of the submodels requires that input data into one submodel be taken from the output of the other submodels in the previous time step. This is appropriate for three reasons.

First, the model is operated with a time step (1 d) over which changes in the computed parameters are small. The error introduced in using, e.g., suspended sediment concentration from the previous day instead of the current day, affects results to a lesser degree (as was verified by adding additional iteration steps to the program) than the uncertainties in estimating model input parameters, e.g., settling velocity of particles and entrainment coefficient of density inflow, surface heat exchange coefficients, albedo, etc.

Second, the coupling between the main submodels in RESQUAL is actually very weak. Most of the inflow into Lake Chicot occurs during the 2 months just prior to the summer stratification period. Outflow is always from the surface mixed layer. Suspended sediment concentration affects mainly the radiation balance of the mixed layer, and suspended sediment concentration in the mixed layer changes only slowly. Dissolved solids concentrations do not control density stratification to any appreciable degree.

Third, the input variables that drive the model most strongly are runoff and weather related. The latter are highly variable in time and have to be obtained from measuring stations remote from the lake. Because of that variability, lake temperatures and mixing "tend" towards constantly changing equilibrium or ultimate temperatures. Therefore, prediction errors are not cumulative.

## B. NUMERICAL SOLUTIONS

The partial differential equations (advection/diffusion equations) for $T$, SS, chl$a$, $P_a$, and $P_n$ are solved by an implicit method.

### 1. Water Temperature (Equation 14)

In the formation of a finite difference scheme, stability and accuracy are the main concern. For the water temperature, a fully implicit central difference scheme was selected. The scheme was formulated for variable layer thickness $\Delta z(i)$. The resulting finite difference equations can be found in Stefan et al.[40] The linear algebraic equations were solved by a tridiagonal matrix algorithm. The boundary conditions applied are

1.    No heat flux to and from the sediment, i.e., $T(j+1, N) = T(j+1, N+1)$, where $N+1$ refers to a dummy layer below the bottom layer of the reservoir.
2.    No diffusive heat flux between the surface mixed layer and the hypolimnion

$$\left[ K_z \frac{\partial T}{\partial z} \right]_{m + \frac{1}{2}} = 0 \tag{49}$$

where $m$ = number of last layer in surface mixed layer.

The numerical solution of the diffusion equation for heat necessary to predict water temperature stratification is accomplished as part of subroutine HEBUG.

### 2. Suspended Sediment (Equation 29)

The subroutine that solves the suspended sediment equation is RESSETL. An implicit, hybrid finite difference scheme was developed to solve the suspended sediment equation. That scheme was described in detail by Dhamotharan et al.[17] It is stable for various combinations of vertical turbulent diffusivities and particle fall velocities.

### 3. Chlorophyll-*a* (Equation 34)

CHLORO is a modified version of the subroutine RESSETL contained in the RESQUAL model. This subroutine solves Equation 34 for all layers of the lake by using an implicit finite difference scheme. The numerical method used to solve this equation is described by Dhamotharan et al.[17] The modification of RESSETL is required to account for the additional source and sink terms that are contained in the chlorophyll-*a* model.

## C. COMPUTATIONAL OPTIONS

One need not use all available submodels each time. Three options are available to select submodels by setting the elements of the integer array MODEL(I) to either 1 or 0. (Zero means a submodel will not be used; one means it will be used.) For example, if all elements of MODEL(I) are set to zero, only the water temperature, suspended solids, and dissolved solids submodels will be selected. If MODEL(I) is set to one and all others zero, the light-limited chlorophyll-*a* submodel will be selected, in addition to the water-temperature, suspended-solids, and dissolved-solids submodels.

## D. MODEL INPUT

Model RESQUAL II requires four types of input data:

## TABLE 1
### Bathymetric Characteristics of Lake Chicot

|  | Lower lake | Upper lake |
|---|---|---|
| Reference water level | 30.48 m above MSL | 30.48 m above MSL |
|  | (100 ft) | (100 ft) |
| Maximum effective | 8.29 m | 4.52 m |
| depth[a] ($h_{ref}$) | (27.2 ft) | (15.5 ft) |
| Mean depth ($d_{ref}$) | 3.8 m | 3.0 m |
|  | (12.5 ft) | (9.9 ft) |
| Surface area | $13.63 \times 10^6$ m² | $3.53 \times 10^6$ m² |
|  | (3369 acres) | (873 acres) |
| Volume ($V_{ref}$) | $51.85 \times 10^6$ m³ | $10.65 \times 10^6$ m³ |
|  | (42036 acre ft) | (8636 acre ft) |

[a]   Effective refers to the deepest bed level, 22.19 m above mean sea level.

- Initial conditions
- Morphometric lake and channel data
- Inflow data
- Weather data

The details of these requirements are given in the instruction manual prepared by Fu.[24] Some of the main points shall be given briefly.

### 1. Initial Conditions
The initial conditions that need to be specified include

1.    The initial number $N$ of layers in the lake, typically 20
2.    $T(i,1)$
3.    $SS(i,1)$
4.    $chla(i,1)$
5.    $P_a(i,1); P_n(i,1)$,

where the number of layers varies, $1 < i < N$. An initial lake stage is also required.

### 2. Lake Morphology
Lake volume (Figure 21) was described by an equation developed by Dhamotharan:[17]

$$\frac{V}{V_{ref}} = \left( \frac{h}{h_{ref}} \right)^m \tag{50}$$

where $V$ = lake volume, $h$ = depth, $V_{ref}$ = reference volume at $h_{ref}$ = reference depth, $m = 2.18$ for lower Lake Chicot, and $m = 1.57$ for upper Lake Chicot (Figure 21). A morphometric map of Lake Chicot was given in Figure 1, and some key bathymetric data are summarized in Table 1.

From the above, the following equations were developed for volume $V$ (in m³):

$$V = 4046.8 \left( 3.28^{m-1} \right) c \, h^m \tag{51}$$

and projected horizontal area $A$ in (m²):

$$A = m\,c\,(3.28h)^{m-1} \tag{52}$$

where $c = 0.305$ m/ft and $h$ = depth in feet.

### 3. Inflow

Inflow rates to Lake Chicot are specified by a relationship of the form:[42]

$$Q_C = 0.85\,Q_M^{0.99}$$
$$Q_M = 140.93\,(S_M - 105.63)^{1.72} \tag{53}$$

where $Q_C$ = discharge into Lake Chicot from Connerly Bayou, $Q_M$ = discharge from Macon Lake, and $S_M$ = stage at Macon Lake in feet above MSL.

Water quality in the inflow (Connerly Bayou) is also specified by correlation functions given by Swain.[42] For inorganic suspended sediment:

$$SS_C = 6.55\,Q_{CAFD}^{0.58} \tag{54}$$

where $Q_{CAFD}$ = inflow from Connerly Bayou in acre ft per day. For chlorophyll-*a* (Cardoni et al. [9,10]):

$$(chla)_C = 317\,(SS_C)^{-0.693} \tag{55}$$

where $(chla)_C$ is in parts per billion and $SS_C$ in parts per million.

Available dissolved orthophosphorus $P_{ac}$ was set constant at 100 ppb. The range of this parameter was $50 < P_{ac} < 150$ ppb.

An equation for nonavailable (absorbed) phosphorus inflow was also developed.[40]

The data from which the above relationships were derived had considerable scatter and are shown in Figures 22 and 23.

### 4. Weather

Required weather data comprise daily total solar radiation, daily mean air temperature, dew point temperature, mean wind velocity, daily precipitation and daily cloud cover.

The weather station nearest Lake Chicot is the Midsouth Agricultural Weather Service Center, NOAA, Stoneville, MS. Air temperature (TA) and total daily solar radiation (RAD) from that station were used in the model. Wind velocity (WIND) and dew point temperature (TD) were the arithmetic means of daily measurements at Memphis, TN, Jackson, MS, and Shreveport, LA. the three stations showed a good correlation and Lake Chicot is located at about the center of a triangle formed by these three stations. Daily precipitation data were from measurements at Stoneville, MS. Additional information on weather stations and data was given by Dhamotharan.[14]

### E. MODEL OUTPUT

Model output is either in the form of tables or graphs, as described by Fu.[24]

## XI. MODEL CALIBRATION

RESQUAL II contains submodels that simulate water temperature, suspended solids, dissolved solids, chlorophyll-*a*, phosphorus and Secchi depth. Coefficients in each of these submodels had to be assigned numerical values. Some of these are physical constants, others were known from previous investigations or could be derived from Lake Chicot data with good reliability. For some coefficients, only order-of-magnitude estimates or ranges of numerical values were known. More precise values of these coefficients that were applicable only to Lake

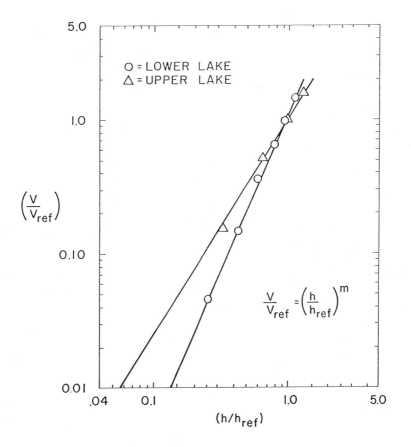

FIGURE 21.  Plot of morphometric equations for upper and lower Lake Chicot.

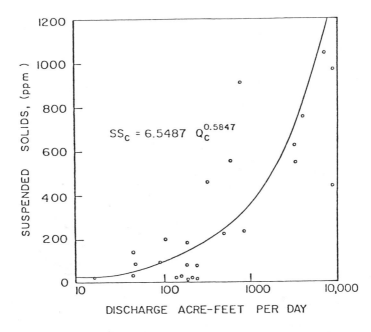

FIGURE 22.  Relationship between suspended solids (sediments) and discharge at Connerly Bayou (Swain, 1980).

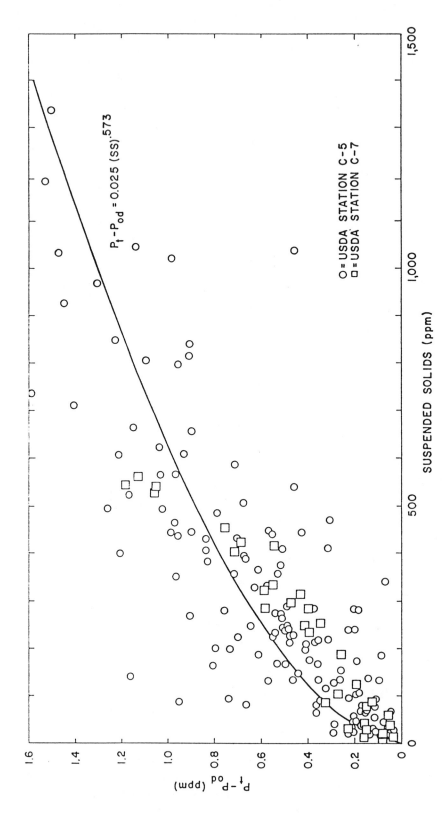

FIGURE 23.    Total phosphorus ($P_t$) minus orthophosphorus ($P_{do}$) versus suspended solids, Lower Lake Chicot.

## TABLE 2
## Coefficients of Submodels

| Submodel | Symbol | Coefficient | Range of values tested | Value used |
|---|---|---|---|---|
| 1. Temperature | $a_m$ | Wind-dependent vertical diffusion coeff. | 28.0 | 28.0 |
| | $b$ | Exponent in the wind-dependent vertical diffusion coeff. | 1.3 | 1.3 |
| | $a_h$ | Wind-dependent vertical diffusion coeff. | 0.1 | 0.1 |
| | $\gamma_{in}$ | Plunging entrainment coefficient | 1.8 – 5.0 | Mixed layer |
| 2. Suspended sediment | | FRAC = fraction of inflow suspended sediment deposited instantly at lake inlet | 0.3 –0.65 (before Mar. 10) 0.8 – 1.0 (after Mar. 10) | 0.35 before March 10 and 1.0 after March 10 |
| | | Particle diameter (DS) | 0.8 – 4.0 µm | 1.0 µm |
| 3. Chlorophyll-*a* | $W_c$ | Fall velocity (FVCHLA) | 0.03 – 1.0 m/d | 0.04 m/d |
| | ♦ | Carbon chlorophyll ratio (YCCHLA) | 25 – 60 | 30 |
| | $K_l$ | Half saturation coeff. for light (HSC) | 75 – 100 µE/m²-s | 100 µE/m²-s |
| | — | Threshold concentration value of chlorophyll-*a* | 1 – 5 ppb | 3 ppb |
| | $K_2$ | Respiration/loss coeff. | 0.003 – 0.006°C⁻¹d⁻¹ | 0.005°C⁻¹d⁻¹ |
| 4. Dissolved ortho-phosphorus (avail. phosph.) | $K_p$ | Half saturation coeff. for available phosphorus (HSCPA) | 0.01 – 0.02 ppm | 0.015 ppm |
| | $r_m$ | Release fraction of phosphorus (RFM) in the mixed layer | 0.6 – 0.8 | 0.8 |
| | $r_h$ | Phosphorus release fraction of phosphorus in the hypolimnion (RFH) | 1.0 | 1.0 |
| | $K_r$ | Bottom release rate (RR) | 0 | 0 |
| | $Y_{ca}$ | Chlorophyll to phosphorus ratio (YCA) | 300 – 700 | 600 (ppb/ppm) |
| | $P_{a\ equil}$ | Equilibrium conc. of avaiable phosphorus (PAEQ) | 0.08 – 0.10 ppm | 0.08 ppm |
| 5. Convertible phosphorus (Nonavail. phosph.) | $\eta$ | Fraction of convertible phosphorus in particulate (PNFRAC) | 0.2 – .25 | 0.2 |
| | $P_{equil}$ | Equilibrium concentration of available (dissolved ortho phos.) | 0.08 – 0.10 ppm | 0.08 ppm |
| | $K_{rr}$ | Resuspension rate of nonavailable (mostly particulate) phosphorus | 0 | 0 g SSm⁻² d⁻¹ |

Chicot were established by model calibration, i.e., comparison of simulated and observed results.

The calibration was made with data collected during water year 1977 (October 1, 1976 to September 30, 1977). The data were collected by the USDA/ARS, mostly at biweekly intervals. Calibrations were made in the major submodels successively and in the order in which they are listed in Table 2.

The mass flow and temperature stratification models were calibrated first. The two parameters that were the least well established and had to be calibrated were the vertical turbulent exchange coefficient, $K_z$, and initial dilution at the plunge point, $\gamma_{in}$.

The temperature model was at first calibrated for mean annual values of $K_z = 400$ m²/d in the surface mixed layer and $K_z = 1$ m²/d in the hypolimnion, respectively.[16] Then, coefficients

$a_m$ = 28 for the mixed layer, $a_h$ = 0.1 for the hypolimnion, and $b$ = 1.3 in Equation 26 were determined by using a mean annual wind in these equations.

In the suspended sediment model, a particle size on the order of 1 μm, as determined by sediment analysis, had to be used. The complete size distribution of particles was never obtained. Calibration confirmed that 1 μm gave loss rates by settling that were in agreement with measurements. Recent analysis using state-of-the-art methodology, however, gave a smaller mean particle size.

The chlorophyll submodel required several coefficients. Literature and Lake Chicot field values of these coefficients were given by Stefan et al.[40] The range of values tested in the model and the calibrated value are given in Table 2. The phosphorus model also had several coefficients to be calibrated. Ranges and values are also shown in Table 2. Simulated results and calibration data are shown in Figures 24 through 29.

RESQUAL II was calibrated by comparing the computed values with available measurements. The root mean square calibration error ($\varepsilon$) for each water quality parameter was computed in the plotting program RESPLT and printed on the output plot. The root mean square calibration error $\varepsilon$ is defined as

$$\varepsilon = \sqrt{\frac{\sum_{i=1}^{n}\left(C_{ic} - C_{im}\right)^2}{n}} \tag{56}$$

where $C_{ic}$ = computed water quality parameter on the $i^{th}$ day, $C_{im}$ = measured water quality parameter on the same day, and $n$ = number of measurements. Calibration coefficients were changed until the value of $\varepsilon$ could not be further reduced.

More systematic schemes to calibrate RESQUAL II, such as the least-squares optimization adopted by Norton,[29] in calibrating the RMA-12 model were not used because of the large computing cost that would have been involved.

## XII. MODEL VERIFICATION

Water quality data sets measured duing the water years 1977/78 and 1978/79 were used for model verification. A comparison between predictions and data was shown by Stefan et al.[40] The root mean square errors between predictions and measurements are reproduced in Table 3. The degree of agreement between predictions and measurements made possible the conclusion that the model was sufficiently verified for application to the exploration of some management alternatives for Lake Chicot.

Contributing to the error are the inaccuracies in inflow rates and the weak correlation between the inflow water quality and inflow rate (see, e.g. Figures 22 and 23), which had to be used in the simulation. The greatest limitation in the model formulation itself is believed to be in the phosphorus model, but this is not crucial, since light rather than phosphorus was the limiting factor most of the time.

## XIII. MODEL APPLICATION

### A. SIMULATION OF 1976-77 CONDITIONS WITH RESQUAL I

Model RESQUAL I, an earlier version of RESQUAL II, was applied to Lake Chicot for the water year 1977.[14] The inflow and outflow of lower Lake Chicot were quite small relative to the lake volume in that year. The bulk hydraulic residence time for the simulation period was 0.3 years.

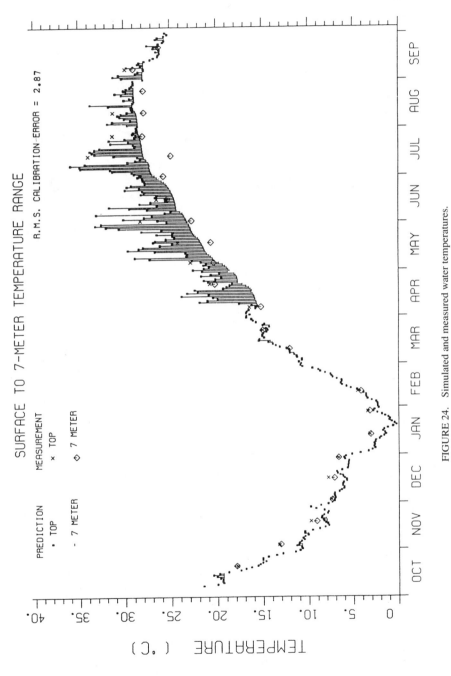

FIGURE 24.  Simulated and measured water temperatures.

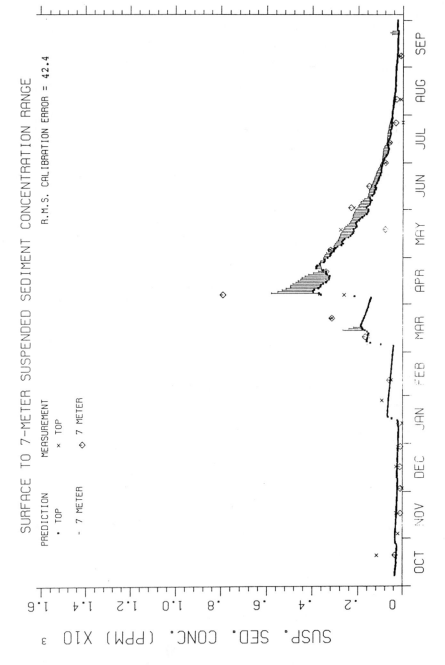

FIGURE 25.   Simulated and measured suspended sediment concentrations.

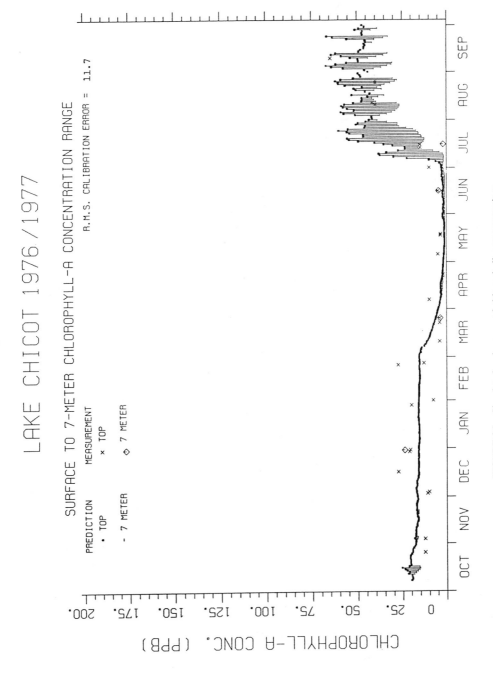

FIGURE 26. Simulated and measured chlorophyll-*a* concentrations.

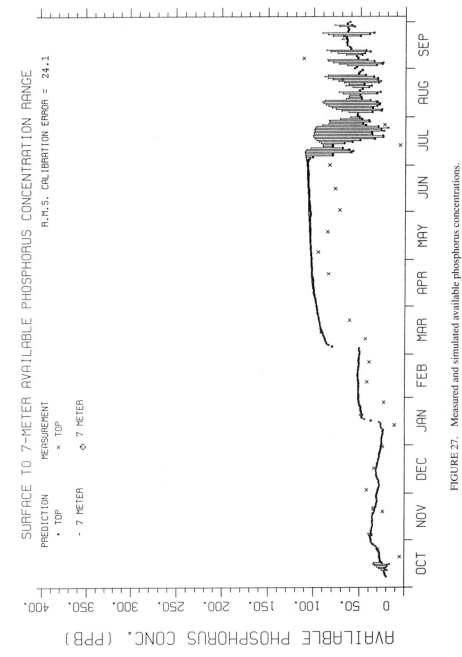

FIGURE 27.  Measured and simulated available phosphorus concentrations.

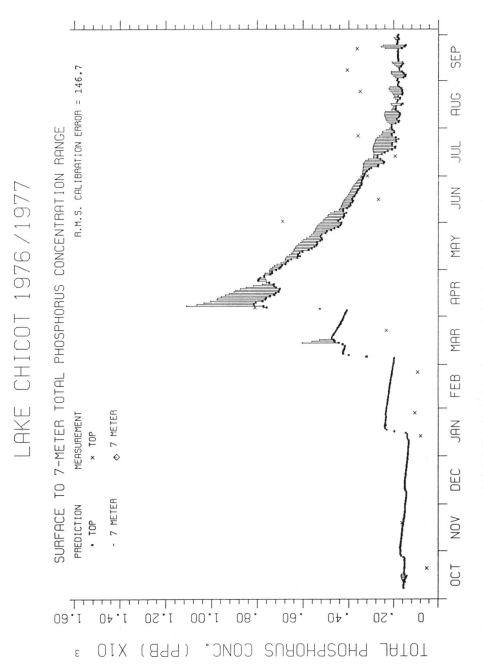

FIGURE 28.    Simulated and measured total phosphorus concentrations.

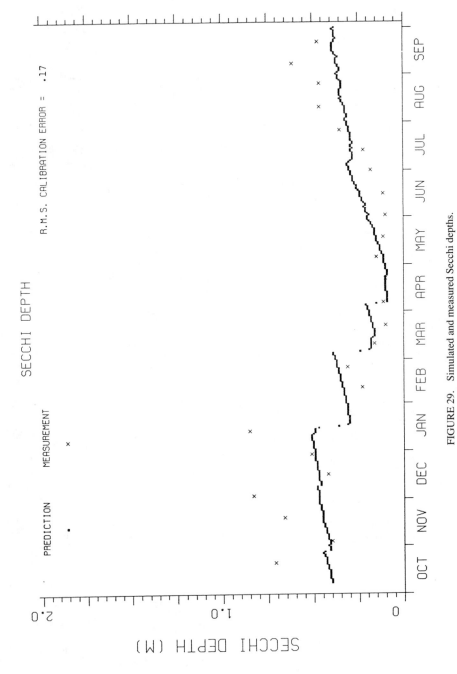

FIGURE 29. Simulated and measured Secchi depths.

**TABLE 3**
**Root Mean Square Errors**

| | Year | | |
|---|---|---|---|
| Parameter | 1976–77 | 1977–78 | 1978–79 |
| Temperature (°C) | 1.7 | 1.2 | 1.2 |
| Susp. solids (ppm) | 42 | 94 | 123 |
| Chlorophyll-*a* (ppb) | 11.7 | 13.9 | 15.4 |
| Available phosphorus (ppb) | 24 | 28 | 31 |
| Total phosphorus (ppb) | 147 | 204 | 199 |
| Secchi depth (m) | 0.17 | 0.10 | 0.13 |

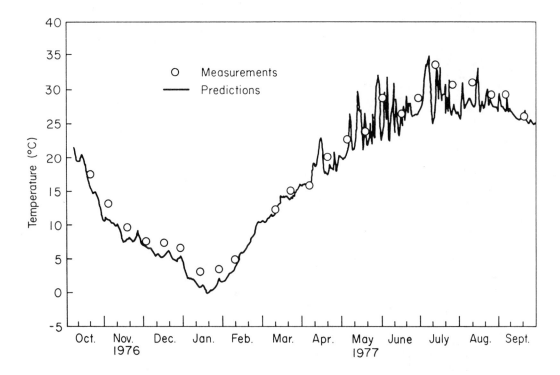

FIGURE 30. Measured and predicted surface water (mixed layer) temperatures in Lake Chicot.

Simulated and measured surface water temperatures are shown in Figure 30. The differences between the measurements and prediction had a standard deviation of 0.34°C. Strong fluctuations in surface water temperatures are shown between April and September, coincident with the period of summer stratification. These fluctuations are related to variations in the depth of the surface mixed layer. When wind velocities are small, the mixed layer becomes temporarily shallow and water temperatures become high.

Water temperatures as a function of depth have been plotted in Figure 31. These simulated results show the summer stratification and surface mixing clearly. The dates of the onset of the summer stratification and of fall overturn were predicted within the observed periods.

The predicted and measured suspended sediment concentration at the water surface in Lake Chicot is shown in Figure 32 (top) and that at 5 m depth is shown in Figure 32 (center). The average suspended sediment concentration over the whole depth, both measured and predicted, is shown in Figure 32 (bottom). The predicted and measured suspended sediment concentration

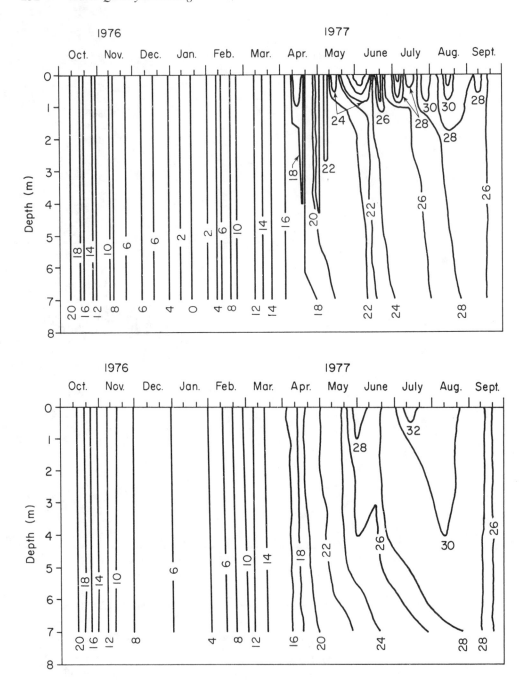

FIGURE 31.   Seasonal temperature structures (Isotherms in °C). Predicted (top) and measured (bottom) water temperature averages at Stations 6, 7, and 8.

vertical profiles for selected dates are shown in Figure 33. The trap efficiency of the reservoir, on a cumulative monthly basis, both measured and predicted, is shown in Figure 34. Comparison of measured and predicted values of temperature, suspended sediment concentration, and trap efficiency indicate good agreement. In the suspended sediment model, the sediment inflow and outflow are treated as source and sink, respectively, in the mixed layer. Previous studies of the Lake Chicot seasonal density structure indicate that this is fairly realistic for most of the year.[18]

Suspended sediment loading of the lake occurs primarily in the spring, when the lake is not

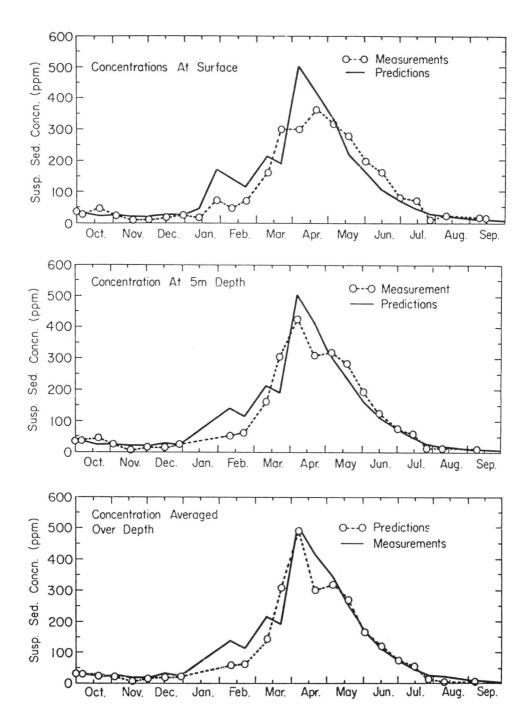

FIGURE 32. Suspended sediment concentration in Lake Chicot.

yet stratified. The differences shown for that period in Figure 32 reflect inaccurate loading information, which had to be based on inflow concentrations measured at weekly intervals. After the onset of stratification (April), the agreement between measurements and predictions is good, indicating that internal mixing and settling mechanisms are well represented in the model. Additional simulation results with RESQUAL I were given by Dhamotharan et al.[18] and by Dhamotharan.[17]

Suspended  Sediment  Concentration  (ppm)

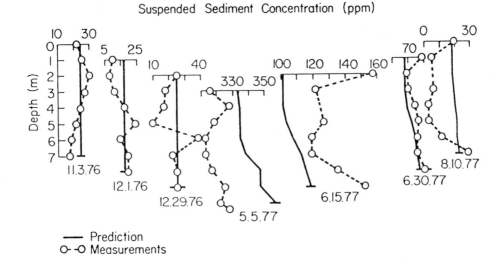

— Prediction
O- -O Measurements

FIGURE 33.    Typical suspended sediment concentration profiles in Lake Chicot.

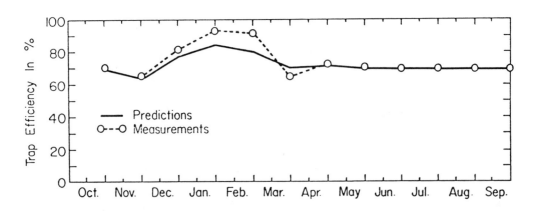

FIGURE 34.    Cumulative trap efficiency of Lake Chicot.

## B. POST-CONSTRUCTION PREDICTION WITH RESQUAL I

Construction of a new outlet weir with increased crest level in lieu of the defective one and/or reduction of inflow so as to decrease suspended sediment load to the lake were the alternatives contemplated for the restoration of the recreational potential of the lake.

Some hypothetical values of weir crest levels and limits on water inflow were selected for study. It was assumed that the weather conditions were the same as those of 1976-77; watershed characteristics and concentrations of the suspended sediments in the inflow remained the same. The outflow over the weir was predicted each day by the standard weir equation:

$$Q_{OUT} = W_c \times L \times H^m \qquad (57)$$

where $Q_{OUT}$ = outflow in cfs, $L$ = weir length = 200 ft (assumed), $W_c$ = weir coefficient = 3.95, $H$ = head of flow over the weir in ft, and $m$ = 1.5. Lake stage was predicted every day by means of the water budget analysis.

Results were obtained for weir crest levels of 106, 108, and 110 ft above MSL, and water inflow was limited to a maximum of 50, 100, and 250 cfs, i.e., the inflow hydrograph peaks were cut off at these values.

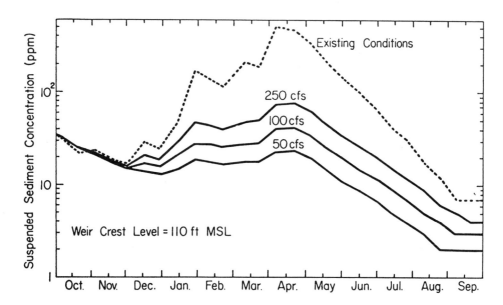

FIGURE 35.    Predicted post-construction suspended sediment concentrations at different residual inflow rates.

For the weir crest level at 110 ft above MSL and for the three flow rate limits, the predicted suspended sediment concentration at the surface of the reservoir is shown in Figure 35. In the same figure, the existing conditions are also shown for comparative purposes. Very similar results were found for the weir crest level at 108 and 106 ft above MSL. In all three cases, the highest predicted postconstruction sediment concentration occurrred during April. The highest April value was 95 ppm for weir crest level 106 ft above MSL and at 250 cfs inflow limit, and 24 ppm for the lake stage of 110 ft and 50 cfs inflow limit, respectively. The preconstruction maximum was 500 ppm. This predicted decrease in suspended sediment concentration after April was very encouraging.

Since the weather conditions remained the same in all situations, the predicted temperature structure remained practically the same as that under the existing conditions. The difference, if any, due to the change in lake volumes was usually less than 1°C.

A sediment mass budget analysis was simulated for the nine postconstruction conditions, existing conditions, and compared with the actually measured data. The results are given in Table 4. The table shows the annual sediment inflow mass, the annual sediment outflow mass, and the sediment mass deposited in the reservoir during the year.

The annual trap efficiency of the reservoir for the various postconstruction predicitions is given in Table 5. The trap efficiency for existing conditions was 68%, both measured and simulated.

It may be pointed out here that the postconstruction predictions were based on the existing measured inflow sediment concentration. Even if the watershed characteristics had not changed, any impoundment of inflow water for the purposes of monitoring the inflow might alter the suspended sediment concentration in the inflow due to either or both (1) settling of sediments in the impoundment and (2) pick up of sediment by clear water downstream of the impoundment before reaching the lake.

Simulation of existing conditions in Lake Chicot showed that about 62,930 t of sediments were deposited in the reservoir during the water year 1976-77. Assuming the specific weight of the incoming coarse clay particles to be about 80 lb/ft$^3$, a lake volume equivalent to 40 acre ft was lost to sediment accumulation during the year. This is about 0.07% of the total lake volume at 105 ft water surface stage.

TABLE 4
Annual Sediment Budget for Lake Chicot (Water Year 1977)

| Inflow limits (cfs) | Annual sediment mass in metric tons | | | |
|---|---|---|---|---|
| | Inflow | Outflow | Deposition | Suspension |
| 250 | 26700 | 4100 | 20100 | 2500 |
| 100 | 14300 | 1400 | 12500 | 400 |
| 50 | 7800 | 500 | 8400 | −1100 |
| Existing conditions (simulated) | 103300 | 32000 | 62900 | 8400 |
| Actual measurements | 103300 | 31200 | — | — |

TABLE 5
Predicted Annual Trap Efficiency of Lake Chicot
in Percentage (Water Year 1977)

| Crest level (ft MSL) | Inflow limit (cfs) | | |
|---|---|---|---|
| | 250 | 100 | 50 |
| 110 | 86 | 91 | 93 |
| 108 | 86 | 90 | 93 |
| 106 | 85 | 90 | 93 |

## C. POST-CONSTRUCTION PREDICTION WITH RESQUAL II

Simulations with reduced inflow rates into Lake Chicot were also made with the expanded model RESQUAL II. In anticipation of the operation of the new pumping station at Macon Lake, inflow rates were truncated at 0.005, 5, 50, 100, 250, and 500 cfs. Flows above these values would be diverted to the Mississippi River. The predicted water quality under the weather conditions encountered in 1978-79 are shown in Figure 36 through 39. 1978-79 was the wettest of the 3 years. It had the largest water run-off into the lake and the highest suspended sediment input. Both of these produced the highest turbidity (lowest Secchi depth) in the lake from December 1978 to September 1979. The effect of inflow diversion on the lake stage is obvious. The suspended sediment concentration, and hence turbidity of the lake, is very much a factor of inflow. As lessened turbidity of the lake permits an increase in light penetration, phytoplankton growth increases as evident in the simulation results.

An estimate of the clarity of the lake at the various flow diversion rates is given in terms of predicted Secchi depths in Figure 39.

Management alternatives were also simulated under different hydrological and weather conditions. The objective was to find an optimum operating rule that would (1) maximize outflow at Ditch Bayou for downstream water use, (2) minimize lake turbidity (mainly suspended sediment concentration) for recreational purposes, and (3) maintain a desirable lake stage (in this case a lake stage between 102 and 106 ft above MSL) during the recreational season (April 1 to September 30).

It was found that by completely diverting inflow into Lake Chicot when the lake stage is higher than 106 ft above MSL, and by withdrawing 75 cfs of outflow through the sluice gates from April 1 to September 30, under natural hydrological conditions, or 25 cfs year round under extreme dry summer conditions, the in-lake water quality and the lake stage could be maintained at satisfactory levels.

How this conclusion was reached shall be explained in more detail. Operating rules were specified as follows:

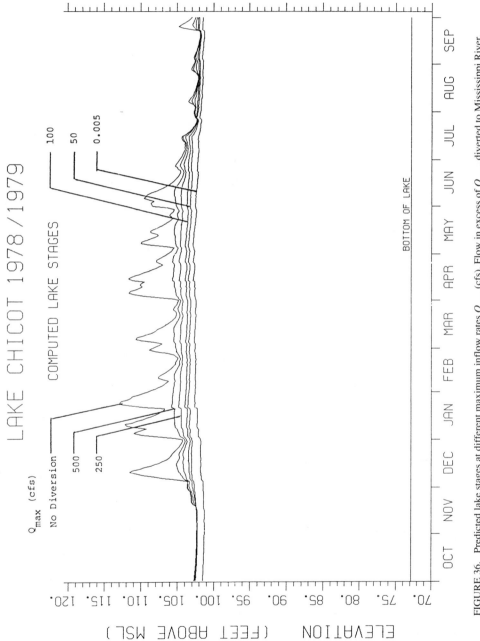

FIGURE 36. Predicted lake stages at different maximum inflow rates $Q_{max}$ (cfs). Flow in excess of $Q_{max}$ diverted to Mississippi River.

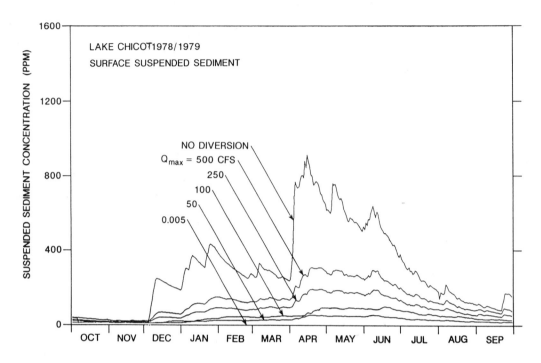

FIGURE 37. Predicted suspended sediment concentration at different maximum inflow rates.

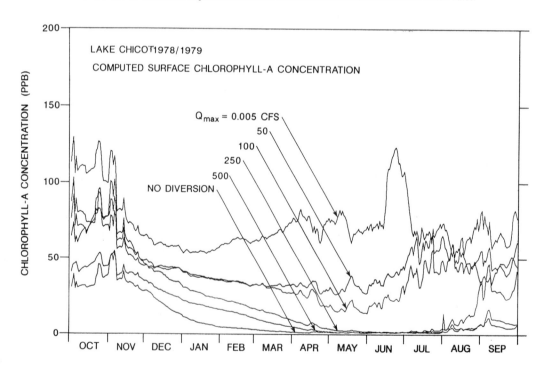

FIGURE 38. Predicted surface chlorophyll-*a* values at different maximum inflow rates.

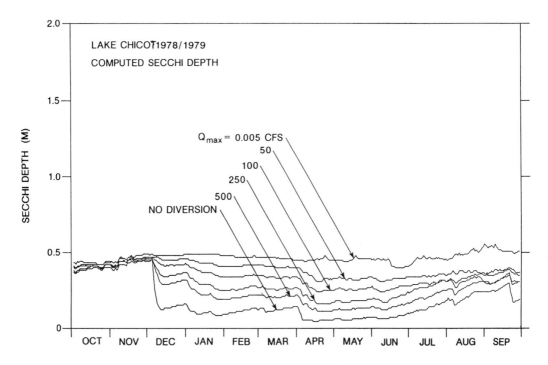

FIGURE 39.   Predicted lake transparency expressed as Secchi depth at different maximum inflow rates.

1.  *Operating Rule No. 1*: If the lake stage is higher than 106 ft, inflow ($Q_{in}$) will be completely diverted, provided it is within the capacity of the pumping station.
2.  *Operating Rule No. 2*: No more than 175 cfs of inflow is allowed into the lake at any time. This inflow corresponds to a lake stage at Macon Lake of 106.92 ft above MSL (before construction).
3.  *Operating Rule No. 3*: If the in-lake and inflow suspended sediment concentrations are higher than 25 ppm, inflow is completely diverted. If the in-lake or inflow suspended sediment concentrations are less than 25 ppm, inflow is allowed into Lake Chicot.

The hydrological and weather conditions considered were

1.  *Natural weather*: No modification of weather data.
2.  *Very dry summer*: Zero precipitation on lake year round. From April 1 to September 30, inflow is a base flow of 7.4 cfs.
3.  *Storm*: A storm occurs from September 1 to September 3, and produces 2 in/d of runoff. The input hydrograph is as follows:

| Date | Inflow (cfs) |
|------|--------------|
| 9/1  | 3143 |
| 9/2  | 13351 |
| 9/3  | 17049 |
| 9/4  | 15159 |
| 9/5  | 5485 |
| 9/6  | 1787 |
| 9/7  | 541 |

**TABLE 6**
**Minimum Outflows through Ditch Bayou (cfs)**

| Hydrological conditions | Inflow operating rules | | |
|---|---|---|---|
| | 1 | 1 + 2 | 1 + 3 |
| Natural | 85 | 75 | 50 |
| Dry summer | 25 | 25 | 25 |
| Storm | | Not needed | |

**TABLE 7**
**Lake Chicot Water Quality Parameters**

| Hydrological conditions | Water quality | Inflow operating rules | | |
|---|---|---|---|---|
| | | 1 | 1 + 2 | 1 + 3 |
| Natural | SD | 31–54 | 39–51 | 36–52 |
| | SS | 5–36 | 5–11 | 5–13 |
| | chl*a* | 26–202 | 28–93 | 31–137 |
| Dry summer | SD | 24–45 | 44–53 | 36–45 |
| | SS | 8–109 | 2–29 | 3–4 |
| | chl*a* | 6 92 | 15–64 | 17–103 |
| After storm | SD | 2–51 | | |
| | SS | 7–1925 | Not needed | |
| | chl*a* | < 3 | | |

*Note:* SD = Secchi depth (cm) range from 4/1 to 9/30, SS = suspended solids (ppm) range from 4/1 to 9/30, and chl*a* = chlorophyll-*a* (ppb) range from 4/1 to 9/30.

The three operating rules were tested using weather data for the water year 1977 (October 1976 to September 1977). Cases that gave promising results were selected and tested for long-term effects under continuous implementation of each operating rule. Weather data for the water years 1977, 1978, 1979, 1980, and 1981 were used for simulation of long-term effects.

Fifteen cases representing different combinations and operating rules and weather were simulated. All cases are listed in detail by Stefan and Fu.[41]

To present the results for the water year 1977 in a readily understandable form, selected cases were summarized under two objectives. The first objective is to maximize outflow, the second is to minimize turbidity (i.e., minimize suspended sediment and maximize Secchi depth). In all cases, lake stage must not fall below 102 ft above MSL. The only results that satisfy these criteria are summarized below.

*Objective 1:* Maximize outflow (cfs) from April 1 to September 30. (See Table 6.)
*Objective 2:* Minimize in-lake turbidity. (See Table 7.)

To access the long-term effects, the seven most promising cases were selected for further study using the weather data of water years 1978, 1979, 1980, and 1981. The results are given by Stefan and Fu.[41]

The best cases under natural hydrologic conditions (1976-81) are summarized in Table 8. According to these results, utilizing operating rule 1 and 2 simultaneously and under normal conditions had the most promise of minimizing in-lake turbidity. In this case, however, the lake stage may fall below 102 ft above MSL. Since a lake stage higher than 102 ft was desirable for recreational purposes, alternative choices were considered.[41]

**TABLE 8**
**Best Operation**

| | Operating outflow (cfs) | | Maximum inflow (cfs) | Lake Stage (ft) | Water Quality |
|---|---|---|---|---|---|
| Case | Rule | Apr–Sept | | Apr–Sept | Apr–Sept |
| A | 1 | 75 | Unlimited | 103–108 | SD: 17–63 cm<br>SS: 5–169 ppm<br>chl*a:* 3–140 ppb |
| B | 1 + 3 | 50 | Unlimited | 103–109 | SD: 13–55 cm<br>SS: 2–222 ppm<br>chl*a:* 25–259 ppb |
| C | 1 + 2 | 75 | 175 | 97–106 | SD: 32–56 cm<br>SS: 2–11 ppm<br>chl*a:* 25–259 ppb |
| D ⎫<br>E ⎭ | 1 + 2<br>1 + 2 | 75 (1976–79)<br>20 (1980–81) | 175 | 105–106 | SD: 32–57 cm<br>SS: 1-11 ppm<br>chl*a:* 25-259 ppb |

The simulation of several water inflow and outflow control alternatives for Lake Chicot under normal and dry weather conditions for up to 5 years led to the following general guidelines:

1. Diversion of inflow from Lake Chicot can improve lake quality significantly if properly operated.
2. Operating rule 1, which requires inflow into Lake Chicot to be completely diverted if the lake stage is higher than 106 ft above MSL, tends to produce considerably improved water quality in Lake Chicot based on a continouous 5-year numerical simulation.
3. Summer outflow rates of 75 cfs, coupled with maximum inflow rates of 175 cfs, will not guarantee lake stages between 102 and 106 ft in all years, but in many years. An operating policy using reductions in outflow rates as a function of lake stage should be made when the lake stage falls below 106 ft.
4. Imposing the above limitations on inflow and outflow appears to result in stable lake levels and the lowest water turbidity when compared to other options tested.
5. Outflow rates of 75 cfs or more appear to draw lake stages considerably below 102 ft on occasion.
6. Admitting water to the lake regardless of the stage appears to increase turbidity.
7. Admitting water at rates considerably higher than 175 cfs appears to have a detrimental effect on water quality.
8. The effects of very major storm events appear to remain detectable for periods as long as 9 months.
9. Further refinements of the desirable inflow and outflow rates may be possible, but those apparent refinements will depend on annual weather cycles. Values for 5 years (October 1976 to September 1981) were used to obtain these guidelines.

## XIV. POST-CONSTRUCTION EXPERIENCE/RECENT DEVELOPMENTS

In March 1985, the U.S. Army Corps of Engineers placed three structures into operation: a new dam equipped with gates on Ditch Bayou to regulate the lake level and discharge, an upstream dam and gate that prevents poor-quality water from entering the lake, and a combination pump-gravity flow facility, which diverts poor quality water to the adjacent

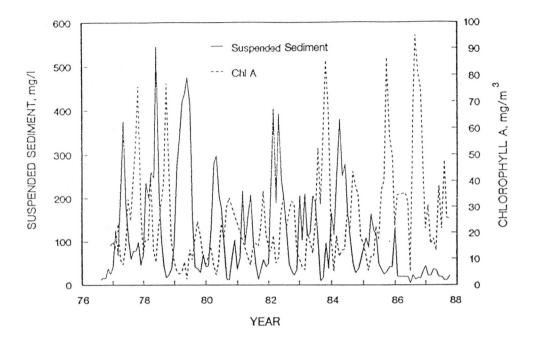

FIGURE 40.    Measured average monthly suspended sediment and chlorophyll-*a* in Lake Chicot 1976 to 1988.

Mississippi River. Operation of these facilities changed the water quality in the lower lake (station 7) after March 1985 significantly (Figure 40). The upper lake remained unaffected, since its inflow cannot be controlled by theses measures. The improvement in water quality in the lower lake is documented in Table 9. In this table, water quality data are presented in two time frames, September 1982 to March 1985 (before changes) and April 1985 to September 1987 (after changes). The upper lake was not affected by the inflow control, and hence water quality was not appreciably changed (Table 10). Because of its much improved water quality, lower Lake Chicot is again attracting visitors, who use the lake for sport-fishing, sightseeing, and swimming. The discharge from the lake is used, in part, for irrigation (50 to 90 cfs in 1987). The new discharge structure has facilitated control of the discharge rate and irrigation water use. Overall the entire project and the predictions made have been successful.

## XV. CONCLUSIONS

The elements and an application of the dynamic, unsteady, spatially one-dimensional numerical model RESQUAL II for the prediction of suspended sediment concentration, water temperature, and dissolved solids in a shallow turbid lake/reservoir have been presented. The model is dynamic because it considers the energy exchange at the air-water interface as the forcing mechanism for lake stratification and mixing; it considers inflow, outflow, and suspended and dissolved material loading by the inflow.

The model has several major submodels, including one for prediction of surface heat transfer and temperature stratification, one for suspended sediment diffusion and settling, one for inflow and outflow in the presence of density stratification, and one for phytoplankton growth and transport. The submodels are uncoupled, resulting in major savings of computational time.

Turbidity caused by clay particles in a shallow lake increases both surface albedo and diffuse radiation attenuation. Relationships for attenuation and albedo based on data from Lake Chicot, AR, were incorporated in the model. Reflectivity and attenuation affect water temperatures and

## TABLE 9
### Secchi Depth and Depth-Averaged Water Quality On Lower Lake Chicot (Station 7) Before and After Restoration

| | Jan | Feb | Mar | Apr | May | Jun | Jul | Aug | Sep | Oct | Nov | Dec |
|---|---|---|---|---|---|---|---|---|---|---|---|---|
| Transparency (m) | | | | | | | | | | | | |
| Sep 1982 - Mar 1985 | 0.17 | 0.08 | 0.11 | 0.10 | 0.08 | 0.10 | 0.12 | 0.32 | 0.45 | 0.38 | 0.31 | 0.15 |
| Apr 1985 - Sep 1987 | 0.53 | 0.78 | 0.80 | 0.51 | 0.75 | 0.70 | 0.50 | 0.56 | 0.51 | 0.46 | 0.57 | 0.51 |
| Suspended sediment (mg/l) | | | | | | | | | | | | |
| Sep 1982 - Mar 1985 | 143.6 | 283.2 | 325.3 | 330.3 | 324.3 | 267.3 | 138.3 | 39.6 | 27.8 | 67.7 | 44.0 | 208.7 |
| Apr 1985 - Sep 1987 | 10.0 | 85.5 | 23.0 | 117.9 | 66.1 | 37.3 | 18.6 | 14.8 | 18.1 | 31.3 | 19.3 | 38.8 |
| Dissolved oxygen (mg/l) | | | | | | | | | | | | |
| Sep 1982 - Mar 1985 | 7.7 | 5.7 | 8.3 | 8.7 | 8.2 | 7.6 | 6.7 | 7.8 | 9.3 | 9.3 | 8.6 | 8.7 |
| Apr 1985 - Sep 1987 | 11.3 | 11.5 | 11.9 | 9.1 | 9.0 | 8.5 | 8.3 | 9.3 | 7.5 | 8.4 | 8.8 | 10.5 |
| Chlorophyll concentration (μg/l) | | | | | | | | | | | | |
| Sep 1982 - Mar 1985 | 1.8 | 4.2 | 1.8 | 2.8 | 1.7 | 20.3 | 16.3 | 13.4 | 39.0 | 24.8 | 56. | 1.6 |
| Apr 1985 - Sep 1987 | 27.7 | 26.3 | 16.8 | 14.3 | 15.1 | 11.0 | 35.7 | 47.6 | 53.5 | 56.5 | 35.2 | 18.3 |
| Total N concentration (mg/l) | | | | | | | | | | | | |
| Sep 1982 - Mar 1985 | 0.8 | 1.2 | 1.3 | 1.5 | 1.2 | 1.2 | 1.2 | 0.9 | 0.8 | 0.8 | 0.6 | 1.1 |
| Apr 1985 - Sep 1987 | 0.7 | 0.4 | 0.5 | 0.7 | 0.7 | 0.7 | 1.0 | 1.3 | 1.3 | 1.0 | 1.0 | 0.7 |
| Total P concentration (mg/l) | | | | | | | | | | | | |
| Sep 1982 - Mar 1985 | 0.246 | 0.219 | 0.354 | 0.342 | 0.593 | 0.389 | 0.202 | 0.106 | 0.097 | 0.149 | 0.118 | 0.314 |
| Apr 1985 - Sep 1987 | 0.044 | 0.035 | 0.035 | 0.193 | 0.128 | 0.105 | 0.096 | 0.140 | 0.335 | 0.138 | 0.072 | 0.069 |

## TABLE 10
### Secchi Depth and Depth-Averaged Water Quality in Upper Lake Chicot (Station 4) Before and After Restoration

| | Jan | Feb | Mar | Apr | May | Jun | Jul | Aug | Sep | Oct | Nov | Dec |
|---|---|---|---|---|---|---|---|---|---|---|---|---|
| Transparency (m) | | | | | | | | | | | | |
| Sep 1982 - Mar 1985 | 0.20 | 0.40 | 0.21 | 0.36 | 0.33 | 0.52 | 0.43 | 0.39 | 0.45 | 0.33 | 0.47 | 0.35 |
| Apr 1985 - Sep 1987 | 0.60 | 0.40 | 0.28 | 0.38 | 0.55 | 0.58 | 0.31 | 0.29 | 0.27 | 0.30 | 0.48 | 0.48 |
| Suspended sediment (mg/l) | | | | | | | | | | | | |
| Sep 1982 - Mar 1985 | 34.7 | 92.2 | 112.3 | 41.7 | 49.8 | 12.3 | 17.0 | 13.0 | 20.2 | 48.0 | 56.2 | 71.3 |
| Apr 1985 - Sep 1987 | 26.0 | 30.0 | 56.0 | 31.4 | 51.9 | 25.5 | 20.7 | 17.0 | 30.5 | 24.3 | 51.2 | 114.2 |
| Dissolved oxygen (mg/l) | | | | | | | | | | | | |
| Sep 1982 - Mar 1985 | 11.7 | 11.6 | 10.2 | 10.2 | 9.6 | 8.0 | 8.7 | 7.2 | 5.2 | 9.2 | 9.9 | 11.3 |
| Apr 1985 - Sep 1987 | 11.2 | 10.2 | 11.4 | 9.5 | 10.1 | 11.5 | 12.0 | 8.7 | 8.8 | 7.7 | 7.9 | 10.6 |
| Chlorophyll concentration (μg/l) | | | | | | | | | | | | |
| Sep 1982 - Mar 1985 | 25.3 | 45.4 | 46.2 | 41.2 | 36.7 | 44.0 | 69.0 | 82.4 | 95.8 | 110.5 | 72.0 | 41.6 |
| Apr 1985 - Sep 1987 | 47.5 | 0.2 | 0.8 | 19.2 | 32.2 | 25.5 | 91.0 | 100.4 | 120.0 | 142.0 | 82.3 | 48.7 |
| Total N concentration (mg/l) | | | | | | | | | | | | |
| Sep 1982 - Mar 1985 | 1.3 | 1.4 | 1.4 | 1.3 | 1.7 | 1.8 | 1.6 | 1.8 | 1.6 | 1.6 | 1.4 | 1.2 |
| Apr 1985 - Sep 1987 | 1.0 | 0.6 | 0.8 | 0.9 | 1.0 | 1.6 | 2.4 | 2.6 | 2.9 | 2.4 | 2.0 | 1.4 |
| Total P concentration (mg/l) | | | | | | | | | | | | |
| Sep 1982 - Mar 1985 | 0.125 | 0.105 | 0.157 | 0.095 | 0.146 | 0.109 | 0.104 | 0.125 | 0.151 | 0.165 | 0.133 | 0.128 |
| Apr 1985 - Sep 1987 | 0.074 | 0.191 | 0.108 | 0.087 | 0.088 | 0.111 | 0.146 | 0.233 | 0.266 | 0.183 | 0.148 | 0.166 |

the strength of the temperature stratification The model was applied to Lake Chicot, AR, which had a severe turbidity problem beginning around 1920 and ending in 1985.

The density stratification structure of the lake was found to be dominated by water temperature; suspended and dissolved solids increased water densities by only small amounts. For this reason, inflowing water, although highly turbid, could spread on the lake surface. Contrary to expectation sinking turbidity currents would not form when turbidity is the highest, but when inflow water temperature was the lowest. Predictions and measurements of water temperatures and of suspended sediment concentrations for 1 year were compared and found to be in good agreement. The model was used to investigate various lake improvement options,[41] and the implementation has resulted in a major reduction in lake turbidity such that it is again used extensively for fishing and recreation.

## ACKNOWLEDGMENTS

This study was conducted for the U.S. Army Corps of Engineers, Vicksburg District, through the U.S. Department of Agriculture Sedimentation Laboratory, Oxford, MS and the Water Quality Laboratory, Durant, OK. The University of Minnesota Computer Center provided a grant for the numerical simulations. The biweekly field data were collected by the USDA. The Corps of Engineers provided data on flow rates.

## REFERENCES

1. **Akiyama, J. and Stefan, H. G.,** Theory of Plunging Flow into a Reservoir, University of Minnesota, St. Anthony Falls Hydraulic Laboratory, Internal Memorandum No. IM-97, December 1981.
2. **Akiyama, J. and Stefan, H. G.,** Gravity Currents in Lakes, Reservoirs, and Coastal Regions: Two-Layer Stratified Flow Analysis, University of Minnesota, St. Anthony Falls Hydraulic Laboratory Project Report No. 253, March 1987.
3. **Bacon, E. J.,** Primary Productivity, Water Quality, and Limiting Factors in Lake Chicot, Arkansas Water Resources Research Center, Fayetteville, AR, Publ. No 56, 1978.
4. **Bacon, E. J.,** Primary Productivity in Lake Chicot, Chicot County, Arkansas, Quarterly Progress Reports, University of Arkansas, Monticello, AR, October 1977–September 1980.
5. **Baker, L.,** Concepts of a Nitrogen Cycling Model of Lake Chicot, Arkansas, University of Minnesota, St. Anthony Falls Hydraulic Laboratory, Internal Memorandum No. 99, July, 1982.
6. **Bowen, I. S.,** The ratio of heat loss by conduction and by evaporation from a water surface, *Phys. Review,* 27, 779–787, 1926.
7. **Brezonik, P. L.,** Effect of organic color and turbidity on Secchi disk transparency, *J. Fish Res. Board Can.,* 35, 1410–1416, 1978.
8. **Cardoni, J. J. and Hanson, M. J.,** Lake Chicot Intensive Field Investigations, June 30 to July 10, 1981, University of Minnesota, St. Anthony Falls Hydraulic Laboratory, External Memorandum No. 176, 92, December 1981.
9. **Cardoni, J. J. and Stefan, H. G.,** A Model for Light and Temperature Limited Primary Productivity in Lake Chicot, University of Minnesota, St. Anthoy Falls Hydraulic Laboratory, External Memorandum 177, May 1982.
10. **Cardoni, J. J., Hanson, M. J., and Stefan, H. G.,** A Model of Phosphorus Available for Phytoplankton Growth in Lake Chicot, University of Minnesota, St. Anthony Falls Hydraulic Laboratory, External Memorandum 178, July, 1982.
11. **Combs, W. S.,** The Measurement and Prediction of Irradiance Available for Photosynthesis by Phytoplankton in Lakes, Ph.D. thesis, University of Minnesota, Minneapolis, 1977.
12. **Cooper, C. M. and Bacon, E. J.,** Effects of suspended sediments on primary productivity in Lake Chicot, Arkansas, in *Proc. Sym. on Surface Water Impoundments,* June 1980, Minneapolis, Minnesota, American Society of Civil Engineers, 1981.
13. **Dake, J. M. K. and Harleman, D. R. F.,** Thermal stratification in lakes: analytical and laboratory studies, *Water Res. Res.,* 5(2), 404–495, 1969.
14. **Dhamotharan, S., Stefan, H. G., and Schiebe, F. R.,** Prediction of post construction turbidity of Lake Chicot, Arkansas, *Proc. Int'l. Symp. on Environmental Effects of Hydraulic Engineering Works,* TVA, IAHR, 1978, 146–159.
15. **Dhamotharan, S. and Stefan, H. G.,** Mathematical model for temperature and turbidity stratification in shallow reservoirs, *Proc. Symp. on Surface Water Impoundments,* ASCE, 1981, 613–623.
16. **Dhamotharan, S., Gulliver, J., and Stefan, H.,** Unsteady one-dimensional settling of suspended sediment, *Water Res. Res.,* 17(4), 1125–1132, 1981.
17. **Dhamotharan, S.,** A Mathematical Model for Temperature and Turbidity Stratification Dynamics in Shallow Reservoirs, Ph.D. Thesis, University of Minnesota, Minneapolis, MN, 319, March 1979.

18. **Dhamotharan, S., Stefan, H. G., and Schiebe, F. R.,** Turbid reservoir stratification modelling, *Proceedings, 26th Annual Hydraulic Div. Specialty Conf.,* ASCE, Univ. of Maryland, College Park, MD, 1978.

19. Environmental Protection Agency, Chicot Lake, Chicot County, Arkansas, EPA Region VI. National Eutrophication Survey, Environmental Monitoring and Support Laboratory, Las Vegas, NV and Corvallis Environmental Research Laboratory, Corvallis, OR, Working Paper No. 484, 1977.

20. **Field, S. D. and Effler, S. W.,** Photosynthesis-light mathematical formulations, *J. Environ. Eng. Div.,* ASCE, 108(EE1), 199–203, 1982.

21. **Filatov, N. N., Rjanzhin, S. V., and Zaycev, L. V.,** Investigation of Turbulence and Langmur Circulation in Lake Ladoga, *J. Great Lakes Res.,* 17(1), 1–6, 1981.

22. **Ford, D. E. and Stefan, H.,** Thermal prediction using integral energy model, ASCE, *J. Hydraul. Div.,* 106(HY1), 39–55, 1980.

23. **Ford, D. E. and Stefan, H.,** Stratification variability in three morphometrically different lakes under identical meteorological forcing, *Water Res. Bull.,* 16 (2), 243–247, 1980.

24. **Fu, A.,** User Instructions for RESQUAL II, A Dynamic Water, Quality Simulation Model for a Shallow Stratified Lake or Reservoir, University of Minnesota, St. Anthony Falls Hydraulic Laboratory, External Memorandum No. 179, July, 1982.

25. **Gibbs, R. J., Mathews, M. D., and Link, D. A.,** The relationship between sphere size and settling velocity, *J. Sediment. Petrolo.,* 41(1), March, 1971.

26. **Jassby, A. D. and Platt, T.,** Mathematical formulation of the relationship between photosynthesis and light for phytoplankton, *Limnol. Oceanogr.,* 21(4)540–547, 1976.

27. **Jorgensen, S. E., Mejer, H., and Friis, M.,** Examination of a lake model, *Ecol. Modelling,* 4, II, 253–278, 1978.

28. **Marciano and Harbeck,** Mass Transfer Studies, Lake Hefner Studies, U.S. Geological Survey Professional Paper No. 267, 1954.

29. **Norton, W. R.,** Final Report for the Upper Mississippi River Basin Model Project, Water Resources Engineers, Walnut Creek, CA, 1974.

30. **O'Connor, D. J., Thomann, R. V., and DiToro, D.,** Dynamic Water Quality Forecasting and Management, Manhattan College, Bronx, NY, EPA-600/3-73-009, 1973.

31. **Pedersen, F. B.,** A monograph on Turbulent Entrainment and Friction in Two-Layer Stratified Flow, Institute of Hydrodynamics and Hydraulic Engineering, Techn. University of Denmark, Lyngby, 1980.

32. **Rothwell, E. D. and Fletcher, B. P.,** Lake Chicot Pumping Plant Outlet Structure, Arkansas, Tech. Report HL-79-10, U.S. Army Waterways Experiment Station, Vicksburg, MS, June, 1979.

33. **Schiebe, F. R.,** Personal communication, U.S.D.A., Durant, OK, 1980.

34. **Schiebe, F. R., Farell, J. O., and McHenry, J. R.,** Water Quality Improvement of Lake Chicot, Arkansas, *Proc. Symp. on Surface Water Impoundments,* ASCE, 1981.

35. **Swain, A.,** Material Budget of Lake Chicot, Ph.D. disserration, University of Mississippi, Oxford, MS, 1980.

36. **Shapiro, J.,** personal communication, 1982.

37. **Stefan, H. and Ford, D.,** Temperature dynamics in dimictic lakes, ASCE, *J. Hydraul. Div.,* 101(HY1), 1975.

38. **Stefan, H. G., Cardoni, J. J., Schiebe, F. R., and Cooper, C. M.,** A model of light penetration in Lake Chicot, University of Minnesota, St. Anthony Falls Hydraulic Laboratory, External Memorandum 174, Feb., 1982.

39. **Stefan, H. G., Dhamotharan, S., and Schiebe, F. R.,** Temperature/sediment model for a shallow lake, ASCE, *J. Environ. Eng. Div.,* 108(EE4), 1982.

40. **Stefan, H. G., Cardoni, J. J., and Fu, A. Y.,** RESQUAL II: A dynamic Water Quality Simulation Program for a Stratified Shallow Lake or Reservoir, University of Minnesota, St. Anthony Falls Hydraulic Laboratory Project Report No. 209, December 1982.

41. **Stefan, H. G. and Fu, A. Y.,** Simulation of Selected Inflow, Outflow and Water Quality Management Options for Lake Chicot, Arkansas, University of Minnesota, St. Anthony Falls Hydraulic Laboratory External Memorancum No. 183, March 1983.

42. **Swain, A.,** Material Budgets of Lake Chicot, Ph.D. dissertation, University of Mississippi, Oxford, 1980.

43. U.S.D.A.-A.R.S. An Assessment and Evaluation of Hydrological, Chemical and Microbiological Regimes of Lake Chicot, Arkansas. Quarterly progress reports, Sedimentation Laboratory, U.S.D.A.-A.R.S., Oxford, MS, October 1977 to September 1980.

44. U.S. Naval Observatory, *The Air Almanac,* 1977 data.

45. **Winter, T. C.,** Numerical simulation of steady state three-dimensional groundwater flow near lakes, *Water Res. Res.,* 14(2), 245–254, 1978.

46. **Wunderlich, W. O.,** The dynamics of density-stratified reservoirs, in Hall, G. E., ed., *Reservoir Fisheries and Limnology,* Spec. Publ. 8, American Fish. Soc., Washington, DC, 1971, 219–231.

Chapter 7

# AN ENVIRONMENTAL MANAGEMENT MODEL OF THE UPPER NILE LAKE SYSTEM

## S. E. Jørgensen

## TABLE OF CONTENTS

# I. INTRODUCTION

During the last decade, a continuous hydrological survey of the Upper Nile Basin has been carried out. The project, called HYDROMET, has eight countries as participants: Burundi, Egypt, Kenya, Ruanda, Sudan, Tanzania, Uganda, and Zaire. The headquarters of the project are in Entebbe, Uganda, where hydrologists from the member countries are all working.

UNDP and WHO agreed to expand the project with a water quality model component, which was developed during 1978 to 1983 by a Danish research team. The model is based on a great number of data, which have been obtained by analysis of water samples from three lakes: Lake Victoria, Lake Kyoga, and Lake Mobuto Sese Seko.

The objectives of the water quality model were

1.   To provide information on the water quality of the lakes. Water quality should be evaluated if the entire water balance is changed due to further development of the water resource, e.g., irrigation, or the discharge of municipal or industrial wastewater to the system. Consideration should be given regarding the use of the water for (1) drinking purposes, (2) industrial purposes, (3) irrigation, (4) stock and wildlife watering, (5) fish and aquatic life, and (6) recreational purposes.
2.   To provide information about the environmental impact on and the ecological changes of the water bodies. This could result from changes in the entire water balance, due to further development of the water resources, e.g., irrigation or discharge of municipal or industrial wastewater to the system.
     This includes
     a.   Eutrophication of the lakes
     b.   Use of water for stock and wildlife watering and for fish and aquatic life
     c.   Ecological balance of the entire ecosystem
     d.   Disease vector agents
     e.   Weed infestation intensity
     f.   Fishstock available for fishery
3.   To provide information on how a suitable water quality could be achieved or maintained to meet present and future requirements. This involves determing the maximum allowable effluent concentrations for wastewater handling methods and the technology to be propagated for use by governments.

# II. GENERAL COMMENTS ON THE MODEL

It would require a very comprehensive model and a lot more time-consuming biological data to meet all of these objectives with high accuracy. Furthermore, industrial planning in the countries involved is not yet known in detail. It was therefore suggested that an ecological model should be developed that will give more detailed and accurate information about the important variables of the system and only rough estimates about those variables that are less important or that require comprehensive biological research and/or information that is available today.

The model was, however, made very flexible and allowed the addition of submodels, when needed, for more detailed information at a later stage or for the time when a more detailed development plan is made available. It was decided to use "intensive" measurements for the parameter estimation that is related to the growth of phytoplankton in the three lakes. This has only been partially possible, due to the quality of the data available. It has, therefore, been necessary to use values from the literature in order to be able to give some reasonably good parameter estimation for the growth of phytoplankton.

Data on forcing functions are available, but unfortunately they are not all from the same period as the water quality measurements. The forcing functions are

1. Water flows
2. Concentration in streams (organic matter, nutrients, conservative species, heavy metals, etc.)
3. Precipitation
4. Concentration in rain water (partially measured, partially taken from previously measured data)
5. Data on predicted water consumption and the urbanization of the area

The quantity and quality of the available data are hardly sufficient to calibrate the model. There is, unfortunately, no independent set of measured values that is available for validation. It must, therefore, be emphasized that the model is not at present validated (as discussed in Chapter 5, for example).

However, the model will give decision-making information. At a later stage, it is recommended that improvements be made in the calibration and that the model be validated based on additional data. The developed model is structured in such a way as to be easy to improve the calibration phase and to validate the model.

## III. DESCRIPTION OF THE MODEL

The total model consists of the following independent submodels:

1. Eutrophication model
2. Fish model
3. Thermocline model
4. Box model describing the exchange of water, elements, and compounds between the box compartments in the lakes
5. DDT model
6. Model describing the interaction of nutrients between sediment and water
7. Heavy metals (giving details for copper, including adsorption of suspended matter and interaction of heavy metals between sediment and water)
8. pH model
9. Bilharzia model

The interrelation between the different forcing functions and the submodels can be seen in Figure 1. An arrow from one box in this figure to another indicates that the data from the first box is used in the second (see also Table 1).

Although no validation has taken place, it is possible to indicate approximately the accuracy of the different submodels. The thermocline model seems to be very accurate, giving temperature profiles when the predicted value is less than 1°C from the measured value.

In temperate lakes, it has been possible to develop eutrophication models with a standard deviation of about 25%[23] relative to measured and predicted values. As this deviation is based on single values, an average annual eutrophication measure will be indicated with a fully acceptable accuracy.

The quality and quantity of the data available for this project are rather limited. The accuracy of predictions on phytoplankton concentrations at certain dates at certain stations are, therefore, limited. However, an overall picture of the entire ecosystem emerges from the model, especially when considering annual average values.

The accuracy of the fishery model is also limited due to the limited number of fishery data available. However, since it seems that the fish biomass in the lakes is, to a certain extent, in balance (the fishery from year to year is relatively constant), it will be possible, based on the fish model, to make predictions on how the fishery will be influenced if significant increases, in the

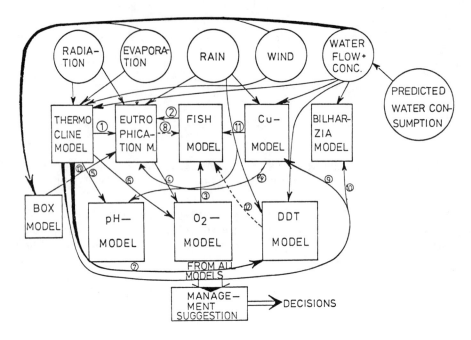

FIGURE 1.   Submodels and their connections.

## TABLE 1
### Transfer of Data from One Submodel to Another

| Number | Explanation |
|---|---|
| 1, 5, 6, 7, 9, 10 | The thermocline model divides the lake in epilomnion and hypolimnion. Under normal conditions, this is only used in the eutrophication model. |
| 2 | Concentration of herbivorous and carnivorous fish. |
| 3 | The mortality of fish is oxygen dependent. |
| 4 | Oxygen consumption is dependent on the mineralization rate for detritus. |
| 8 | The phytoplankton concentration averaged over 6 months indicates the feed concentration for herbivorous fish species. |
| 11, 12 | The mortality of fish is dependent on the copper concentration. |
| 13 | The box model is generally applicable to all submodels, but is mainly used for the eutrophication model. |
| 14 | The growth rate and mortality of phytoplankton is dependent on the copper concentration. |

fish catch take place. It is, furthermore, estimated that the relationship between pollution and fishery conditions is rater well described in the model. This relationship is based on quite well-known facts. In addition, a literature review has given great support to the parameter estimations of the fish model. All in all, it must be estimated that the fishery model is a useful model for developing a future fishery strategy, although additional data might considerably improve the reliability of this submodel.

It is not possible today to set up very accurate heavy-metals submodels. However, the required accuracy is limited, as it is necessary, due to unknown sublethal effects, to reduce heavy-metals contamination to at least a magnitude less than the concentration corresponding to lethal effects.

The oxygen submodel and the pH submodel are both relatively accurate. The experience with such submodels is extensive. If additional data are to be provided, it will be rather easy to improve the paramter estimation of these two submodels.

The DDT submodel is based on a very small amount of data. However, the processing of DDT in the environment is quite well known from other studies, and this knowledge has been taken into account in the DDT submodel for the Upper Nile Lake system. It is, furthermore, not necessary to have high accuracy in the DDT model in order to apply the model to practical environmental management. Acceptable concentrations of DDT will be at least one order of magnitude less than that which produces harmful effects. Therefore, it can be concluded, in spite of the scarce data used for the calibration of this submodel, that the submodel is applicable for practical management purposes in the region.

The Bilharzia model is only semi-quantitative. This submodel will indicate whether the conditions for the Bilharzia snail are changed relative to what is known to be the present condition. However, Bilharzia disease is mainly dependent on the quality of the water supply, an individual's overall health condition, and other factors that, under no circumstances, can be taken into account in the model.

In summary, it can be concluded that the submodels that are a part of the total model of the Upper Nile Lake system can be used for practical management, but that the output from the model should be treated with caution. Although the model is applicable as it stands today, it is strongly recommended that the model be improved and maintained in the coming years. It must especially be recommended that additional data be provided to improve the calibration and validation of the submodels. This is of particular importance for the pH, $O_2$, fishery, and eutrophication models. All four of these submodels provide a level of accuracy that can only be used to give an overall picture of the water quality in the lake system. This means that the predicted value must be taken as an average value over a certain period or combined for a number of stations in order for it to be reliable; but if these precautions are taken into account, the model can be of great use for environmental management in the region.

## IV. MANAGEMENT ISSUES

The model can be used to explore a number of management issues. We have selected four to illustrate applications of the model as a management tool.

1. Urbanization of the area will approximately triple the discharge of wastewater into the Equatorial Lakes between 1980 and 2000. What effects will this have on water quality in the lakes in terms of eutrophication and the oxygen concentration?
2. Copper mining activity is expected to increase in the future. What will be the implications of the increased wastewater discharge?
3. What effect will the steady use of DDT have on environmental quality?
4. Will it be possible to increase the sizes of fisheries without producing any long-term harm to the fish population, which would result in a decrease in fisheries size in the future?

## V. THE EUTROPHICATION MODEL

Independent cycles of nitrogen, phosphorus, and carbon are included in the model. The three elements are allowed to vary between a minimum and maximum value in phytoplankton, and these variations are also reflected in the upper trophic levels due to the system of equations, which is based upon conservation of mass.

The state variables in the eutrophication model for phosphorus are shown in Figure 2. Similar

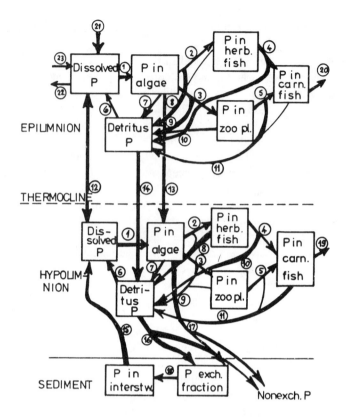

FIGURE 2. Eutrophication model illustrated by use of phosphorus-cycling.

flow diagrams can be set up for nitrogen and carbon. The state variables are: dissolved phosphorus, algal phosphorus, detrital phosphorus, phosphorus in zooplankton, phosphorus in herbivorous fish (two species are modeled independently in the fish model: haplochromis and tilapia; they are lumped in the eutrophication model), and phosphorus in carnivorous fish. These six state variables are applied in the epilimnion as well as the the hypolimnion. In addition, there are two phosphorus-state variables in the sediment: phosphorus-exchangeable in the sediment and phosphorus in interstitial water in total fourteen phosphorus-state variables.

The arrows between the boxes in Figure 2 are processes. Descriptions of the following processes are included in the model:

1. Uptake of phosphorus by algae
2. Grazing by herbivorus fish
3. Grazing by zooplankton
4, 5. Predation of fish and zooplankton, respectively, by carnivorous fish
6. Mineralization
7. Mortality of algae
8-11. Grazing and predation loss
12. Exchange of phosphorus between the epilimnion and hypolimnion
13. Settling of algae (epilimnion → hypolimnion)
14. Settling of detritus (epilimnion → hypolimnion)
15. Diffusion of phosphorus from interstitial → to lake water
16. Settling of detritus (hypolimnion → sediment) (a part goes to the nonexchangeable fraction)

17. Settling of algae (hypolimnion → sediment) (a part goes to the nonexchangeable fraction)
18. Mineralization of phosphorus in exchangeable fraction
19, 20. Fishery
21. Precipiation
22. Inflows (tributaries)

It is characteristic of the model that it describes algal growth as a two-step process: the uptake of a nutrient (a Michaelis-Menten expression is used) and growth are determined by the intracellular concentration of the nutrient relative to a minimum and maximum value. It is our experience[21,22] that this description in some cases gives a better agreement between the model and observations. The growth is described by the use of a growth rate coefficient, CDR, which is limited by four factors:

1. A temperature factor

$$FTI = \exp\left[-2.3\left|t - 17.5\right|/15\right]$$

An alternative and sometimes better factor is

$$\exp\left[A\left(t - T_{opt}\right)\right]\left[\left(T_{max} - T\right)\middle/\left(T_{max} - T_{opt}\right)\right]A\left(T_{max} - T_{opt}\right)$$

where $A, T_{opt}$, and $T_{max}$ are species-dependent constants. $t$ is the temperature in both expressions.

2. A factor for intracellular nitrogen, NC:

$$FN3 = 1 - NC_{min}/NC$$

A parallel factor for intracellular phosphorus:

$$FP3 = 1 - PC_{min}/PC$$

Intracellular carbon:

$$FC3 = 1 - CC_{min}/CC$$

NC, PC, and CC are determined by nutrient uptake rates:

$$UC = UC_{max} \bullet FC1 \bullet FC2 \bullet FRAD$$

$$UN = UN_{max} \bullet FN1 \bullet FN2$$

$$UP = UP_{max} \bullet FP1 \bullet FP2$$

where $UC_{max}$, $UN_{max}$, and $UP_{max}$ are species-dependent constants (maximum uptake rate). Generally the smaller the size of the considered phytoplankton, the greater is the $UC_{max}$.

FC1, FN1, and FP1 are expressions that give the limitations in uptake:

$$FC1 = \left(FCA_{max} - FCA\right)\big/\left(FCA_{max} - FCA\right)$$

$$FN1 = \left(FNA_{max} - FNA\right)\big/\left(FNA_{max} - FNA_{min}\right)$$

$$FP1 = \left(FPA_{max} - FPA\right)\big/\left(FPA_{max} - FPA_{min}\right)$$

$FCA_{max}$, $FCA_{min}$, $FNA_{max}$, $FNA_{min}$, $FPA_{max}$, and $FPA_{min}$ are constants that indicate the maximum respectively, minimum content of nutrient in phytoplankton.

FC2, FN2, and FP2 give the limitations in uptake caused by the nutrient level in the lake water:

$$FC2 = C/KC + C$$

$$FN2 = NS/NS + KN$$

$$FP2 = PS/PS + KP$$

Thus, these expressions are in accordance with the Michaelis-Menten formulation.

FRAD is a complex expression, regarding the influence of solar radiation. The influence is integrated over depth and the self-shading effect is included.

The intracellular nitrogen, phosphorus, and carbon can now be determined by differential equations (see Table 2, level I). FCA, FNA, and FPA are determined as

$$FCA = CC/PHYT, \; FNA = NC/PHYT, \; and \; FPA = PC/PHYT$$

The eutrophication model is presented by the use of symbols taken from Park et al.[33] Figure 3 gives the flow chart of the phytoplankton model of Jørgensen[21] and Jørgensen et al.[22] Table 2 summarizes the equations, Table 3 the parameter values, Table 4 the forcing functions, and Table 2 the differential equations using simpler symbols.

The crucial parameters are determined by the use of an intensive measuring program (sampling every second day over 2 to 3 weeks). This technique gives a considerably better parameter estimation, as many parameters (e.g., maximum growth rate of algae) reflect dynamic variability requiring a high frequency of measurement. The technique is described in detail in Jørgensen et al.[23]

Oxygen depletion should only be a problem at the sediment surface. The oxygen equation included in our model is as follows. With $O_2$ equal to the oxygen concentration at the sediment surface,

$$dO_2/dt = -\left(\text{oxygen consumption in sediment and on the surface}\right)$$
$$+ \text{diffusion from hypolimnion}$$

Oxygen consumption is paralleled by the release of phosphorus from the sediment and at the surface. This mineralization, in accordance with generally valid stoichiometry, requires 135 times more oxygen than the amount of phosphorus released. The diffusion can be computed from the oxygen concentration gradient. The diffusion coefficient is calibrated using 1.6 cm²/d as the first estimation.

Application of the model to the projected situation in the year 200 yields the following results:

1.    The central parts of Lake Victoria, Lake Kyoga, and Lake Mobuto Sese Seko would not be affected. It is easy to understand from the model why this is the case: input of water to

## TABLE 2

The model has 17 state variables:

| | |
|---|---|
| CC | Carbon in algal cells (g/m$^3$) |
| FNF | Proportion of nitrogen in fish |
| FNZ | Proportion of nitrogen in zooplankton |
| FPF | Proportion of phosphorus in fish |
| FPZ | Proportion of phosphorus in zooplankton |
| NC | Nitrogen in algal cells (g/m$^3$) |
| ND | Nitrogen in detritus (g/m$^3$) |
| NSED | Nitrogen in sediment (g/m$^3$) |
| NS | Soluble nitrogen (g/m$^3$) |
| PB | Phosphorus released biologically from sediment (g/m$^3$) |
| PC | Phosphorus in algal cells (g/m$^3$) |
| PD | Phosphorus in detritus (g/m$^3$) |
| PE | Exchangeable phosphorus in sediment (g/m$^3$) |
| PHYT | Phytoplankton in biomass (g/m$^3$) |
| PI | Phosphorus in interstitial water (g/m$^3$) |
| PS | Soluble phosphorus (g/m$^3$) |
| ZOO | Zooplankton biomass (g/m$^3$) |

Level I. The Differential Equations:

$$dCC/dt = (UC - RC) \cdot PHYT - (SA + GZ/Y + Q/V) \cdot CC$$
$$dFNF/dt = (PRED/Y) \cdot (FNZ - FNF)$$
$$dFNZ/dt = MYZ \cdot (FNA - FNZ)$$
$$dFPF/dt = (PRED/Y)(FPZ - FPF)$$
$$dFPZ/dt = MYZ \cdot (FPA - FPZ)$$
$$dNC/dt = UN \cdot PHYT - (SA + GZ/Y + Q/V) \cdot NC$$
$$dND/dt = L \cdot GZ \cdot NC + MZ \cdot NZOO + L \cdot PRED \cdot NFISH - (KDN + SD + Q/V) \cdot ND + QNDIN$$
$$dNSED/dt = (SA \cdot NC - SD \cdot ND - NREL)/AE$$
$$dNS/dt = KDN \cdot ND + RZ + NZOO + PRED \cdot NFISH + NREL - UN \cdot PHYT + QNSIN - (Q/V + DENIT) \cdot NS$$
$$dPB/dt = QSED/AB - QBIO - QDSORP$$
$$dPC/dt = UP \cdot PHYT - (SA + GZ/Y + Q/V) \cdot PC$$
$$dPD/dt = L \cdot GZ \cdot PC + MZ \cdot PZOO + L \cdot PRED \cdot PFISH - (KDP - SD + Q/V) \cdot PD + QPDIN$$
$$dPE/dt = [\tfrac{12}{29} \cdot SA \cdot PC - QSED + SD \cdot PD]/AE - KE \cdot PE$$
$$dPHYT/dt = (CDR - SA - GZ/Y - Z/Q) \cdot PHYT$$
$$dPI/dt = (AE/AI) \cdot KE \cdot PE - QDIFF \cdot AL$$
$$dPS/dt = KDP \cdot PD + RZ \cdot PZOO + PRED \cdot PFISH - UP \cdot PHYT - QDIFF + QPSIN - (Q/V) \cdot PS + AB \cdot (QBIO + QDSORP)$$
$$dZOO/dt = (MYZ - RZ - MZ - Q/V) \cdot ZOO - PRED \cdot FISH/Y$$

QNDIN (QPDIN) and QNSIN (QPSIN) respresent the flows of detrital and soluble nitrogen (phosphorus)

Level II. Rates:

| | |
|---|---|
| CDR | $= CDR_{max} \cdot FT1 \cdot FN3 \cdot FC3 \cdot FP3$ |
| GZ | $= MYZ \cdot FZP$ |
| KDN | $= KDN_{10} \cdot FT3$ |
| KDP | $= KDP_{10} \cdot FT3$ |
| KE | $= KE_{20} \cdot FT2$ |
| MYZ | $= MYZ_{max} \cdot FPH \cdot FTI$ |
| NREL | $= FTS \cdot (KREL \cdot NSED + 0.08)/(1000D)$ |
| PRED | $= PRED_{max} \cdot FT1 \cdot FZ$ |
| QBIO | $= 0.563 \cdot FT6 \cdot (PB/1800)/(1000 \ DB)$ |
| QDIFF | $= FT4 \cdot [1.21 \cdot (P1 - P2) - 1.701] (1000D)$ |
| QDSORP | $= (0.60 \lg PS - 227)(1000 \cdot DB)$ |
| QSED | $= \min(SA \cdot PC, 5.06 \cdot 10^{-3})$ |
| RC | $= RC_{max} \ FC4 \ FT1$ |

## TABLE 2 (continued)

| | |
|---|---|
| RZ | $= RZ_{max} FT1$ |
| SA | $= (SVS\ D)\ (FT2)^{1/2}$ |
| SD | $= (SVD\ D)\ (FT2)^{1/2}$ |
| UC | $= UC_{max}\ FC1 \cdot FC2 \cdot FRAD$ |
| UN | $= UN_{max}\ FN1\ FN2$ |
| UP | $= UP_{max}\ FP1\ FP2$ |

Level III. Limiting Factors:

| | |
|---|---|
| FCA | $= CCPHYT$ |
| FC1 | $= (FCA_{max} - FCA)/(FCA_{max} - FCA_{min})$ |
| FC2 | $= C/(KC + C)$ |
| FC3 | $= I - CC_{min}/CC$ |
| FC4 | $= (CC/CC_{min})^{2/3}$ |
| FNA | $= NC/PHYT$ |
| FN1 | $= (FNA_{max} - FNA)/(FNA_{max} - FNA_{min})$ |
| FN2 | $= NS/(NS + KN)$ |
| FN3 | $= I - NC_{min}/NC$ |
| FPA | $= PC/PHYT$ |
| FPH | $= max\ 0, (PHYT - 0.5)/(PHYT + KA)$ |
| FP1 | $= (FPA_{max} - FPA)/(FPA_{max} - FPA_{min})$ |
| FP2 | $= PS/(PS + KP)$ |
| FP3 | $= I - PC_{min}/PC$ |
| FRAD | $= lg[(RAD + KL)/(RAD\ BEER + KL)]\ \Omega = (\xi + \beta\ PHYT)D$ |
| FT1 | $= exp[-2.3|t - 16.5|/15]$ |
| FT2 | $= \theta^{t-20}$ |
| FT3 | $= \phi^{t-10}$ |
| FT4 | $= (t + 273)/280$ |
| FT5 | $= exp(0.151t)$ |
| FT6 | $= exp(0.203t)$ |
| FZ | $= max(0, (ZOO - KS)/(ZOO + KZ))$ |
| FZP | $= ZOO/PHYT$ |

Level IV. Other Equations:

| | |
|---|---|
| FISH | $= FISH_0\ \{1 + 0.8\ sin\ [0.017453\ (DAY + 150)]\}$ |
| NWAT | $= NC + ND + NS + NZOO$ |
| PWAT | $= PC + PD + PS + PZOO$ |
| NTOT | $= NC + ND + NS + NZOO + NFISH + AE\ NSED$ |
| PTOT | $= PC + PD + PS + PZOO + PFISH + AE\ PE + AI\ PI + AB\ PB$ |
| NZOO | $= FN2 \cdot 200$ |
| PFISH | $= FNF\ FISH$ |
| PROD | $= CDR\ PHYT$ |
| AB | $= (DB/D)\ DMUU$ |
| AE | $= LUS\ DMU/D$ |
| AI | $= LUL\ (I - DMU)/D$ |

Lake Victoria comes mainly (about 90%) from precipitation. (This implies that the outflow from Lake Victoria would likewise be unaffected). Lake Kyoga is more eutrophic than Lake Victoria andwould be affected if tributaries north of Lake Victoria received significantly more nutrients, though it would not be affected by the outflow from Lake Victoria. Because major development will take place north and west of Lake Victoria, Lake Kyoga would be only slightly more eutrophic in 2000 than it is today. Lake Mobuto Sese Seko would not be affected at all, because water quality there is entirely dependent on the water quality of its southern tributaries.

2.   If a long period of little or no wind were to occur, the sediment might become anaerobic, but that is also true at a much lower wastewater loading. In other words, climatic conditions are more important factors in the determination of the oxygen conditions of Lake Victoria.

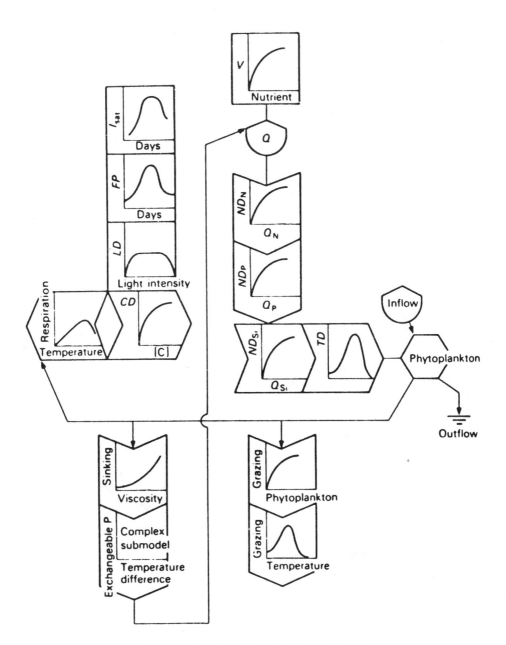

FIGURE 3.    Flow chart of the phytoplankton model.

3.    Kampala and Kisumu Bays would be affected by the increased discharge of wastewater, mainly because development will take place in adjacent areas. Even now, the discharge of nutrients in the two bays is at unacceptable levels and has already caused increasing eutrophication. In fact, according to model projections, the nutrient loading should be decreased to about 50% of its *present* level.

## VI. THE THERMOCLINE MODEL

Our thermocline model is based upon Ryan and Harleman.[35] It considers the following heat fluxes:

## TABLE 3
## Parameter Values

| Symbol | Definition | Unit | Values Model I | Model II | Values based upon |
|--------|-----------|------|---------|----------|-------------------|
| $n$ | Extinction coefficient of water | $m^{-1}$ | 0.27 | 0.27 | Chen and Orlob (1975) |
| $\beta$ | Specific extinction coefficient of phytoplankton | $m^2/g$ | 0.18 | 0.18 | Chen and Orlob (1975) |
| $C$ | Concentration of inorganic carbon | mg/1 | 100 | 100 | Measurements |
| $CDR_{max}$ | Maximum growth rate of phytoplankton | $d^{-1}$ | 2.3 | 2.35 | Calibration |
| $D$ | Depth | m | 1.8 | 1.8 | Measurements |
| DB | Depth of biologically very active layer | m | — | $2 \cdot 10^{-3}$ | Measurements |
| DENIT | Denitrification coefficient | $d^{-1}$ | 0.02 | 0.03 | Nitrogen balance |
| DMU | Dry matter in sediment | — | 0.07 | 0.07 | Measurements |
| $FISH_0$ | Concentration of fish | mg/1 | 0.55 | 0.3 | Calibration |
| KA | Michaelis constant for zooplankton grazing on phytoplankton | mg/1 | 0.5 | 2.0 | Chen and Orlob (1975) |
| KC | Michaelis constant for carbon uptake | mg/1 | 0.5 | 0.5 | Chen and Orlob (1975) |
| $KDN_{10}$ | Decompositon rate of detritus nitrogen at 10°C | $d^{-1}$ | 0.1 | 0.1 | Calibration |
| $KDP_{10}$ | Decompositon rate of detritus phosphorus at 10°C | $d^{-1}$ | 0.25 | 0.4 | Calibration |
| $KE_{20}$ | Decompositon rate of PE at 20°C | $d^{-1}$ | $2.5 \cdot 10^{-4}$ | $2.2 \cdot 10^{-4}$ | Calibration |
| KL | Michaelis constant for light | $kcal/m^2/d$ | 400 | 400 | Gargas (1975) |
| KN | Michaelis constant for nitrogen uptake | mg/1 | 0.2 | 0.2 | Lehman et al. (1975), Chen and Orlob (1975) |
| KP | Michaelis constant for phosphorus uptake | mg/1 | 0.02 | 0.02 | Lehman et al. (1975), Chen and Orlob (1975) |
| KREL | Rate constant for release of nitrogen | $d^{-1}$ | 0.0040 | 0.0040 | Jacobsen and Jørgensen (1975) |
| KS | Threshold zooplankton biomass | mg/1 | 0.75 | 0.75 | Steele (1974) |
| KZ | Michaelis constant for fish feeding on zooplankton | mg/1 | 0.35 | 0.35 | Calibration |
| LUL | Unstable layer of sediment | m | 0.1 | 0.1 | Measurements |
| DS | LUL (1-DMU) | m | — | — | — |
| MA | Mortality of phytoplankton | $d^{-1}$ | 0.09 | — | Calibration |
| $MYZ_{max}$ | Maximum growth rate of zooplankton | $d^{-1}$ | 0.2 | 0.188 | Calibration |
| MZ | Mortality of zooplankton | $d^{-1}$ | 0.025 | 0.033 | Calibration |
| $NH_4P$ | Ammonia concentration in rainwater | mg/1 | 0.2 | 0.2 | V Jørgensen (1972) |
| $NO_3P$ | Nitrate concentration in rainwater | mg/1 | 0.16 | 0.16 | V Jørgensen (1972) |
| $\phi$ | Temperature coefficient for degredation of detritus | — | 1.072 | 1.072 | |

## TABLE 3 (continued)
## Parameter Values

| Symbol | Definition | Unit | Values Model I | Values Model II | Values based upon |
|--------|------------|------|---------|----------|-------------------|
| $PRED_{max}$ | Maximum feeding rate of fish on zooplankton | $d^{-1}$ | 0.06 | 0.012 | Calibration |
| PP | Phosphorus concentration in rainwater | mg/1 | 0.0015 | 0.0015 | V Jørgensen (1972) |
| $RC_{max}$ | Maximum respiration rate of phytoplankton | $d^{-1}$ | 0.13 | 0.088 | Calibration |
| $RZ_{max}$ | Maximum respiration rate of zooplankton | $d^{-1}$ | 0.035 | 0.028 | Calbration |
| SVD | Settling rate of detritus | m/d | 0.002 | 0.0019 | Jørgensen (1976) |
| SVS | Settling rate of *Scenedesmus* | m/d | 0.06 | 0.19 | |
| $\theta$ | Temperature coefficient for decomposition of PE | $d^{-1}$ | 1.03 | 1.03 | Chen and Orlob (1975) |
| $UC_{max}$ | Maximum rate of carbon uptake | $d^{-1}$ | 0.65 | 0.55 | Calibration |
| $UN_{max}$ | Maximum rate of nitrogen uptake | $d^{-1}$ | 0.03 | 0.015 | Calibration |
| $UP_{min}$ | Maximum rate of phosphorus uptake | $d^{-1}$ | 0.003 | 0.0014 | Calibration |
| $FCA_{min}$ | Maximum kg $C$/kg phytoplankton biomass | — | 0.15 | 0.15 | |
| $FCA_{max}$ | Maximum kg $C$/kg phytoplankton biomass | — | 0.6 | 0.6 | |
| $FNA_{min}$ | Minimum kg $N$/kg phytoplankton biomass | — | 0.15 | 0.15 | |
| $FNA_{max}$ | Maximum kg $N$/kg phytoplankton biomass | — | 0.10 | 0.10 | Jørgensen (1976) |
| $FPA_{min}$ | Minimum kg $P$/kg phytoplankton biomass | — | 0.001 | 0.001 | |
| $FPA_{max}$ | Maximum kg $P$/kg phytoplankton biomass | — | 0.013 | 0.013 | |
| V | Volume of the lake | $m^3$ | 420,000 | 420,000 | |
| Y | Yield of feeding zooplankton and fish | — | 0.63 | 0.63 | Chen and Orlob (1975) |
| L | $1/Y - 1$ | — | 0.59 | 0.59 | — |

## TABLE 4
## Forcing Functions (All Given as Tables)

| | |
|---|---|
| QTRI | $m^3$/d tributaries |
| NTOTRI | Total $N$ (mg/l) tributaries |
| $NH_4TRI$ | mg $NH_4^+$ N/l tributaries |
| $NO_3TRI$ | mg $NO_3^-$ N/l tributaries |
| PTOTRI | Total $P$ (mg/l) tributaries |
| PTRI | mg $PO_4^3$ P/l tributaries |
| $T$ | Temperature of lake water |
| QWAS | $m^3$/d wastewater |
| NTOWAS | Total N mg/l wastewater |
| $NH_4WAS$ | mg $NH_4^+$N/l wastewater |
| $NO_3WAS$ | mg $NO_3^-$ N/l wastewater |
| PWAS | Total $P$ (mg/l) wastewater |
| RAD | Global irradiance (kcal/$m^2$/d) |
| $Q$ | Outflow ($m^3$/d) |
| QPREC | Precipitation ($m^3$/d) |

1. Incident solar radiation
2. Reflected solar radiation
3. Incident long-wave radiation (atmospheric variation) and the reflection of long-wave radiation (atmospheric radiation)
4. Long-wave radiation from the water surface
5. Evaporative heat flux
6. Conduction (sensible heat flux)

The heat content of rain and evaporated water is neglected; they are small and tend to balance each other. The components in the heat fluxes are discussed below.

## A. INCIDENT SOLAR RADIATION

Net solar radiation is emitted by the sun. It reaches the lake surface after passing through the atmosphere, where it undergoes scattering, reflection and adsorption by clouds, air, and dust. The radiation reaching the surface is then a mixture of long- and short-wave components and of direct and diffuse components. The short-wave radiation can be estimated by:

1. Measurements with a pyrheliometer
2. Empirical equations (cf. Chapters 1 and 2)

Direct measurements are important when an accuracy greater than 15% is required.

## B. REFLECTION

The reflection of the incident solar radiation is a function of the solar elevation and of the type of radiation—direct or diffuse. Typical values for a solar elevation of 50° are 2% reflection of direct sunlight and 7% of diffuse light.[14] In this model, a reflection of 6% has been applied.

## C. INCIDENT LONG-WAVE RADIATION

This radiation, is created by the different internal fluxes in the atmosphere. Some of these are radiation by components in the atmosphere (water vapor, ozone, and carbon dioxide). No precise analytical expressions are available for these factors. The atmosphere could be considered as black body, following the Stefan-Boltzmann law; the total amount of energy emitted, integrated over all wavelengths, is proportional to the fourth power of the body's absolute temperature, $T$, where the constant of proportionality, $\sigma$, is the Stefan-Boltzmann constant. Since the atmosphere is not a black body, a constant of emission, $\varepsilon$, is applied to the equation ($\varepsilon$ is dimensionless). Then

$$\phi_l = \varepsilon \bullet \sigma \bullet T^4$$

is the equation describing this flux.

Several formulae for $\varepsilon$ have been proposed, containing relations to vapor pressure, $e$, air temperature (or different combined functions of these), and clouds. A combined formula proposed by Brutsant[3] is used, giving the following equation for net atmospheric radiation, $\phi_N$:

$$\phi_N = 5.9 \bullet 10^{-3} \bullet \left(e/T_a\right)^{1/7} \bullet T^4 \bullet \left(1.0 + 0.17 \bullet c^2\right)$$

where $e$ is the vapor pressure in millibars and $c$ is the fraction of the sky covered by clouds. This formula includes a reflectance of the water surface of 3% for long-wave radiation:

$$\phi_W = 0.97 \bullet \phi_N$$

Back radiation from the water surface is $\phi_w$. The emissivity of a material is equal to the absorptivity (Kirchhoff's Law). This radiation is then

$$\phi_W = 0.97 \bullet \sigma \bullet T_W^4$$

when $T_W$ is the water surface temperature in degrees Kelvin.

## D. EVAPORATIVE HEAT FLUX, $\phi_e$

This is the latent heat flux. The evaporation from a lake is driven by the difference between the saturated vapor pressure ($e_S$) and the actual vapor pressure ($e_z$) at height $z$. The evaporation heat flux is

$$\phi_e = L_v \bullet EV$$

where $L_v$ is the latent heat of evaporation.

$$L_V = \left(2493 - 2.26 \bullet T_s\right) \bullet 10^3 \ J/kg$$

where $T_s$ is the surface temperature in °C; $EV$ is the mass flux of evaporation based on the law of mass transfer:

$$EV = \rho \bullet F\left(W_z\right) \bullet \left(e_s - e_z\right)$$

where $\rho$ is the density of water and $F(W_z)$ is a wind-speed function for mass flux, including both free (wind-driven) and forced (buoyancy-driven) convection effects. This is given as

$$F\left(W_z\right) = 3.08 \bullet 10^{-4} + 1.85 \bullet 10^{-4} \bullet W_z$$

where $W_z$ is the windspeed at height $z$ and the units of $F(W_z)$, are m/day$^{-1}$(mm Hg)$^{-1}$. A zero displacement of 6 inches has been assumed. This is the height at which the horizontal wind disappears, also known as the roughness parameter. In total this means that

$$E_e = \left(3.08 \bullet 10^{-4} + 1.85 \bullet 10^{-4} \ W_z\right) \bullet \left(e_s - e_z\right)\left(2493 - 2.26 \bullet T_s\right)10^3$$

with units of $J/(m^2 \bullet d)$.

## E. CONDUCTIVE HEAT FLUX $\phi_c$

The conductive heat flux, $\phi_c$, can be found from the evaporative heat flux, $EV$. The ratio between these is the Bowen ratio:

$$B = \frac{\phi_c}{EV}$$

The Bowen ratio can be calculated as

$$B = \left(C_P/L_v\right) \bullet \left(T_s - T_z\right)/\left(e_s - e_z\right)$$

where $C_p$ is specific heat of water and $L_v$ is the latent heat of evaporation of water. The ratio $(C_p/L_v) = 0.24/590 = 269.1$ and the conductive heat flux can be found by

$$\phi_c = 269.1\left(T_s - T_z\right)\big/\left(e_s - e_z\right) \bullet \mathrm{EV}$$

where $T_s$ is the water surface temperature in °C and $T_z$ is the air temperature at height, $z$.

## F. INTERNAL HEAT TRANSFER

The absorption of incident radiation is not solely a surface phenomenon. The insolation at any depth is

$$\phi_z = (1 - K) \bullet \phi_s \bullet \exp(-n \bullet z)$$

where $\phi_z$ is the insolation at depth $z$; $K$ is a constant, typically from 0.4 to 0.5; $\phi_s$ is the surface insolation; and $n$ is a local parameter. It can be found from field measurements, or it can be related to the Secchi disc depth, $D$, with $n \simeq 1.7/D$.

This equation is not valid close to the surface. However, for the present purpose, these equations are sufficiently accurate (more detailed discussion in Chapter 6). If the primary production of a lake is considered, a more accurate description of the absorption should be used.

The temperature profile in a lake is also affected by advection rivers are "placed" in the lake water depth according to density calculation.

Seiching is not included in the model. Seiching occurs when a lake is tilted by the wind and an oscillatory motion is created. This would require a three-dimensional model.

Convection caused by instability of density differences is included together with molecular diffusion. This latter is small and is included for computational convenience.

## G. INTERNAL MIXING CALCULATIONS

The basic transport equation for an internal element is

$$\frac{\partial T}{\partial t} = \frac{1}{A}\frac{\partial (Q_v T)}{\partial z} + \frac{K_z}{A}\frac{\partial}{\partial z}\left(A\frac{\partial T}{\partial z}\right) + \frac{BU_i T_i}{A} - \frac{BU_o T}{A} - \frac{1}{\rho c A}\frac{\partial}{\partial z}\left(A\,\phi_s\right)$$

where $T$ = the temperature at $z$, $A$ = area of the element considered, $B$ = width of the element, $U_i$ = horizontal inflow velocity of river water, $T_i$ = temperature of inflow, $U_o$ = horizontal flow velocity, $Q_v$ = vertical flow velocity, $t$ = time, $\phi_s$ = internal short-wave radiation, $K_z$ = vertical turbulent diffusion coefficient, $c$ = heat capacity of water, and $\rho$ = density of water.

The quantity $\rho \bullet c \bullet T$ represents the heat per unit volume, and it is assumed that $\rho \bullet c$ is constant. The equation is expressed by the computed by means of the temperature at the previous time step $T(z, t{-}1)$.

$$\partial T\big/\partial z = f\left(T(z,t-1), \phi, A, U_i, T_i, B\right)$$

By applying the explicit scheme for the time-derivative term:

$$\left(T(z,t) - T(z,t-1)\right)\big/\Delta t = f\left[T(z,t-1), K_z, A, U_i, T_i, B\right]$$

where $\Delta t$ = the time step. This gives a distribution of the temperature at time, $z$:

$$T(z,t) = T(z,t-1) + f\left(T(z,t-1), K_z, A, U_i, T_i, B\right) \bullet \Delta t$$
$$= T(z,t-1) + \Delta T(z,t-1)$$

This equation gives the temperature distribution $T(z,t)$ at time $t$ by using the temperature distribution $T(z, t-1)$ at time $t-1$.

## H. WIND

There is a wind-mixing routine applied. The wind affects the surface of a lake and creates a sheer stress at the surface, which can be characterized by a shear stress coefficient, $C_z$. This shear stress at the surface gives a kinetic input of energy, KE

$$KE = \rho_0 \bullet U \bullet A_{surface}$$

where $U$ = the friction velocity and

$$\tau_o = C_z \bullet \rho_{air} \bullet W_z^2$$

where $\rho_{air}$ = the density of the air.

$C_z$ can be related to the fetch length based on Froude's scaling:

$$\frac{1}{\sqrt{C_z}} = \frac{1}{k} \ln\left( \frac{1}{0.11 \bullet C_z \bullet F^2} \right)$$

where $F = W_z / \sqrt{g\ z_m}$ ; $z_m$ = the wind measurement height, $W_z$ = the wind speed at $z_m$, and $K$ = von Kármán's constant, equal to 0.41. The factor 0.011 is obtained from the relationship between the shear velocity and zero displacement.

## I. POTENTIAL ENERGY

The potential energy of the $i^{th}$ element above the $j^{th}$ element is defined as

$$PE = g \bullet \Sigma A(i) \bullet \Delta z [\rho(j,t) - \rho(i,t)] \bullet D(i,t)$$

where $A(i)$ = the area of the $i^{th}$ element; $\rho(i,t)$ = the density of the $i^{th}$ element at time $t$; $\rho(j,t)$ = the denity of the element immediately below the mixed layer at time $t$; $z$ = thickness of an element; $D(i,t)$ = the distance between the $i^{th}$ and the $j^{th}$ element at time $t$, and $g$ = gravitational acceleration.

The upper layer of the lake (epilimnion) is assumed to be well mixed, so that the potential energy can be rewritten as

$$PE = g \bullet \Delta\rho \bullet V_m \bullet \Delta z \bullet m/2 = g \bullet \Delta\rho \bullet \overline{A} \bullet H^2$$

where $m$ = number of elements in the mixed layer; $m \bullet z$ = depth of the mixed layer = $H$; $V$ = volume of mixed layer = $a \bullet H$, where $A$ = the average cross-sectional area, and $\Delta\rho$ = the density difference between the mixed layer and the elements immediately below.

A critical ratio can now be calculated. The ratio KE/PE is equivalent to the Richardson number $R_i$.

The mixing algorithm calculates the Richardson number.

$$f(R_i) = KE/PE$$

If the calculated value of the Richardson number is less than KE/PE, mixing will be applied between the two elements of the water body.

The Richardson number is calculated as

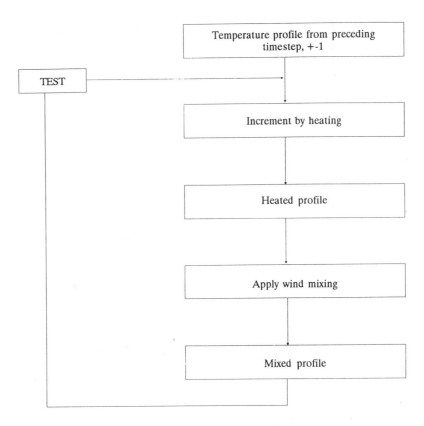

FIGURE 4.   Temperature profile.

$$R_i = g \bullet H \bullet \Delta\rho \Big/ \big(U^2 \bullet \rho_a\big)$$

where $\Delta\rho$ = the density difference and $\rho_a$ = the density.

The ratio KE/PE will, by the wind-mixing algorithm, be evaluated against a function $R_i$. Different functions have been used for different $R_i$ ranges, i.e.,

$$f\big(R_i\big) = 1,$$

$$f1\big(R_i\big) = \frac{0.57\, R_i \big(29.55 - \sqrt{R_i}\,\big)}{14.2 + R_i} \text{ and}$$

$$f2\big(R_i\big) = \frac{R_i}{14.2 + R_i}$$

When KE/PE is less than the calculated $R_i$, the two elements considered will be mixed and a new temperature will arise in the $i^{th}$ element. Therefore, the order of calculation is important.

The procedure shown in Figure 4 has been used. The test is done by calculation of the total outgoing heat flux and then comparing the new total heat flux with the old one. Should the numeric value be small, calculation of a new time step is made. Should the difference be larger, the heating and mixing procedure is carried out again.

Two characteristic plots resulting from the model are shown in Figures 5 and 6. They are three-dimensional plots giving temperature as a function of depth and date.

NO WIND MIXING
D=0.125 m2/day

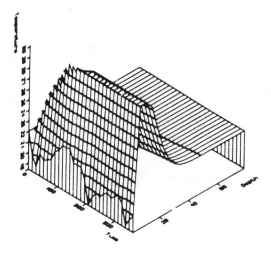

FIGURE 5.   Three-dimensional plots.

WIND   MIXING
if dPE/dKE LT f2(R$_i$)

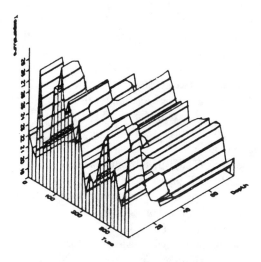

FIGURE 6.   Three-dimensional plots.

## Cu - submodel

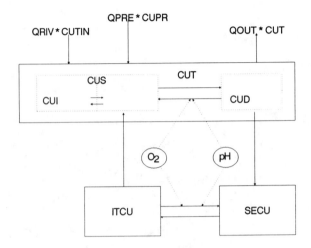

FIGURE 7.   Conceptual model for the distribution of copper. CUT is the total copper; CUS is the dissolved copper; CUD is the particulate copper; CUI is the ionic copper; CUS-CUI is the complex, bound dissolved copper; SECU is the particulate copper in sediment; and ITCI is the dissolved copper in interstitial water.

# VII. THE HEAVY-METAL MODEL

At present, heavy-metal pollution does not appear to create any serious problems in the area, since only a few external sources exists. However, the effluents from the Kilembe copper mines north of Lake George may contribute some copper via the Kazinga Channel through Lake Idi Amin (Lake Edward) and further on through the Semiliki River to Lake Mobuto Sese Seko.[4] Also, copper-processing plants, near Jinja north of Lake Victoria, and increasing industrialization in the Kisumu area in the northeastern and the Mwanza area in the southern part of the Lake Victoria drainage basin are potential sources of heavy-metal pollution.

Since the study area is very large, the distribution of heavy metals is only a small part of the whole project, and the data are limited, a rather simple model was chosen. Some of the input data for the heavy-metal submodels are created as output from the other submodels in the project (Figure 1) and some are determined experimentally. In the present discussion, only the submodel for copper distribution in Lake Victoria is shown. For information on the other metals and the other lakes, see Jørgensen et al.[24]

The bulk part of the loading of heavy metals into the lakes occurs as suspended solids carried by the tributaries. Similarly, the bulk part of the heavy-metals in the lakes is present in particulate form, either as sediments or as seston. Since the toxic effects of the metals are most likely to be associated with the levels of the ionic and other dissolved forms, the model and the experiments are focused on the distribution between dissolved and particulate forms of the metals.

The copper submodel shown in Figure 7 considers the loading to the lake via suspended solids from the tributaries and via precipitation. The distribution between dissolved and particulate copper in the water phase is described by sorption equilibria governed by pH and oxygen conditions. The dissolved copper is separated in ionic and complexed copper by an assumed partition coefficient. Particulate copper sediments out, bound to inorganic particles and detritus. In the sediment, the copper separates as dissolved, interstitial copper and particulate copper. The dissolved copper diffuses back to the water phase by an ordinary Fickian diffusion, corrected for sediment porosity. The model complexity, and the considerations behind it, are similar to

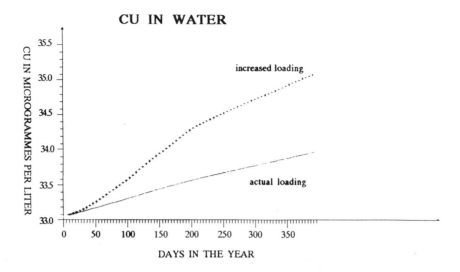

FIGURE 8.    A 1-year simulation of total copper in the water phase in Lake Victoria.

those previously described by Kamp-Nielsen et al.[26] for phosphorus in the study area. The results of simulations in which the model was applied are shown in Figures 8 and 9.

## VIII. THE DDT MODEL

Should DDT be conservative, it would of course be very easy to set up a DDT balance following the water balance. However, DDT is degraded by photodegradation. There is an equilibrium between DDT in water and in suspended matter (adsorption), and there is direct uptake from the water to fish and also from fish food directly to fish. The concentration factors giving the ratio of concentrations in fish and water are known from several investigations. As well as those giving the ratio between DDT in food items and in the water. When fish have a certain concentration of DDT, they are able to very slowly excrete a certain amount of DDT. The excretion rate can be considered to be first-order reaction. The excretion rate is also known from several investigations.

The model has been used to track the DDT concentration in the water of Lake Victoria, based on mass balances and first-order photodegration. Furthermore, the DDT concentration in carnivorous species is modeled, as is the highest concentration due to bioaccumulation (see the food web in Figure 10). Figure 11 shows the predicted concentrations in water and fish (1) at normal (present) input, (2) at a tenfold concentration of the river input, and (3) at a tenfold concentration of precipitation. Output levels of DDT in the model can be controlled with data from Kokman and Pennings.[27] The fish analyses used here were sampled in Homa Bay and in Mirunda Bay, Kenya; concentrations of DDT in fish were measured at 0.010 to 0.025 ppm, dieldrin concentrations at 0.009 to 0.086 ppm.

The projections shown in Figure 11 indicate that increased use of DDT is unacceptable. Its concentration in fish today is already too high.[27] Further control of this problem should be very seriously considered in order to prevent bioaccumulation of DDT in the regional population.

## IX. FISHERY MANAGEMENT

Fisheries in Lake Victoria have for many years been approximately balanced by a steady fish population.[2,6,10,12] However, the human population in the area is increasing, and it is of great

TOTAL CU IN SEDIMENT

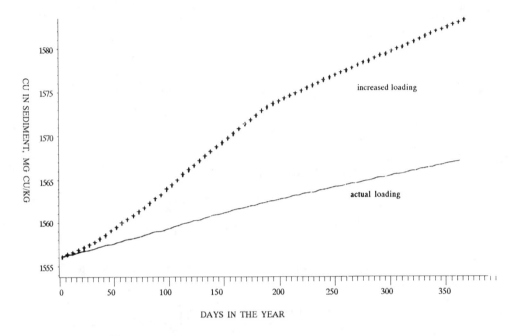

FIGURE 9.    A 1-year simulation of total copper in the sediment in Lake Victoria.

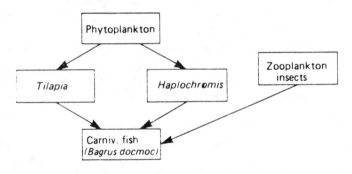

FIGURE 10.    Food web in Lake Victoria.

importance to obtain an optimum yield in the years to come and yet to avoid overfishing. It has been discussed whether it would be advantageous to concentrate in fisheries in herbivorous stock by eliminating carnivorous species, including the Nile perch, which was introduced into Lake Victoria a number of decades ago.[17]

A simplified food web for the lake is shown in Figure 10. The two herbivorous species, *Haplochromis* and *Tilapia*, are dominant in concentration as well as in the current fishery. The carnivorous species are treated together here, as they are of less importance for fisheries (about 12% of the current catch) and are dominated by only two species: *Bagrus* and Nile perch.

Zooplankton play a minor role in Lake Victoria compared to lakes in temperate or subtropical regions. However, the fisheries model is a submodel of a more comprehensive model that provides information about phytoplankton and zooplankton concentrations as a function of time—a eutrophication model that basically follows principles of the model published by Jørgensen, et al.[22] and Jørgensen.[21]

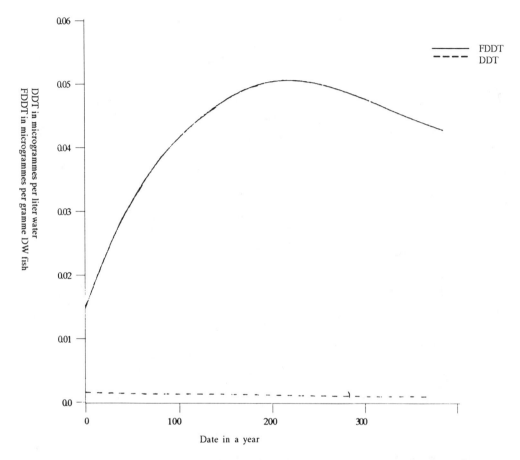

FIGURE 11. Modeled DDT concentrations in carnivorous fish. (A) At present DDT input; (B) With tenfold DDT input. The simulation results of A are in accordance with measurements (see text).

The state variables in the fish submodel are number per cubic meter and weight for each age class (1 age class = 1 ring year = 6 months) of *Haplochromis*, *Tilapia*, and carnivorous fish averaged over 1 ring year. Intensive studies over recent decades have provided further information about the characteristics of the fish species in Lake Victoria; these data are included in the model as table functions.

When the fisheries model was applied to evaluate present the fisheries policy, results showed that fish catch could be increased 20 to 25% over the present level without any decrease in the fish population—provided that the net size is not decreased (it even seems slightly favorable to increase the net size). Thus, the optimum harvest in Lake Victoria would require an increase in fisheries above present efforts.

# X. DISCUSSION

The total model is based upon a partial calibration and literature survey on the lake system and specific parameter values. As the Global Environmental Modeling System (GEMS) continues the data sampling in the area, it would be possible to get further data for an improved calibration in the future. The complexity of the model will require further data if the advantages

**DDT in Lake Victoria
Normal Input**

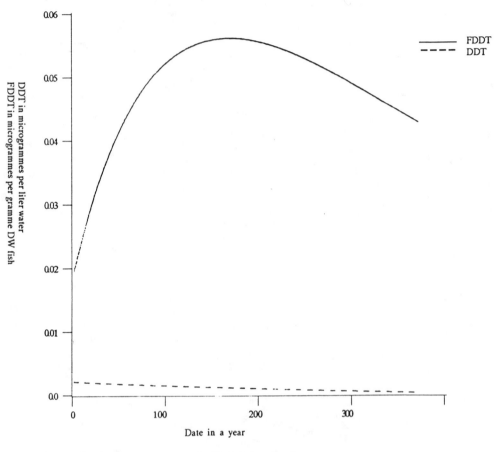

FIGURE 11 (continued).

of the model are to be fully available for environmental management. Later it may also be possible to obtain enough data to validate the model.

In its present form, the model is already useful for managers of the region, and the use of this tool for planning in this region must be strongly recommended.

Predictions based upon forecasts of water consumption have been undertaken, but currently new regional plans should be used as forcing functions to the model to supply the managers with a prediction on the consequences of pollution of the Upper Nile Lake system.

Considering that the model is only partially calibrated and that meteorological data cannot be forecast years ahead, the model output will give ony typical average consequences for an annual situation, but that will be sufficient for environmental planning of the region. The accuracy of the predictions will, however, be improved currently by the addition of new data and modeling developments, which are recommended to be incorporated into the model. Therefore, continuing model maintenance and improvement must be carried out.

# XI. CONCLUSIONS

The project demonstrates that it is possible to set up a workable environmental model for a region as large as the Upper Nile Lake system. The model reflects of course the available data and the present knowledge of environmental modeling. The model is, therefore, only able to deal with the processes that are incorporated in the model with an accuracy corresponding to the available data. Therefore, we have only incorporated the environmental problems, that are, or might be, of interest for the region.

We have presented four applications that demonstrate the use of a comprehensive model for management decision making on future environmental and ecological policy in the region of the Upper Nile lakes. Although the data currently available to the model are of insufficient quality to allow precise predictions, we believe they are sound enough to permit provisional and approximate management decisions. We should be able to improve the database in the near future, and we already plan to make the model more accessible for use on microcomputers.

# REFERENCES

1. **Beavington, F. and Cawse, P. A.,** The deposition of trace elements and major nutrients in dust and rainwater in Northern Nigeria, *Sci. Total Environ.*, 13, 263–274, 1979.
2. **Bergstrand, E. and Cordone, A. L.,** Exploratory bottom trawling in Lake Victoria, *Afr. J. Hydrobiol. Fish.*, 1(1), 13–23, 1971.
3. **Brutsaert, W.,** On a derivable formula for long-wave radiation from clear skies, *Water Resour. Res.*, 11(5), 42–56, 1975.
4. **Bugeny, F. W. B.,** Copper ion distribution in the surface waters of lakes George and Idi Amin, *Hydrobiologia*, 64, 9–15, 1979.
5. **Cridland, C. C.,** Laboratory experiment on the growth of *Tilapia* spp., *Hydrobiologia*, 15, 135–160, 1960.
6. **Cridland, C. C.,** Laboratory experiment on the growth of *Tilapia* spp., *Hydrobiologia*, 20, 155–166, 1962.
7. **Crosby, N. T.,** The determination of nitrate in water using Cleves acid, 1-naphthylamine-7-sulphonic acid, in *Proc. Soc. Wat. Treatment Exam.*, 16, 51, 1967.
8. **Davey, E. W., Morgan, M. J., and Erikson, S. J.,** A biological measurement of the copper complexing capacity of seawater, *Limnol. Oceanogr.*, 18, 993–997, 1973.
9. **Deutsches Einheitsverfahren,** Verlag Chemie, GmbH., *Weinheim*, E 3, E 4, 1960.
10. **EAFRO** East African Fishery Research Organization (Jinji).
11. **Frissel, M. J. and Reininger, P.,** in *Simulation of Accumulation and Leaching in Soils*, Wageningen, The Netherlands, 1974.
12. **Garrod, D. J.,** A review of Lake Victoria fishery service records 1954-1959, *East Afr. Agric. Forest. J.*, 26, 42–48, 1960.
13. **Golterman, H. L.,** *Methods for Chemical Analysis of Fresh Waters*, IBP Handbook No. 8, Blackwell Scientific Publications, Oxford, 1969.
14. **Golterman, H. L.,** *Physiological Limnology*, Elsevier, Amsterdam, 1975.
15. **Greenwood, P. H.,** Two new species of *Haplochromis* (Pisces: Cichlidae) from Lake Victoria, *Ann. Mag. Natur. Hist.*, 8, 303–318, 1965.
16. **Gächter, R., Davis, J. S., and Marés, A.,** Regulation of copper availability to phytoplankton by macromolecules in lake water, *Environ. Sci. Tech.*, 12, 1416–1421, 1978.
17. **Hamblyn, E. L.,** The food and feeding habits of Nile perch *Lates niloticus* (Linné) (Pisces: Centropomidae), 1966.
18. **Harwood, J. E. and Kühn, A. L.,** A colorimetric method for ammonia in natural waters, *Water Res.*, 4(21), 805–811, 1970.
19. **Hydromet,** Hydrometerological Survey of the Catchments of Lakes Victoria, Kyoga and Albert, UNDP and WHO report, Vol. 1, Nairobi, Kenya, 1975.
20. **Jackson, P. B. N.,** The African Great Lakes fisheries: past present and future, *Afr. J. Hydrobiol. Fish.*, 35–49, 1970.

21. **Jørgensen, S. E.,** A eutrophication model for a lake, *Ecol. Modeling*, 2, 147–165, 1976.

22. **Jørgensen, S. E., Mejer, H., and Friis, M.,** Examination of a lake model, *Ecol. Modeling*, 4, 253–278, 1978.

23. **Jørgensen, S. E., Jørgensen, L. A., Kamp-Nielsen, L., and Mejer, H. F.,** *Ecol. Modeling*, 13, 111–129, 1981.

24. **Jørgensen, S. E., Jørgensen, L. A., Kamp-Nielsen, and Mejer, H. F.,** Water Quality Model for the Upper Nile Lake Basin, Final Report, Copenhagen, Denmark, 1982.

25. **Kamp-Nielsen, L.,** Some comments on the determination of copper fractions in natural waters, *Deep-Sea Res.*, 19, 899–902, 1972.

26. **Kamp-Nielsen, L., Jørgensen, L. A., and Jørgensen, S. E.,** A sediment-water exchange model for lakes in the Upper Nile Basin, in *Progress in Ecological Engineering and Management by Mathematical Modeling*, Editions Cebedoc, Liege, Belgium, 1980.

27. **Koeman, J. H. and Pennings, J. H.,** An orientational survey on the side-effects and environmental distribution of insecticides used in tsetse-control in Africa, *Bull. Environ. Cont. Toxicol.*, 5(2), 164–170, 1970.

28. *Limnological Methods*, University of Copenhagen, Hillerød, 1977.

29. **Lu, J. C. S. and Chen, K. Y.,** Migration of trace metals in interface of sea-water and polluted surficial sediments, *Environ. Sci. Tech.*, 1, 174–182, 1977.

30. **Morel, F. and Morgan, J. J.,** A numerical method for solution of chemical equilibria in aqueous systems, California Institute of Technology, Pasadena, CA, 1968.

31. **Mullin, J. B. and Riley, J. P.,** The colorimetric determination of silicate with special reference to sea and natural waters, *Anal. Chim. Acta*, 12, 162–176, 1966.

32. **Murphy, J. and Riley, J. P.,** A modified single solution method for the determination of phosphate in natural waters, *Anal. Chim. Acta*, 27, 21–26, 1962.

33. **Park, R. A., Groden, T. W., and Desormeau, C. J.,** Modification to model CLEANER, requiring further research, in *Perspectives on Lake Ecosystem Modeling*, Scavia, D. and Robertson, A., Eds., Ann Arbor Science, Ann Arbor, MI, 1979, 87–108.

34. **Roberts, T. R.,** Geographical distribution of African freshwater fishes, *Zool. J. Limn. Soc.*, 57, 249–319, 1974.

35. **Ryan, P. J. and Harleman, D. R. F.,** An analytical and experimental study of transient cooling pond behavior, MIT Dept. of Civil Eng., R. M. Parsons Laboratory for Water Resources and Hydrodynamics Tech. Report, 161, 1, 1973.

36. **Schindler, D. W.,** Two useful decides for vertical plankton and water sampling, *J. Fish. Res. Board. Can.*, 26, pp. 1948–1955, 1969.

37. **Ssentongo, G. W.,** Yield isopleths of *Tilapia esculenta* Graham 1928 in Lake Victoria and *Tilapia nilotica* (Linnaeus) 1957 in Lake Albert, *Afr. J. Trop. Hydrobiol. Fish.*, 2(2), 121–128, 1976.

38. **Steeman Nielsen, E.,** The use of radioactive carbon for measuring organic production in the sea, *J. Cons. Perm. Int. Explor. Mer.*, 18, 117–140, 1952.

39. **Strickland, J. D. H. and Parson, T. R.,** A practical handbook of sea-water analysis, *J. Fish. Res. Board. Can.*, 167, 167–202, 1968.

40. **Stumm, W. and Morgan, J. J.,** *Aquatic Chemistry*, Wiley-Interscience, New York, 1970.

41. **Welcomme, R. L.,** Studies of the effect of abnormally high water levels on the ecology of fish in certain shallow regions of Lake Victoria, *J. Zool. London*, 160, 405–436, 1970.

42. **Yanni, M.,** Biochemical studies on some Nile fish 1. Fat and water contents of *Anguilla vulgaries*, *Synodontis schall* and *Clarias Lazfra*, *Zool. Soc. Egypt. Bull.*, 23, 90–101, 1970/71.

Chapter 8

# APPLICATION OF A DECISION SUPPORT SYSTEM TO DEVELOP PHOSPHORUS CONTROL STRATEGIES FOR SOUTH AFRICAN RESERVOIRS

**D. C. Grobler and J. N. Rossouw**

## TABLE OF CONTENTS

# I. INTRODUCTION

Like many other countries in the world,[1] South Africa is experiencing serious water quality problems as a result of eutrophication. Phosphorus is the growth-limiting plant nutrient in most freshwater systems and is considered to be the easiest to control.[2]

The South African authorities responsible for water resources management adopted a policy whereby preference is given to controlling the causes, rather than the consequences of eutrophication. The assumption was that between 80 and 90% in phosphorus loads on the receiving water bodies in the so-called sensitive drainage basins originated from point sources.[3] As a result, all wastewater effluents that are returned to water bodies in sensitive drainage basins are legally required to comply to a 1 mg P/l phosphorus standard. This standard was arrived at from an assessment of the technological and economical feasibility of phosphorus removal from the effluent of typical wastewater treatment plants.

The assimilative capacity of individual water bodies to absorb phosphorus loads without experiencing undesirable changes in eutrophication-related water quality was ignored in setting a uniform effluent standard. Therefore, enforcing it could not be expected to always be cost effective. In many cases, it is likely to have little impact, e.g. if the receiving water body has a large assimilative capacity. In other cases, the uniform standard could be expected not to have enough impact, e.g., if the assimilative capacity of the receiving water is grossly exceeded.

It was realized that a phosphorus control strategy based on enforcing a uniform standard would be inappropriate. It would lead either to unnecessary expenditure or to inadequate protection of water quality. Fortunately, the 1 mg P/l standard does not have to be applied uniformly. The laws governing water quality protection allow for standards to be relaxed or to be made stricter if it is required. This flexibility allows the authorities to take into account the assimilative capacity of receiving waters for phosphorus pollution in designing and implementing phosphorus control strategies for point sources.

If the assimilative capacity of receiving water bodies for phosphorus pollution is to be considered in decisions on the implementation of a phosphorus standard, the impacts of these decisions on water quality must be assessed. One of the best ways to do this is to use water quality models to simulate eutrophication-related water quality in the receiving water bodies in response to different point-source control strategies. However, the final decision on which control strategy to adopt depends not only on water quality issues, but also on financial, social, and political factors. Decision-making models, such as the analytical hierarchy process (AHP) developed by Saaty,[4] are useful tools for incorporating quantitative information and qualitative value judgments into the same decision framework.

In this chapter we briefly describe the models used for simulating the eutrophication process, the decision support system (DSS) in which they were incorporated, some case studies in which the DSS was applied, and how the analytical hierarchy process was used to help decide on appropriate eutrophication control measures.

# II. MODELING THE EUTROPHICATION PROCESS

## A. COMPONENTS OF THE EUTROPHICATION PROCESS

Assessing the impacts of different phosphorus concentrations in point-source effluents on water quality in the receiving water bodies required simulation of the three main components of the eutrophication process. These are the export of nutrients from the drainage basin, their fate in river reservoir systems, and the eutrophication-related water quality resulting from the ambient nutrient concentrations. Caswell[5] distinguished between two types of models that could be developed for simulating the behavior of ecosystems, i.e., research- and management-oriented models. Research models are used mainly as aids to understand the functioning of ecosystems, whereas management-oriented models are used to predict the response of systems to alternative management options. The requirements of management-oriented models are that

**THE EUTROPHICATION PROCESS**

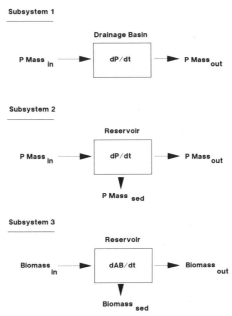

FIGURE 1. Three different subsystems considered in modeling the eutrophication process.

they should be conceptually simple, easy to implement, and compatible with the available data.[6] Separate models, conforming to the requirements of management-oriented models, were developed for simulating each of the three main components of the eutrophication processes in South African reservoirs.[7] These models were later incorporated into a DSS which was used to predict the impacts of alternative phosphorus control measures on water quality in reservoirs in South Africa.

In the analysis of lake restoration by phosphorus reduction, it is recommended that analysis of the lake recovery behavior be performed by conceptually dividing the ecosystem into two compartments. These are a subsystem for describing the fate of phosphorus in lakes and a subsystem for describing the relationship between biological response variables and in-lake phosphorus concentrations.[8] The two subsystems require different modeling approaches. We used a similar framework in our analysis of eutrophication of reservoirs and the effect of control measures, but used three instead of two subsystems.

## Subsystem 1

Subsystem 1 describes the relationship between phosphorus control measures in a drainage basin and the external phosphorus loads received by lakes or reservoirs (Figure 1). In this case we are primarily interested in the amount of phosphorus leaving the drainage basin. The processes causing changes in the amount of phosphorus leaving the system are the phosphorus entering the system (interbasin water transfers and point sources) and the changes with time in phosphorus content of the system ($dP/dt$). The processes affecting $dP/dt$ in a drainage basin (erosion, storage, remobilization, etc.) are highly complex, and it is very difficult to quantify $dP/dt$. If $dP/dt$ cannot be estimated independently, the principle of conservation of mass cannot be used to simulate the mass of phosphorus leaving the system (phosphorus export) simply as a function of phosphorus entering the system and changes in the phosphorus content of the system. Consequently a statistical approach, i.e., a regression model, was developed to simulate phosphorus export as a function of phosphorus entering the system and drainage basin properties.

**Subsystem 2**

Subsystem 2 describes the relation between the external phosphorus load and phosphorus concentrations in a lake or reservoir (Figure 1). The dominant processes causing changes in phosphorus mass with time ($dP/dt$) in the lake or reservoir are the mass of phosphorus entering and leaving the system, and the mass lost through sedimentation. $P mass_{in}$, $P mass_{out}$, and $P mass_{sed}$ can be estimated independently. Therefore, the principle of conservation of mass can be used to simulate $dP/dt$. Because we are ultimately interested in the phosphorus concentration in the water body, changes in the volume of water stored also need to be simulated. Therefore, a water balance must be done simultaneously with the phosphorus mass balance.

**Subsystem 3**

Subsystem 3 describes the relation between in-lake phosphorus concentration and indicators of algal production, e.g., standing biomass (chlorophyll-*a* concentration) (Figure 1). The processes causing changes in standing algal biomass ($AB$) with time ($dAB/dt$) in the lake or reservoir are the biomass entering and leaving the system, i.e., $biomass_{in}$, $biomass_{out}$ and $biomass_{sed}$, and the growth (as a function of phosphorus concentration and other factors) and decay (respiration, grazing, etc.) of biomass in the system. $Biomass_{in}$, $biomass_{out}$, and $biomass_{sed}$ can be estimated independently. However, these are generally not the dominant processes determining $dAB/dt$. Growth and decay dominate the temporal changes in standing algal biomass in most lakes and reservoirs and are generally very difficult to quantify. Therefore, the principle of conservation of mass cannot be used to simulate the temporal changes in biomass ($dAB/dt$) simply as a function of biomass entering and leaving the system, as was the case with phosphorus. Consequently, a statistical approach, i.e., a regression model, is generally used to simulate temporal changes in algal biomass (as chlorophyll) as a function of ambient phosphorus concentrations.

The combination of models developed for the different subsystems required for simulating the eutrophication process are referred to as the reservoir eutrophication model (REM). Although REM is conceptually similar to the Organization for Economic Co-operation and Development (OECD) eutrophication modeling approach,[9,10] it represents some major departures from the OECD approach[7]:

1.    REM is a dynamic model, whereas the OECD eutrophication modeling approach assumes water bodies to be in steady state. REM accepts time-varying inputs of hydraulic and phosphorus loads to simulate the time-varying output of state variables such as the volume of water stored and the phosphorus concentrations in a water body. This departure from the OECD modeling approach was required because the very large hydrological variability characteristic of semiarid regions, such as South Africa, makes the assumption of steady state untenable. To be able to simulate the eutrophication process for a representative range of hydrological conditions in regions experiencing highly variable hydrology requires that long time series of hydrological data be used. In South Africa, 60 years and longer time series of hydrological data, e.g., monthly runoff, are typically used in investigations done for water resources development.
2.    Parameter estimation and verification for the OECD modeling approach was done with data from a cross-section of lakes.[11] In the case of REM, it is done with time-series data from a specific water body.[7] The departure from the cross-sectional to a time-series approach is required because in the semiarid regions, for which REM was developed, temporal variation usually exceeds spatial variation.

**B. PHOSPHORUS EXPORT**

We use the term *load* to describe the phosphorus flux, in units of mass/time, entering a water body in the form of an effluent or passing a sampling point in a river. The term *export*, which has units of mass/area/time, is used to describe the total phosphorus load from a drainage basin that was normalized through dividing it by the area of the drainage basin. Similarly we use the term *discharge*, which has units of volume/time, to describe the flow of water past a point and the term *runoff*, which has units of volume/area/time(depth/time), for discharge normalized per unit of the area of a drainage basin.

The sources of phosphorus in drainage basins are divided for convenience into point sources, i.e., those that can be easily identified, quantified, and controlled, and nonpoint sources, i.e., all the remaining ones. The contributions from point and nonpoint sources to the total phosphorus load are usually estimated separately.

## 1. Point Sources

Monthly phosphorus loads derived from point source effluents are estimated as the product of the monthly means of effluent discharges and phosphorus concentrations in the effluents, i.e.,

$$PL_{ij} = PQ_{ij} * C_{ij} \qquad (1)$$

where $PL_{ij}$ is the phosphorus load, $PQ_{ij}$ is the average effluent volume, and $C_{ij}$ is the average phosphorus concentration during month $i$ in the effluent of point source $j$. The total point–source–derived phosphorus load in a drainage basin is calculated as the sum of the individual point-source loads, modified by a loss rate ($LR$), which takes into account losses of phosphorus (self-purification) between a specific point source and the receiving water body:

$$PL_i = \sum_{j=1,n} \left( PL_{ij} * LR_j \right) \qquad (2)$$

Presently, $LR_j$, the specific loss rate for point source $j$, is estimated on the basis of historical data and an understanding of the behavior of the stretch of river between a point source and what is considered to be the receiving water body. $LR_j$ can be varied to establish how sensitive the model output is to assuming different values for it.

Projections of future phosphorus loads from point sources in a drainage basin are obtained as the product of mean projected effluent volumes and assumed mean phosphorus concentrations in the effluent, which reflect the particular phosphorus control option being evaluated. Mean projected effluent volumes are estimated as a function of assumed effluent growth rates, which are based on projected growth rates for the population and industrial and mining activities in the drainage basin under consideration.

We use the means of projected effluent volumes and phosphorus concentrations in effluents to estimate mean point–source-derived phosphorus loads instead of time-series data. There is so little temporal variation in point-source phosphorus loads that it does not justify the extra effort to develop time series of projected effluent volumes and phosphorus concentrations in effluents.

## 2. Nonpoint Sources

In contrast to point sources, nonpoint-source-derived phosphorus loads are characterized by large temporal and spatial variability. Consequently, they are difficult to estimate and predict and require the use of mathematical models, which relate phosphorus export to hydrological factors and drainage basin characteristics.[12,13] Runoff, land form, and land use are considered to be the most important factors affecting the quantity of nonpoint-source-derived phosphorus export from drainage basins.[14-16]

Of all the factors affecting nonpoint-source phosphorus export, runoff has the dominating effect. It explained more than 80% of the variance in phosphorus export from drainage basins in South Africa and the United States.[17,7] It was only after the variance explained by runoff had been removed that the effects of land form and land use on phosphorus export became apparent. Contrary to what is often believed, of these two factors, land form is the dominating one.[14] Possible reasons for the confusion about the effects of land use and land form on nonpoint-source phosphorus export are the multicolinearity between land use and physiography of drainage basins,[18] or, as Walker[19] has shown, that regional effects (differences in climate, general agricultural practices, and soils) may be greater than land-use effects.

Because runoff has such a dominating effect on nonpoint–source-derived phosphorus export from drainage basins, we used a nonlinear regression model of monthly phosphorus export as a function of monthly runoff to simulate it, i.e.,

$$P \text{ export}_i = a * \text{runoff}_i^b \tag{3}$$

Land form and land-use effects are taken into account by estimating the regression model parameters $a$ and $b$ separately for specific drainage basins.

The following procedure[20] was used to calibrate the model parameters:

1. Select a drainage basin, or part of it, that contains no significant point sources for calibrating the parameters of the model. No point sources must be present, because it is impossible to separate the contributions from point and nonpoint sources in the total amount of phosphorus exported from a drainage basin. The drainage basin must have been adequately gauged over a long enough period to result in discharge and phosphorus concentration records that are both accurate and representative.
2. Estimate the daily phosphorus flux from continuously recorded daily discharge and phosphorus concentration records obtained from grab sampling. We used the methods described in Walker[21] to estimate the daily phosphorus flux.
3. Calculate monthly flows and phosphorus loads from daily discharge and phosphorus flux data, and convert these to monthly runoff and phosphorus export through dividing flows and loads by the area of the drainage basin.
4. Use a nonlinear regression procedure to estimate the parameters for the nonlinear regression model of phosphorus export as a function of runoff.

The calibrated model is used to simulate phosphorus export, which is converted to monthly nonpoint source phosphorus loads from a drainage basin:

$$\text{NPL}_i = P \text{ export}_i * A \tag{4}$$

where $\text{NPL}_i$ is the nonpoint source derived phosphorus load, $P \text{ export}_i$ is the phosphorus export from the drainage basin during month $i$, and $A$ is the area of the drainage basin.

The time series of monthly total phosphorus loads ($TL_i$) entering the receiving water body is obtained by adding, for each month, the point- and nonpoint-source derived phosphorus loads:

$$TL_i = PL_i + \text{NPL}_i \tag{5}$$

## C. FATE OF PHOSPHORUS IN RESERVOIRS

The fate of phosphorus in reservoirs is modeled with a phosphorus budget model, which is based on the principle of conservation of mass,[7] i.e.,

$$dP/dt = P_{in} - P_{out} - P_{sed} \tag{6}$$

in which $P$ is the mass of phosphorus stored in the water in a reservoir, $P_{in}$ is the mass of phosphorus entering the reservoir, $P_{out}$ the mass of phosphorus leaving the reservoir, and $P_{sed}$ the amount of phosphorus lost by sedimentation, during $dt$.

$P_{in}$ and $P_{out}$ are obtained as either measured or simulated time series of phosphorus loads entering or leaving a reservoir. $P_{sed}$ is simulated as a function of the mass of phosphorus and the ambient phosphorus concentration in the reservoir, i.e.,

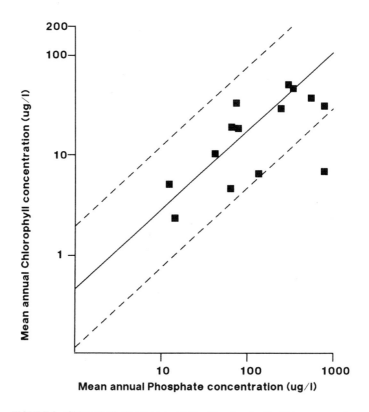

FIGURE 2. Chlorophyll-phosphorus relationships for South African reservoirs showing the OECD chlorophyll-phosphorus regression line with its 95% confidence

$$P_{sed} = s * P \tag{7}$$

$$s = k[P]^2 \tag{8}$$

Where $s$ is the phosphorus-concentration-dependent sedimentation rate, $k$ the sedimentation parameter, and $[P]$ the ambient phosphorus concentration during $dt$. The parameter $k$ is calibrated for a specific reservoir by solving the mass budget equation using time series of observed $dP/dt$, $P_{in}$, and $P_{out}$.

The ambient phosphorus concentration during $dt$ is calculated by dividing the mass of phosphorus by the volume of water stored in the reservoir during the time interval $dt$.

## D. CONVERSION OF PHOSPHORUS TO CHLOROPHYLL

The final step in simulating the eutrophication process relates the ambient phosphorus concentration in a reservoir to the water quality variable of concern. Water quality problems associated with eutrophication are usually caused by the occurrence of undesirable quantities and/or species of algae. Therefore, the algal standing crop in a reservoir, usually expressed as a chlorophyll concentration, is most often selected as the water quality variable of concern.

To account for the large differences in light penetration, as a consequence of differences in mineral turbidity in South African reservoirs, different regression models for simulated chlorophyll concentrations as a function of ambient phosphorus concentrations are used. For clear water reservoirs, i.e., reservoirs in which light extinction is dominated by algae, the OECD chlorophyll:phosphorus regression model[22] is used (Figure 2):

$$\text{chl} = 0.45[P]^{0.79} \tag{9}$$

**TABLE 1**
**Turbidity Classes for Reservoirs Used in Selecting the Appropriate**
**Chlorophyll:P Model**

| Turbidity class | Secchi Depth (m) | Turbidity (NTU) | Inorganic suspended solids (mg/l) |
|---|---|---|---|
| Highly turbid | 0.0–0.1 | 260 | 419 |
| | 0.1–0.2 | 87 | 142 |
| Turbid | 0.2–0.4 | 44 | 72 |
| Moderately turbid | 0.4–1.0 | 19 | 32 |
| Clear | > 1.0 | < 10 | < 10 |

where chl and $[P]$ are respectively the mean chlorophyll and phosphorus concentrations in the reservoir. For turbid reservoirs, i.e., ones in which light extinction is dominated by mineral turbidity, a chlorophyll:phosphorus:inorganic suspended solids regression model[23] is used:

$$\log \text{chl} = 1.13 \log [P] - 1.03 [\text{ISS}]/[P] - 0.47 \qquad (10)$$

where [ISS] is the inorganic suspended solids concentration in the reservoir. Suspended solids concentrations are not generally available for South African reservoirs, but turbidity and Secchi disk depth are commonly used as measures of water transparency in most reservoirs. To allow Equation 10 to be applied in cases where only turbidity or Secchi depth measurements are available, the classification system in Table 1 was developed to help select the appropriate model for converting phosphorus to chlorophyll.

Based on experience and available information, a reservoir must be placed in one of the turbidity classes, which will then automatically lead to the selection of the appropriate chlorophyll:P model.

From a management point of view, the frequency at which algal concentrations resulting in nuisance conditions are experienced are of concern. Walmsley[24] developed an esthetic and use-impairment classification system that was based upon mean annual chlorophyll concentrations observed in reservoirs:

| Chlorophyll (µg/l) | Rating |
|---|---|
| 0–10 | No problems encountered |
| 10–20 | Algal scum present |
| 20–30 | Nuisance conditions |
| > 30 | Severe nuisance conditions |

A procedure was also developed to predict, from the mean annual chlorophyll concentration, the frequency at which each of these use-impairment classes will occur in a reservoir. For example, the percentage time that chlorophyll concentrations can be expected to exceed 30 µg/l and therefore result in severe nuisance conditions being experienced is calculated as

$$F = 1.19 \text{ chl} - 5.36, \qquad (11)$$

where $F$ is the frequency of occurrence of severe nuisance conditions, expressed as a percentage of a year.

Based on experience in South Africa, and elsewhere in the world, the Department of Water Affairs provisionally adopted a general eutrophication management objective of ensuring that severe nuisance conditions in reservoirs occur less than 20% of the time. Using Equation 11, this

**TABLE 2**
**Main Physical and Hydrological Characteristics of Hartbeespoort Dam**

| | |
|---|---|
| Drainage basin area | 4112 km$^2$ |
| Full supply volume | 194.63 x 10$^6$ m$^3$ |
| Full supply area | 2034 ha |
| Full supply max. depth | 32.5 m |
| Full supply mean depth | 9.6 m |
| Mean annual runoff (natural) | 163 x 10$^6$ m$^3$ |
| Mean annual precipitation | 703 mm |
| Mean annual evaporation (S pan) | 1684 mm |

translates into maintaining a mean chlorophyll concentration of less than 21 µg/l, which in turn translates into maintaining mean phosphorus concentrations (Equation 9) at below 130 µg/l in a clear water reservoir.

# III. DECISION SUPPORT SYSTEM

The models described in the previous sections were incorporated into a user friendly DSS. A DSS is defined as an interactive computer-based system that utilizes data and models to help solve decision problems. A DSS usually consists of three components, i.e., a modelbase, a database, and a dialogue system, which is a computer program that links the model and database and provides the user interface (Figure 3).

The modelbase contains the different REM submodels that were described in the previous sections. The database contains the hydrological, point source, drainage basin, and reservoir data for the reservoir under investigation. The DSS that was developed for REM helps the user to prepare the input data for the models, runs the models, and then allows the user to display the results of the simulation in various formats. It is responsible for storing and retrieving input data and interim results. The user interface relies on a system of menus to interact with the user. It also does error checking, allows for corrections to be made, and provides help facilities to inexperienced users.

The DSS was developed to run on IBM-compatible personal computers under the MS-DOS operating system. The output can be displayed on screen, a printer, or a plotter. The DSS was developed in TURBO PASCAL (Borland International). Other supporting software were used for developing the dialogue system.

# IV. APPLICATION OF THE DECISION SUPPORT SYSTEM

## A. BACKGROUND

The reservoir, Hartbeespoort Dam, is situated about 37 km west of Pretoria on the Crocodile River (Figure 4). It was initially constructed to provide water for irrigation, but it is presently also used for many purposes such as potable water supply, recreation, waterfront housing development, and flood control.The main physical and hydrological characteristics of Hartbeespoort Dam are given in Table 2.

The Crocodile River accounts for over 90% of the inflow into the reservoir. The upper reaches of the Crocodile River drain large parts of the Witwatersrand metropolitan area, which covers about 12% of the drainage basin. The remaining 88% are rural areas, of which only 2% is under irrigation and the rest is principally used for grazing and low-density residential purposes.

Rietvlei Dam is situated upstream of Hartbeespoort Dam on the Jukskei River tributary of the Crocodile River (Figure 4) and serves as a source of potable water to the city of Pretoria. Rietvlei Dam is a highly eutrophied, nitrogen-limited reservoir, with TP concentrations in the order of 1500 µg/l. More than half its annual inflow is treated sewage effluent.

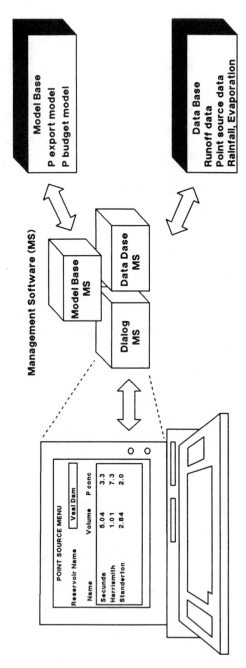

FIGURE 3.    Diagram of the main components of the decision support system for eutrophication control.

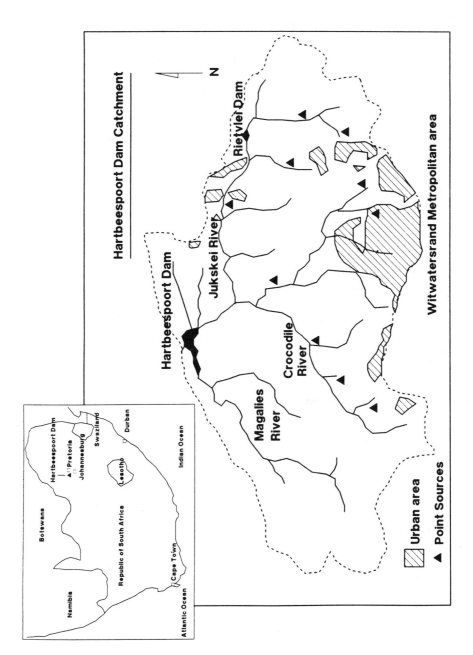

FIGURE 4.　Map of the drainage basin of Hartbeespoort Dam showing the main inflowing rivers, as well as the location of sewage treatment plants (point sources).

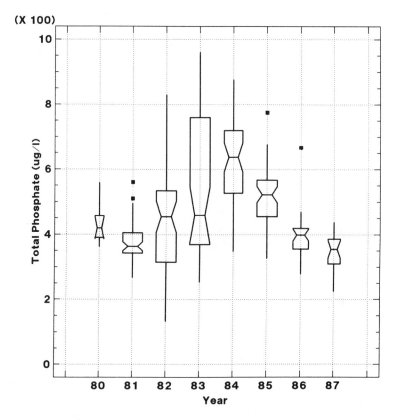

FIGURE 5.   Boxplots of phosphorus concentrations (μg/l) observed in Hartbeespoort Dam from 1980 to 1986. Extreme points beyond 1.5 times the interquartile range (box length) are plotted as individual points.

Hartbeespoort Dam is a hypertrophic reservoir.[25,26] Since 1980, the median total phosphorus concentrations varied from about 470 to 770 μg/l (Figure 5), and total nitrogen concentrations remained fairly constant at about 2000 μg/l. The *N/P* ratio in the reservoir was about 4:1, but is expected to increase to a point where phosphorus will become the limiting nutrient as a consequence of a reduction in the phosphorus input to the reservoir.

Hartbeespoort Dam is characterized by dense algal populations due to warm water, high solar radiation, and high nutrient concentrations.[26] The reservoir is dominated by a single blue-green alga, *Microcystis aeroginosa*. This species has a buoyancy mechanism that allows it to float to the surface and form algal mats (scum) during calm periods. The mean chlorophyll concentration in Hartbeespoort Dam in 1986 (Figure 6), integrated over the light penetration zone, was 135 μg/l, which is very high compared to other reservoirs in the region. Due to the buoyant nature of the algae, surface chlorophyll concentrations higher than 1000 μg/l have often been observed. During prolonged calm periods, *Microcystis* colonies have accumulated in wind-protected bays, forming a foul smelling mass of decomposing algae which may be up to a meter thick and exceed a hectare in area.[26,27]

## B. WATER RESOURCE MANAGEMENT SCENARIOS

The analysis of water resource management scenarios require alternative development options to be considered. Each simulation with the DSS is done for a specific water resource scenario, which is a combination of water quantity and quality management options for a given time horizon. Generating a scenario for a time horizon, say for the year 2000, involves three major steps (Figure 7):

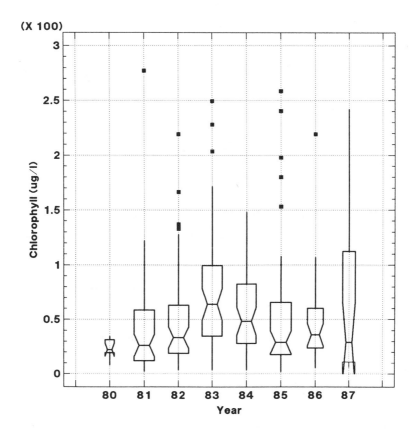

FIGURE 6. Boxplots of chlorophyll concentrations (μg/l) observed in Hartbeespoort Dam from 1980 to 1986. Extreme points beyond 1.5 times the interquartile range (box length) are plotted as individual points.

1.  Determine the total phosphorus load to the reservoir (subsystem 1)
2.  Determine the fate of phosphorus in the reservoir (subsystem 2)
3.  Determine the biological response to the ambient phosphorus concentration (subsystem 3).

Three water resource management scenarios, which reflect a particular point source control option, were evaluated with the DSS for eutrophication control to assess the impact of eutrophication control measures on Hartbeespoort Dam.

### 1. Scenario 1: "Do Nothing" Scenario

For the "do nothing" option, it was assumed that no action was taken and that effluent volumes would increase in the future, but the phosphorus concentrations in the effluents would remain at their present levels. It was assumed that there would be no phosphorus losses in the river between the point sources and Hartbeespoort Dam, and that a 1 mg P/l standard would be implemented in the Rietvlei Dam drainage basin.

### 2. Scenario 2: 1 mg P/l Standard Scenario

This scenario required that all the point sources in the Hartbeespoort Dam drainage basin comply to the 1.0 mg P/l effluent standard. It was assumed that there were no phosphorus losses in the river between the point sources and the Hartbeespoort Dam, and that a 1.0 mg P/l standard was enforced in the Rietvlei Dam drainage basin.

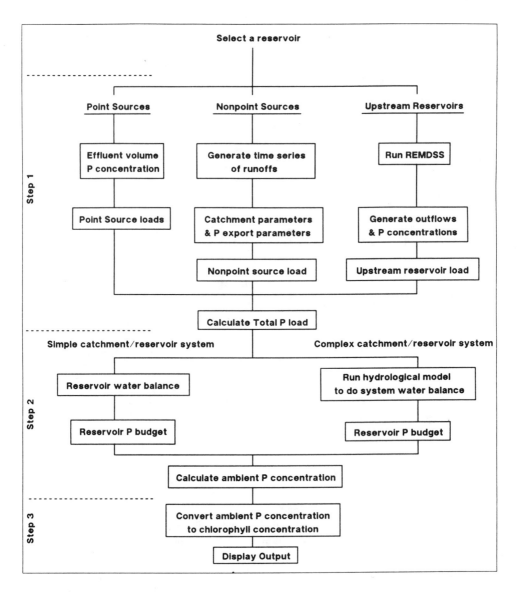

FIGURE 7. Flow diagram of the application of the decision support system for eutrophication control.

### 3. Scenario 3: 0.1 mg P/l Standard Scenario

For this scenario, the impact of implementing a stricter standard of 0.1 mg P/l on all the point sources in the Hartbeespoort Dam drainage basin was evaluated. It was assumed that there were no phosphorus losses in the river between the point sources and the Hartbeespoort Dam and that a 0.1 mg P/l standard was also implemented in the Rietvlei Dam drainage basin.

### C. PREPARATION OF INPUT DATA

The following input data to the DSS was prepared to assess the impact of the three water resource management scenarios.

### 1. Input Data Preparation for Subsystem 1

To estimate the total load to the reservoir from the drainage basin, given a point-source management scenario, the contribution from point and nonpoint sources must be quantified.

### a. Point Sources

The point-source contribution to the total load into the reservoir usually reflects the particular management scenario being evaluated. There are nine sewage treatment plants in the Hartbeespoort Dam drainage basin (Figure 4), and it was estimated that there will be about a 5% growth in future effluent discharges. This increase will be mainly due to an increase in the Witwatersrand metropolitan area. The total effluent volumes and loads for the different time horizons were calculated, given the three water resource management options as follows:

| Time horizon | 1990 | 2000 | 2020 |
|---|---|---|---|
| Total effluent volume ($10^6$ m³/a) | 140.3 | 227.5 | 518.0 |
| Phosphorus loads (tons/a) | | | |
| Scenario 1 - Do nothing | 857.6 | 1385.1 | 3160.8 |
| Scenario 2 - 1 mg/l P standard | 141.0 | 227.5 | 518.0 |
| Scenario 3 - 0.1 mg/l P standard | 14.1 | 22.7 | 51.8 |

### b. Natural Runoff

A representative time series of "natural" monthly runoffs is required to estimate the nonpoint-source phosphorus load. This is done with a hydrological model, which take the changes in land use into account for the given time horizon.

A time series of runoffs from the Hartbeespoort Dam drainage basin, excluding the Rietvlei Dam drainage basin, was simulated with the Pitman hydrological model[28] and was adjusted for the projected changes in land use in the drainage basin. The increase in runoff is mainly due to increased runoff from the expanding urban areas of the Witwatersrand metropolitan area.

| Time horizon | 1984 | 1990 | 2000 | 2020 |
|---|---|---|---|---|
| Natural MAR ($10^6$ m³/a) | 147.0 | 147.0 | 147.0 | 147.0 |
| Irrigation runoff ($10^6$ m³/a) | 37.5 | 38.4 | 40.2 | 43.9 |
| Urban runoff ($10^6$ m³/a) | 30.7 | 41.8 | 67.0 | 185.1 |
| Total runoff ($10^6$ m³/a) | 142.2 | 150.4 | 173.8 | 288.2 |

### c. Nonpoint Sources

The phosphorus export parameter values ($a = 0.163$ and $b = 1.288$) were used in Equation 3 to calculate the monthly nonpoint-source TP export as a function of runoff. These parameters were calibrated on load and flow data for 1979 to 1986 collected in the Magalies River drainage basin, which forms part of the Hartbeespoort Dam drainage basin.[20] There are no significant point sources in the Magalies drainage basin.

### d. Upstream Reservoirs —Rietvlei Dam

For the purpose of the investigation, water levels and TP concentrations of Rietvlei Dam were simulated,[30] and the outflow volumes and TP concentrations were used to calculate the phosphorus loads from Rietvlei Dam to Hartbeespoort Dam. The median monthly Rietvlei Dam outflow volumes and loads were

| Time horizon | 1990 | 2000 | 2020 |
|---|---|---|---|
| Outflow volume ($10^6$ m³) | 120.0 | 121.6 | 121.6 |
| Phosphorus load (tons/a) | 104.0 | 214.4 | 608.4 |

## 2. Input Data Preparation for Subsystem 2

To describe the relation between external phosphorus loads and phosphorus concentrations in the lake, a phosphorus mass balance (Equation 6) is undertaken. The phosphorus lost to the sediment (Equation 7) is characterized by a sedimentation parameter calibrated for the reservoir.

**TABLE 3**

**Summarized Results of the Simulated TP and Chlorophyll Concentrations in Hartbeespoort Dam for the Three Time Horizons**

| | Time horizons | | |
|---|---|---|---|
| | **1990** | **2000** | **2020** |
| **Median Nonpoint Source TP Load (tons/annum)** | | | |
| Scenario 1 | 15.92 | 21.54 | 50.92 |
| Scenario 2 | 15.92 | 21.54 | 50.92 |
| Scenario 3 | 15.92 | 21.54 | 50.92 |
| **Median Rietvlei Dam TP Load (tons/annum)** | | | |
| Scenario 1 | 6.67 | 15.31 | 51.90 |
| Scenario 2 | 6.67 | 15.31 | 51.90 |
| Scenario 3 | 2.89 | 6.04 | 17.61 |
| **Total Point Source TP Loads (tons/annum)** | | | |
| Scenario 1 | 857.6 | 1388.1 | 3160.8 |
| Scenario 2 | 141.0 | 227.5 | 518.0 |
| Scenario 3 | 14.1 | 22.8 | 51.8 |
| **Average Predicted TP Concentration (µg/l)** | | | |
| Scenario 1 | 568 | 659 | 856 |
| Scenario 2 | 297 | 334 | 418 |
| Scenario 3 | 144 | 146 | 158 |
| **Average Predicted Chlorophyll Concentration (µg/l)** | | | |
| Scenario 1 | 67 | 76 | 93 |
| Scenario 2 | 41 | 44 | 53 |
| Scenario 3 | 23 | 23 | 25 |

Scenario 1 = "Do nothing".
Scenario 2 = 1 mg/l P standard.
Scenario 3 = 0.1 mg/l P standard.

## a. Sedimentation

A concentration-dependent sedimentation parameter ($k$) of 2.0 x $10^{-6}$/month/(mg/m$^3$)$^2$ was used in Equation 8 to simulate the monthly sedimentation losses of TP in Hartbeespoort Dam. This parameter was calibrated on Hartbeespoort Dam using observed inflow, outflow, and reservoir TP concentrations.[7]

A water balance for the reservoir must be done simultaneously with the phosphorus mass balance using characteristics such as the full supply volume, dead storage volume, and volume/area relationship, and average monthly rainfall and evaporation must also be entered. The projected water demands (abstractions) must be estimated, and provision is also made for entering rules for applying water restrictions during drought periods.

## b. Abstractions

It is projected that the maximum yield of Hartbeespoort Dam will be utilized in the future. Demands are expected to increase as a result of an increase in the demand for urban and industrial users in the Hartbeespoort Dam area of supply.

| Time horizon | 1990 | 2000 | 2020 |
|---|---|---|---|
| Agricultural demands ($10^6$ m$^3$) | 120.0 | 121.6 | 121.6 |
| Urban & industrial demands ($10^6$ m$^3$) | 104.0 | 214.4 | 608.4 |

### c. Rainfall and Evaporation

To calculate a water balance for Hartbeespoort Dam, the average rainfall (mm) and symonspan evaporation (mm) data for 1925/26 to 1979/80 were used.[29]

| Month | Oct | Nov | Dec | Jan | Feb | Mch | Apr | May | Jun | Jul | Aug | Sep |
|---|---|---|---|---|---|---|---|---|---|---|---|---|
| Rainfall | 59 | 113 | 112 | 130 | 99 | 85 | 48 | 21 | 8 | 7 | 6 | 15 |
| Evaporation | 194 | 193 | 200 | 189 | 155 | 143 | 107 | 85 | 67 | 78 | 114 | 158 |

### 3. Input Data Preparation for Subsystem 3
### a. Chlorophyll/phosphorus Relationship

Hartbeespoort Dam is classified as a clear water reservoir, because the extinction of light in the water is mainly due to suspended algae and not suspended sediment.[26] The OECD chlorophyll/phosphorus relationship (Equation 9) was therefore used to relate the mean seasonal TP concentrations to the mean seasonal chlorophyll-*a* concentration.

### D. SIMULATED WATER QUALITY RESPONSE

The input data described in the previous sections were used as input to the DSS for eutrophication control. The TP concentrations in Hartbeespoort Dam were simulated, subject to three point source phosphorus control scenarios (Table 3), for the three time horizons.

Hartbeespoort Dam is a point source dominated system. For the "do nothing" option, the point-source loads are about 50 times larger than the median nonpoint source loads. The implementation of a 1 mg/l P effluent standard reduce point–source loads to about nine times larger than median nonpoint source loads. It is only when a 0.1 mg/l P effluent standard is implemented that the predicted point source loads are reduced to about equal to the median nonpoint source loads. The predicted phosphorus concentrations in the reservoir in response to the alternative options are summarized in Figure 8. From these it is obvious that for rehabilitation, Hartbeespoort Dam will require the enforcement of at least a 0.1 mg P/l standard.

# V. SELECTION OF A EUTROPHICATION CONTROL OPTION

### A. BACKGROUND

In the selection of appropriate eutrophication control strategies, decision makers are faced with a multiplicity of conflicting objectives of technological, financial, environmental, social, institutional, and political nature. Application of the DSS described above only yields part of the required information, namely, that related to some of the technological and environmental objectives. Information on the financial, social, institutional, and political objectives, and the remainder of the information on technological and environmental objectives have to be obtained from other sources and must be sought in different ways.

Once all the relevant information has been gathered, it has to be integrated into the same decision framework to arrive at an optimal eutrophication control strategy. Benefit/cost analysis could be used to arrive at an optimal strategy. However, severe reservations have been expressed about traditional benefit/cost analysis procedures in which all noneconomic values have to be expressed in monetary terms.[31–33] It was suggested that in cases where noneconomic and intangible factors play an important role, more realistic benefit/cost analysis can be done by using a decision-making process such as the AHP,[4,34] which allows for quantitative and qualitative information to be incorporated in the same decision framework.

The AHP is a multiobjective, multicriterion decision-making system that employs a pairwise comparison procedure to arrive at a scale of preferences amongst sets of alternatives. It is applied by conceptualizing the decision problem as a hierarchy of decision elements. The AHP can cope with intuitive, rational, and irrational judgments of measurable quantities and immeasurable qualities. This feature of the AHP is a major advantage in decision problems in which

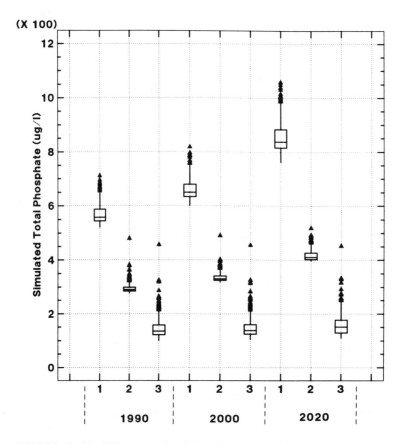

FIGURE 8.   Predicted TP concentrations in Hartbeespoort Dam for the no control (scenario 1), 1 mg/l P effluent standard (scenario 2) and the 0.1 mg/l P effluent standard (scenario 3) as eutrophication control options.

quantitative, as well as qualitative information has to be considered. A commercially available software package, EXPERT CHOICE (Decision Support Software, Inc.), was developed to assist with the application of the AHP.

Grobler[35] used the AHP in a hypothetical analysis to show how it could be used to select the most appropriate eutrophication control option from several alternatives to improve the water quality in Hartbeespoort Dam. Since then, the Department of Water Affairs actually selected, from several alternatives, a eutrophication control strategy for Hartbeespoort Dam.[36] Its strategy involves enforcing a 1 mg P/l standard on point source effluents in the reservoir's drainage basin and to introduce aeration/destratification of the reservoir to treat some of the symptoms of eutrophication.

We decided to use the AHP to do a retrospective analysis of the decision, which resulted in this particular strategy being selected. We hoped that such an analysis would illuminate the factors, and their weights, which influenced the department's decision. We believe that such an analysis would provide valuable information for decisions concerning the future development of the DSS for eutrophication control. The information required for the analysis was acquired through interviewing those officials in the Department of Water Affairs involved in arriving at the decision that was made.

## B. ALTERNATIVE EUTROPHICATION CONTROL STRATEGIES

Four alternative eutrophication control strategies were considered:

1.  No control (NO CONT): No action is taken, i.e., the 1 mg P/l phosphorus standard is not enforced and no in-lake management procedures are introduced to treat the symptoms of eutrophication in the reservoir. The rationale behind this option was either that the reservoir is so badly eutrophied that it is unlikely to respond to control measures or that the cost of eutrophication control will not be justified by the benefits to be derived.

2.  1 mg P/l standard (1 MG/L): Require all point-source effluents in the drainage basin of the reservoir to comply to a 1 mg P/l effluent standard. The rationale behind this option was that the technology for achieving this standard in effluents is well established in South Africa; that implementation of the standard will cause a notable improvement in eutrophication-related water quality; and, therefore, some material and nonmaterial benefits will be derived from selecting this option.

3.  0.1 mg P/l standard (0.1 MG/L): Require all point source effluents to comply to a 0.1 mg P/l standard. The rationale for this option is that a much more stringent standard than 1 mg P/l will be required to lower the phosphorus load on the reservoir sufficiently to cause a notable improvement in water quality. It recognizes that the technology for treating effluents to achieve phosphorus concentrations of 0.1 mg P/l is available, although it has not been tested locally. It acknowledges that in the medium term, this option would be considerably more expensive for polluters to implement than the 1 mg P/l standard.

4.  1 mg P/l standard plus aeration (1 MG/L+): This option requires that a 1 mg P/l standard be enforced on all point sources in the drainage basin of the reservoir, while simultaneously an aeration/destratification system is installed in the reservoir to alleviate the symptoms of eutrophication in the reservoir. The rationale for this option is that enforcing a 1 mg P/l standard alone is not expected to result in the desired water quality in the reservoir. It has therefore to backed up by additional measures. The severity of eutrophication-related water quality problems experienced in Hartbeespoort Dam is largely due to *Microcystis* being the dominant algal species. It is believed that aeration/destratification is a practical and economical in-lake management operation, which could cause a shift from *Microcystis* dominance to less obnoxious algal species. It was established that the cost of aeration/destratification would be considerably less than that of introducing a 0.1 mg P/l standard.

## C. FACTORS THAT AFFECTED THE CHOICE OF A CONTROL OPTION

The major factors that affected the choice of a eutrophication control option for Hartbeespoort Dam were determined to be:

1.  Predicted water quality response (PRED RES): The eutrophication-related water quality in Hartbeespoort Dam in response to the various options was thought to have played an important role in the final selection of a particular option. The responses to options 1 to 3 were simulated with the DSS as described above. The predicted water quality in response to option 4 was obtained partly by simulation and partly in the form of an expert opinion expressed by a group of experienced researchers who studied the limnology of the reservoir over several years.[26]

2.  The cost of each option (COST): The estimated financial costs to polluters, water users, and water resource managers associated with each of the options was considered to be an important factor in the selection of a control option.

3.  The feasibility of each option (FEASABIL): Although each of the options on the list was considered feasible, there are major differences in the degree of feasibility. For example, the effluent treatment technology for achieving a 1 mg P/l standard is well established with the polluters involved, whereas the technology for achieving a 0.1 mg P/l standard is not. Similarly, aeration/destratification has never been tried anywhere in South Africa.

4.    Conservatism inherent in government institutions (CONS): Government institutions, like the Department of Water Affairs, which deal with public funds and are responsible to Parliament for their actions, are usually very conservative. In the selection of options, they tend to show a strong preference for options that are proven, standard, and non-controversial.

5.    The image of the Department of Water Affairs (IMAGE): It was realized in the early 1970s that eutrophication is becoming a serious water quality problem in South Africa. The 1 mg P/l standard for sensitive drainage basins was promulgated in 1980, and polluters were given until 1985 to gear up for complying to the standard. During this time the eutrophication-related water quality in Hartbeespoort Dam deteriorated rapidly to the point where the reservoir achieved the reputation of being one of the most eutrophied water bodies in world. The Department's eutrophication control policies and strategies have previously often been criticized in the public media and in scientific circles. On the one hand, it was criticized for allowing water quality in Hartbeespoort Dam to deteriorate to its present status and for its lack of action to improve the situation. On the other hand, the Department's decision to enforce effluent phosphorus standards have been criticized as costly measures that are not justified by the benefits derived from them.

However, officials in the Department are genuinely concerned about the poor water quality in the reservoir and are serious about fulfilling their obligation to protect the quality of South Africa's water resources. They are aware of the damaging effect the present water quality is having on the Department's image and would therefore prefer an option that is likely to cause a notable improvement in water quality in the reservoir. However, they are also sensitive about the Department's image amongst polluters as being strict but fair, strict in the sense that it has the will to enforce effluent standards if it believes these are necessary to protect the water environment; fair in the sense that it will not impose unnecessary strict eutrophication control measures. Therefore, from this perspective, the least costly option that will produce the desired results would be preferred.

## D. STRUCTURING AND ANALYSIS OF THE DECISION PROBLEM

Those involved in the decision agreed that the decision problem was to "select the best option from the four alternatives being considered" and that it could be structured in a hierarchy, as shown in Figure 9. The pairwise comparison procedure of the AHP was used to first determine the relative preferences for each of the options under each of the factors (each time ignoring all the other factors). As an example, the comparisons of the options, considering only the predicted water quality responses to the various options, are described below. To obtain a bench/mark, the two options with the greatest difference between their predicted responses were compared first.

The 0.1 mg P/l option was very strongly preferred above the no-control option. This large difference in preference resulted from the knowledge that with no control the reservoir developed severe eutrophication-related water quality problems. In contrast, a 0.1 mg P/l standard was predicted to largely solve the eutrophication problems in the reservoir.

The 1 mg P/l option was preferred moderately above the no-control option. The relatively small difference in preference between these two options is mainly because the 1 mg P/l standard is not expected to improve water quality much compared to what it would be with no control.

The 1 mg P/l plus aeration/destratification option was strongly preferred above the no control option, because it is believed that aeration/destratification, in combination with 1 mg P/l standard, will considerably improve the water quality above what it is now.

The same pairwise comparison procedure was used to rate the various options in terms of preference under each of the remaining factors. The relative preferences are converted in the \HP to weights that are summarized in Table 4.

It is interesting to note the opposite trends in the relative weights associated with the different

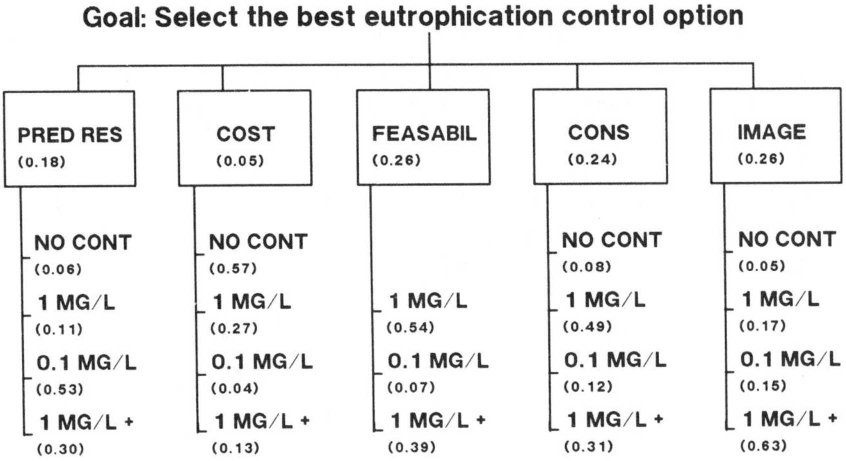

## Goal: Select the best eutrophication control option

FIGURE 9. The decision problem of selecting the best eutrophication control option for Hartbeespoort Dam structured into a hierarchy. The relative weights associated with each factor or option under each factor are given in parentheses.

**TABLE 4**
**The Weights Assigned to Each Option for a Given Factor**

| Factor | Relative Weight of Options (Total = 100%) | | | |
|---|---|---|---|---|
| | NO CONT | 1 MG/L | 1 MG/L+ | 0.1 MG/L |
| Predicted Response | 6 | 11 | 30 | 53 |
| Cost | 56 | 27 | 13 | 4 |
| Feasibility | — | 54 | 39 | 7 |
| Conservatism | 8 | 49 | 31 | 12 |
| Image | 6 | 17 | 63 | 14 |

options when comparing predicted response and cost as factors. The relative weights under feasibility, as expected, showed the same trend as under cost. The inherent conservatism of the Department of Water Affairs resulted in most of the weight under this factor being given to the 1 mg P/l option, followed by the 1 mg P/l plus aeration/destratification option. Both the no-control and 0.1 mg P/l options received low weights. When the image of the department was considered, the 1 mg P/l plus aeration/destratification option was by far the most preferred.

Finally, the relative weights associated with the factors in the first level of the decision hierarchy had to be determined such that in this retrospective analysis we would arrive at the same decision as the department did earlier. That meant that the weight associated with the main factors must result in (1) the 1 mg P/l plus aeration/destratification option being the most preferred options and (2) the options being ranked in the same order of preference as in the original decision by the department. The relative weights derived for each of the factors, in decreasing order of magnitude, were image (27%), feasibility (26%), conservatism (24%), predicted response (18%), and cost (5%). These weights associated with each of the factors resulted in the following overall preferences for the various options. The weights shown in parenthesis are those calculated in the AHP: The most preferred option was 1 mg P/l plus aeration/destratification (40%), followed by 1 mg P/l (33%), 0.1 mg P/l (19%), and no-control (7%).

## E. DISCUSSION

The weights calculated for the various options were regarded by those involved in the actual decision making to be a fairly realistic reflection of the relative preference associated with each of the options when the actual decision was made. However, we and those responsible for the actual decisions were surprised at the weights that had to be given to the different factors to achieve these weights and the particular rank order of the options. It was initially anticipated that the factors predicted response and cost would have had much higher weights than they actually received in this analysis. However, after careful consideration it was agreed that these weights are not an unreasonable reflection of the relative importance attached to these factors when the decision was made.

It must be stressed that the weights associated with these factors depend very much on the particular circumstances. At a different site, where different alternatives are being considered, or at a different time, the weights associated with the factors, or even the factors themselves, could be very different from what they were during this analysis.

This post-mortem analysis of the decision to select a particular eutrophication control option proved to be valuable from different perspectives:

1. The two factors that dealt with quantitative information, i.e., predicted response and cost, together carried less weight (23%) than any one of the remaining factors. Rather than suggesting that predicted response and cost are not important factors, we believe it reemphasized that many factors played an important role in the selection of eutrophication control strategies.

2. For the 1 mg P/l plus aeration/destratification option, the predicted response consisted of quantitative information obtained as output from the DSS (response to the 1 mg P/l standard) and qualitative information obtained as expert opinion (response in terms of algal species composition to aeration/destratification). In selection of an appropriate control option, equal value was attached to each of these sources of information by those involved in making the decision.

   Obtaining the information required by factors other than predicted response, i.e., information that had to be obtained from sources other than our models, required as much effort and time as was required to apply our models.

These perspectives provided us, as modelers, with a healthy dose of modesty about the impact the output from our models are having on the actual decisions that are made about eutrophication control.

# REFERENCES

1. **Vollenweider, R. A.,** Eutrophication: A global problem, *Water Qual. Bull.,* 6, 59, 1981.
2. **Toerien, D. F.,** A review of eutrophication and guidelines for its control in South Africa. Special Report WAT, 48, Division of Water Technology, CSIR Pretoria, South Africa, 1977.
3. **Taylor, R., Best, H. J., and Wiechers, H. N. S.,** The effluent phosphate standard in perspective: Part 1. Impact, control and management of eutrophication, *IMIESA,* 9, 43, 1984.
4. **Saaty, T. L.,** *Decision-Making for Leaders: The Analytical Hierarchy Process for Decisions in a Complex World,* Lifetime Learning Publications, Belmont, CA, 1982.
5. **Caswell, H.,** The validation problem, in *Systems Analysis and Simulation in Ecology,* Vol. 3, Patten, B. C., Ed., Academic Press, New York, 1976.
6. **Skopp, J. and Daniel, T. C.,** A review of sediment predictive techniques as viewed from the perspective of nonpoint pollution management, *Environ. Manag.,* 2, 39, 1978.

7. **Grobler, D. C.,** Phosphorus budget models for simulating the fate of phosphorus in South African reservoirs, *Water SA,* 11, 219, 1985.

8. **Sas, H.,** Introduction to the framework adopted for case study analysis. Guidelines for participants in the International Symposium: Lake Restoration by Reduction of Phosphorus Loading: Experiences, Expectations and Future Problems, 17–19 April 1988, Leeuwenhorst Congress Center, The Netherlands, 1988.

9. **Vollenweider, R. A. and Kerekes, J. J.,** The loading concept as a basis for controlling eutrophication. Philosophy and preliminary results of the OECD programme on eutrophication, *Prog. Water Technol.,* 12, 5, 1981.

10. **OECD,** Eutrophication of Waters: Monitoring Assessment and Control, OECD, 2, Paris, France, 1980.

11. **Reckhow, K. H. and Chapra, S. C.,** *Engineering Approaches for Lake Management. Volume 1: Data Analysis and Empirical Modeling,* Ann Arbor Science, Butterworth, Boston. 340, 1983.

12. **Dillon, P. J. and Kirchner, W. B.,** The effects of geology and land use on the export of phosphorus from watersheds, *Water Res.,* 9, 135, 1975.

13. **Gilliom, R. J.,** Estimation of background loadings and concentrations of phosphorus for lakes in the Puget Sound region, Washington, *Water Resour. Res.,* 17, 410, 1981.

14. **Sonzoni, W. C., Chesters, G., Coote, D. R., Jeffs, D. N., Konrad, J. C., Ostry, R. C., and Robinson, J. B.,** Pollution from land runoff, *Environ. Sci. Technol.,* 14, 148, 1980.

15. **Beaulac, M. N. and Reckhow, K. H.,** An examination of land use-nutrient export relationships, *Water Resour. Bull.,* 18, 1013, 1982.

16. **Jolankai, G.,** Modelling of nonpoint source pollution, in *Application of Ecological Modeling in Environmental Management,* Jørgensen, S. E., Ed., Elsivier, Amsterdam, 1983.

17. **Grobler, D. C. and Silberbauer, M. J.,** The combined effect of geology, phosphate sources and runoff on phosphate export from drainage basins, *Water Res.,* 19, 975, 1984.

18. **Hill, A. R.,** Factors affecting the export of nitrate-nitrogen from drainage basins in southern Ontario, *Water Res.,* 12, 1045, 1978.

19. **Walker, W. W.,** Land-use-nutrient export relationships in the Northeast, in *A Preliminary Analysis of the Potential Impacts of Watershed Development on the Eutrophication of Lake Chamberlain,* New Haven Water Company, New Haven, CT, 1978.

20. **Grobler, D. C. and Rossouw, J. N.,** Nonpoint Source Derived Phosphorus Export from Sensitive Catchments in South Africa, Report no. E0000/00/0188, Department of Water Affairs, Pretoria, South Africa, 1988.

21. **Walker, W. W.,** Empirical Methods for Predicting Eutrophication in Impoundments: Report 4, Phase 3: Applications Manual, Technical Report E-81-9, U.S. Army Corps of Engineer Water Ways Experiment Station, CE, Vicksburg, Ms, 1987.

22. **Jones, R. A. and Lee, G. F.,** Recent advances in assessing impact of phosphorus loads on eutrophication-related water quality, *Water Res.,* 16, 503, 1982.

23. **Hoyer, M. V. and Jones, J. R.,** Factors affecting the relation between phosphorus and chlorophyll *a* in midwestern reservoirs, *Can. J. Fish. Aquat. Sci.,* 40, 192, 1983.

24. **Walmsley, R. D.,** A chlorophyll *a* trophic status classification system for South African impoundments, *J. environ. Qual.,* 13, 97, 1984.

25. **Scott, W. E., Ashton, P. J., Walmsley, R. D., and Seaman, M. T.,** Hartbeespoort Dam: a case study of a hypertrophic, warm, monomictic impoundment, in *Hypertrophic Ecosystems,* Barica, J. and Mur, L. R., Eds., Developments in Hydrobiology, Vol. 2. Dr. W. Junk, The Hague, Netherlands, 1980.

26. **National Institute for Water Research,** The Limnology of Hartbeespoort Dam, South African National Scientific Programmes Report No. 110, 1985.

27. **Zohary, T.,** Hyperscums of the cyanobacterium *Microcystis aeruginosa* in a hypertrophic lake (Hartbeespoort Dam, South Africa), *J. Plankton Res.,* 7, 399, 1985.

28. **Pitman, W. V.,** Hydrology of the Hartbeespoort Dam catchment, report No. PC000/00/5486, Vaal River Systems Analysis, Directorate of Planning, Department of Water Affairs, Pretoria, South Africa, 1986.

29. **Department of Water Affairs,** Evaporation and Precipitation Records, Monthly Data up to September 1980, Hydrological Information Publication No. 13, Department of Water Affairs, Pretoria, South Africa, 1985.

30. **Rossouw, J. N.,** Evaluation of the Impact of Eutrophication Control Measures on Water Quality in Rietvlei Dam, Contract report to the Directorate of Water Pollution Control, Department of Water Affairs, Pretoria, South Africa, 1988.

31. **Schumacher, E. F.,** *Small is Beautiful: Economics as if People Mattered,* Harper and Row, New York, 1973.

32. **Fedra, K.,** A modular interactive simulation system for eutrophication and regional development, *Water Resour. Res.,* 21, 143, 1985.

33. **Christie, W. J., Becker, M., Cowden, J. W., and Valentyne, J. R.,** Managing the great lakes as a home, *J. Great Lakes Res.,* 12, 2, 1986.

34. **Saaty, T. L.,** Priority setting in complex problems, *IEEE Trans. Eng. Manag.,* EM 30a, 140, 1983.

35. **Grobler, D. C., Rossouw, J. N., van Eeden, P., and Oliveira, M.,** Decision support system for selecting eutrophication control strategies, in *Systems Analysis in Water Quality Management,* Beck, M. B., Ed., Pergamon Press, Oxford, 1987.

36. **Department of Water Affairs,** Executive Summary: Evaluation of the Impacts of Phosphate Control Measures on Eutrophication-Related Water Quality in Sensitive Catchments, Department of Water Affairs, Pretoria, South Africa, 1988.

Chapter 9

# FUTURE DIRECTIONS FOR WATER QUALITY MODELING

**B. Henderson-Sellers**

## TABLE OF CONTENTS

# I. INTRODUCTION

In this final chapter we will summarize the range of mathematical and computer-based tools available (or likely to become available in the near future) for assisting in the development of water quality simulation models and environmental decision support systems (EDSS). In both scientific and managerial areas, the new software development concept of object-oriented programing will be explained and presented in the context of prototype software modules intended to provide decision support for the water manager.

In the final section, it is noted that traditional water resource management models are based on a historical analysis of records for streamflow, rainfall, etc. Both quantity and quality assessments presume that these records can be equally applied to future conditions. However, atmospheric scientists over the last decade have become increasingly convinced of the reality of climate change.[1] A key consequence of these studies for water management is the likely shift of location of rainbelts, intensity of rainfall events, and warmer temperatures (which, for instance, control evaporation rates and algal growth rates). Some of the water management areas likely to be impacted by climate change are outlined. Little is known about the potential magnitudes of the impacts on water quality and water quantity. It is, however, vitally important that water management policies being implemented now for the next few decades take into account such possible climate trends and variability, and that further evaluation of the impacts on water supply be undertaken immediately.

# II. SOFTWARE MODELING TOOLS

Problem solving in applied mathematics has a long history. It is only recently that mathematical modeling has assumed a new guise with the advent of the digital computer. The concept that a model was useful only if it had an analytical solution is being rapidly displaced by the use of numerical techniques in problem solving. In addition, the availability of increasingly powerful computers provides an opportunity for the development of numerical models for the solution of problems that had previously been dismissed as being intractable using only analytical modeling tools.

In the water industry, software modeling tools include those for scientific evaluation (e.g., the degree of eutrophication expected in a reservoir), for technical management (e.g., real-time control of a filter plant), and for information systems similar to those used in most businesses today (e.g., payroll, maintenance scheduling, decision support systems[DSS]).

## A. TECHNICAL AND SCIENTIFIC MODELING

Water quality models for lakes, reservoirs, and rivers have tended to remain in the domain of the model developer and to be applied in a consultancy framework. Only recently have some of these packages been released in a more general sense. For example, the reservoir water quality models of the U.S. Army Corps of Engineers have been available to the district offices of the Corps for some time, and some of the aquatic pesticide models have had a relatively high profile in the marketplace (see also specific examples in earlier chapters of this book). Thermal stratification models and their counterparts of ecosystem/eutrophication models have provided many successful applications in various parts of the world, but again usually as part of international research projects in collaboration between the industrial water quality manager and the researcher himself or herself; although in some instances (for example, the U.S. Environmental Protection Agency Water Quality Modeling Research Center[2]) a government agency provides model support for released versions of the code.

Problems of nontransferability of models have become evident (see further discussion in Chapter 5), especially when the motive for the original model development was very strictly oriented (probably because of the source of funding) towards one particular case study. Selection of appropriate models for use in problem solving presents a difficulty to the water quality manager who does not have the detailed knowledge of the limits of applicability of various simulation models. Two possible solution methodologies are becoming possible with new software tools: EDSS[3] and expert systems (ES). Indeed, Guariso and Werthner[3] state that an EDSS could contain an ES as an extra component, in addition to the three shown in Figure 1 of Chapter 1. Other authors maintain a clear distinction between these two software tools inasmuch as (1) an expert system is intended to *replace* an expert and relieve the expert for more difficult tasks and (2) a DSS is intended to *assist* the expert in making a more focused management decision.[4]

DSSs have been discussed in Chapter 1, and further details can be found in the book by Guariso and Werthner[3] and a forecast for the future in Keen.[5] In addition to the simulation model component, a DSS contains a database. A database can be considered as a storehouse of information on a given topic. It is analogous to a hardcopy-based filing system. A database management system (DBMS) is a software system interposed between the data repository and the user in such a way that only a single copy of each datum exists so that changes in the data format can be accomplished in a way that is "transparent" to all users of those data. Database management software permits data from different origins to be concatenated so that multiple-component enquiries can be satisfied without resort to a set of individual enquiries of the several departments concerned. The DBMS must be capable not only of locating the requested information (possibly using a 4GL as an interface to the user), but also of presenting it to the manager in the detail required for effective use of the information by the manager.

In contrast, an ES or intelligent knowledge-based system (IKBS) is intended to encapsulate the knowledge that a human practitioner of some years experience has in applying the *rules* of his or her profession. Experts' knowledge is encapsulated in the expert system (this encapsulation being perhaps the hardest part of designing an ES, giving rise to the new discipline of knowledge engineering) and the user interrogates the system, being prompted by a set of questions that will differ depending upon previous answers. In essence, the program follows links through information stored in a treelike structure, but is also able to back-chain through the structure in order to explore all the viable possibilities. In addition to delivering an answer, an ES is also able to state a line of reasoning that led the system to that answer. Additionally, decisions need not be made solely in terms of a yes/no response, but can elicit answers on a grey scale, for example, from 1 to 10 or from "not at all" to "to a large extent".

ES have been implemented successfully in medical areas, where the answers to a set of questions yields the prognosis, with an associated degree of probability, for any particular patient. In the water industry environment, it has become clear that the water supply manager has difficulty in choosing between several water quality models for application to the particular case study. Seeking the advice of a consultant is often the only recourse; yet encapsulation of the knowledge of a number of such "experts" into a software package could make available on a wider scale the information pertaining to the different domains of applicability of the large number of water quality models now available worldwide and would permit the "tailored choice" of the model most appropriate for the particular problem of concern.

As part of the Water Quality Operational Studies (WOTS) of the U.S. Army Corps of Engineers, a knowledge base on streambank protection, levée projects, and flood control channel projects has been developed as a PROLOG-based expert system, known as ENDOW (= **EN**vironmental **D**esign **O**f **W**aterways),[6] which will permit the rapid identification of environmental alternatives for inclusion in project plans, designs, or operational procedures. A software package, RAISON, developed in Canada[7] helps the user select catchment acidification models. This system also has several of the elements of an EDSS as envisaged by Guariso and Werthner.[3] In the area of water quality models in rivers and reservoirs, technical reviews, such as that of Henderson-Sellers et al.,[8] should provide sufficient scientific information from which an ES-oriented management information system could be developed.

Other software engineering and information system tools that are being applied to water management include geographic information systems and computer graphics. Indeed, Jamieson and Fedra[9] attribute much of the increased usage of computer information systems in water management to the recent availability of interactive computing and color graphics.

## B. NEW PARADIGMS FOR SOFTWARE RELIABILITY AND MAINTAINABILITY

Research development of standard procedural languages is being rapidly displaced by a strong and growing interest in functional programming (FP) languages and object-oriented programming (OOP) languages, which promise greatly increased software reliability, maintainability, and portability. Both paradigms have developed from the mathematical ideas of "category theory"; in FP, the emphasis is on the functions that relate the objects (data structures), whereas in OOP, the object is the focus of attention being stimulated into action by "message passing" (analogous to the function of FP). These types of new programming methodologies are establishing a user base, but at present are not well established in commerce and industry.[10]

### 1. Functional Programing

In a functional programing language[11] such as MIRANDA* or LISP, there are four components: (1) a set of primitive functions (predefined by the language), (2) a set of functional forms, (3) the application operation, and (4) a set of data objects.[12] The programer develops the functional forms using the predefined functions and the built-in mechanism of the application operation. Predefined (or primitive) functions may include selection operations (such as LAST, FIRST, TAIL), structuring operations (such as ROTATE RIGHT, LENGTH), arithmetic operations, predicate functions (i.e., those having a true/false value), logical operations (such as AND, OR, NOT), and identity functions. Assignments of procedural languages are replaced by function definitions. This all means that less emphasis is placed on the ideas of assignment of individual values to specific, named memory cells and instead the use of uniform data objects, such as sequences, lists, and arrays, permits the construction of more complicated data structures.

### 2. Object-Oriented Programing

In the object-oriented approach, both for software design and implementation (program coding), the system to be modeled is regarded not as a set of nested procedures (derived by a top-down functional decomposition of the system), but as a conglomerate of interacting objects that are identifiable in part with the substantive objects of the real system. Thus, for example, in a water supply system, objects would include reservoirs, valves, and pipes. Having identified the objects in a system, their interaction is then evaluated, and not until a fairly late stage in program development would the algorithm describing their internal procedures be finalized. This view of the world corresponds more closely with our own view, in which often the procedure is hidden from view and is of no interest. For example, in driving a car we wish to be assured that the system has a working set of brakes but we do not care how that braking system is implemented (hydraulic or mechanical, drum or disc) so long as it functions correctly when we put our foot on the brake pedal. In other words, the functional implementation is hidden from the user's view, since knowledge of its details are unnecessary in order to be able to utilize the object. Essentially, then, the object-oriented paradigm requires the emphasis to be shifted from PROCEDURES (+data) to DATA/OBJECTS (+procedures). Although an apparently small shift in mindset, it has extensive ramifications for software engineering, many of which are still being evaluated in the context of a wide range of potential application areas (including water resources).

In using modularization techniques based on objects and classes of objects, rather than procedures, as a focus for systems design, it has been shown[13,14] that software can be written using an OOP language, which requires less maintenance than that written in standard

---

* MIRANDA is the trademark of Research Software Limited.

procedural languages. Since maintenance can account for about 70% of overall software costs, the use of object-oriented code promises large decreases in overall software development costs.

OOP centers on classes of objects, each class defining a new type (an extension to types REAL, INTEGER, BOOLEAN, etc.). In a pure object-oriented language, each class is itself a module.[14] Each class has a small interface to the outside world, such that implementation details are hidden from the user of the class. Thus, for example, a class (or module) that evaluated an employee's weekly wage would provide only that information upon demand. The calculation itself, in terms of rates of hourly pay, overtime rate, bonus, etc. would be hidden within the module. Consequently, changes to the *implementation* of "calculate weekly wage" could be done in a single location without disruption to users of that module. This leads to the possibility of much more reliable maintenance and extendibility of existing object classes. New services can easily be added to a module, since all information pertaining to a given class of objects is highly (and tightly) localized within the one module.

Other key attributes of OOP are (1) *inheritance,* by which attributes of a class can be acquired from a parent (or super) class. This means that common characteristics do not require duplicate code, with the resultant decreased risk of mistakes when alterations are required to that code in more than one place (as normally occurs with procedural coding techniques). (2) *Polymorphism and dynamic binding,* which permit classes to refer to different data types at run time. So, for example, a class representing a *list* could sometimes represent and handle a list of animals and at the same time handle a list of people. The decision on type is deferred to runtime, not compile time. Other advanced features, not available in some languages, are automatic garbage collection (i.e., reallocation of memory space no longer required) and multiple and repeated inheritance. Meyer[14] suggests that only languages satisfying all these criteria deserve the appellation of *object-oriented.* The language EIFFEL* satisfies these criteria and can therefore be called fully object-oriented. The language C++ also fulfils most of the requirements. Its pedigree is very different, being essentially a superset of the language C, which itself is a cross between an assembly-level and a high-level language. C++ is probably a wise choice for experienced C programers, whilst EIFFEL or Smalltalk** would present the novice objected-oriented programer with less problems. Water quality models written in a C++ environment are beginning to be disseminated[15]; and projects in this domain are under way at the University of New South Wales using EIFFEL.

One potentially huge benefit with object-oriented systems development depends on the availability of base classes from which larger systems can be constructed "bottom-up". Such library classes are sometimes provided with the compiler (e.g., EIFFEL), whilst in other cases sharing of such software is less formalized, and potentially therefore less reliable and secure. Trading in such software modules has the potential of relieving the programer of "reinventing the wheel" for every new, usually slightly different, application, and consequently of providing a high degree of flexibility and significant increases in software productivity once the learning period has been passed. In addition, the highly modular structure, coupled with the software libraries, should make prototyping not only rapid, but also the basis for the real system.

As a final thought regarding the emergence of new mathematical and software ideas and paradigms, mathematical modelers should be at least cognizant of (and hopefully will utilize) new *mathematical* theories, concepts, and tools. In water quality modeling, there has been little discussion of the potential use of (1) *catastrophe theory* (but cf. ref 16), which has the potential of elucidating the phenomenon of an algal bloom, and (2) the concept of *chaos,*[17] which would appear to have direct application in modeling and understanding turbulence in fluid flows, including lake circulation, jetted inflows, etc. These tools are now available to the modeler. How can they best be used to improve both our understanding and ability to simulate reservoir ecosystems?

---

\*    EIFFEL is the trademark of Inductive Software Engineering, Inc.

\*\*  Smalltalk is the trademark of Xerox.

## III. POTENTIAL FUTURE CLIMATIC IMPACTS

In this book, the authors have essentially followed the traditional experimental design prevailing in hydrology and water supply; that is, when using values for meteorological data such as rainfall, which are the forcing variable for the water supply system, interpretation of these data is undertaken with the implicit assumption that the set of hydrometeorological statistics describing the atmospheric climate are stationary (in a statistical sense). In other words, although there are year-to-year variations, the long-term average (say over *n* years) will be the same independent of the particular set of *n* years that is chosen. Water engineering designs for water storage reservoirs, drought attenuation impoundments, sea defenses and river levées are all built traditionally on the design principle of a stationary climate. However, over the last few years, it is becoming evident that climate is likely to change over the lifetime of these civil engineering projects. For example, in a climate where precipitation may, on average, increase, impoundments will flood more frequently, levées will be overtopped, etc. On the other hand, in such a warmer climate, more water is likely to evaporate, thus decreasing stored water levels and tending to make undersupply problems potentially more frequent.

It is now realized that climatic change, historically the domain of the atmospheric scientist, may impinge upon decisions made *now* by water industry management, especially in areas of strategic planning.[18] In the report of the meeting held in Villach, Austria, organized by the World Meteorological Organization (WMO), the UN Environment Programme (UNEP) and the International Council of Scientific Unions (ICSU) in October 1985, it was concluded that: "Many important economic and social decisions are being made today on . . . major water resource management activities such as irrigation and hydropower; drought relief; agricultural land use; structural designs and coastal engineering projects; and energy planning — all based on the assumption that past climatic data, without modification, are a reliable guide to the future. This is no longer a good assumption."[19]

Not only may average (i.e., the first statistic) atmospheric conditions change as a result of increasing atmospheric loading of pollutant gases, such as carbon dioxide, but also the variance (the second statistic) is likely to increase. In other words, climatic forecasts are that the mean atmospheric temperature will increase by somewhere between 2 and 5°C over the next half century or so, and that the day-to-day, month-to-month, and year-to-year variations will also be larger. Since much of hydrological forecasting depends upon extreme values statistics, this could have severe repercussions on water supply strategic planning (and the design of the decision support tools used for this exercise). A changed atmospheric climate (especially increased air temperatures) is likely to impact snow lines, spring melt periods, and rates (and hence seasonal runoff patterns). Changes in seasonal water quantity distributions may impact not only public water supply (especially in semi-arid regions), but also water-based recreation and hydropower generation.[18]

There are essentially two basic areas of concern for water quality managers and water quality modelers: water quantity (and its effects on lake levels, flow channel volumes) and water quality. At the time of writing, few studies have been undertaken specifically for lake and reservoir applications in either of these areas, so the discussion will of necessity be brief.

Modeling studies of the potential climatic impact on water bodies have, to date, been focused largely on the thermodynamic subcomponent and largely in oceanographic, rather than limnological, applications. This concern pertains largely to the role of the oceans in affecting the global climate.[20-22] Typically the surface energy budget of thermodynamic models is perturbed in order to evaluate both the feedback effect on atmospheric temperatures and also the impact of increasing atmospheric temperatures on water temperatures.

An additional factor is the significance of water level changes due to increasing atmospheric greenhouse gases (such as carbon dioxide). In the oceanic environment, there is an increasing interest in the assessment of the magnitude of the likely sea-level rise, whereas in the lacustrine environment, few studies have been undertaken. Hostetler[23] is using thermal stratification models in an evaluation of lake level changes in the central United States over historical and archaeological time periods. In an initial forecast for the impact of doubled carbon dioxide (a typical scenario used in atmospheric modeling), Henderson-Sellers[24] simulated changes in reservoir levels and found that, after a 50-year simulation, the water level had changed significantly.

The second area of concern is the impact on such hydrological changes on water quality parameters. Wigley and Jones[25] evaluated possible runoff changes for a doubled carbon dioxide scenario. Since the lake inflow rates are directly related to runoff, this permits an assessment of the change of the influent water quantity. Then, making some reasonable assumptions about the associated sediment load, a change in nutrient loading can be calculated. If runoff increases, the nutrients (nitrogen and phosphorus especially) leached from the catchment are likely to increase roughly in parallel. This would cause ambient lake concentrations to increase, further compounded if lake water levels were to fall. An order of magnitude analysis[24] suggests that nutrient concentration changes could be significant and that the trophic status of the lake could change on the same time scale as the climatic perturbation.

At this stage, it is difficult to make a precise evaluation of the significance of forecast climatic change on water quality. Software described in the early part of this chapter may at least give the water quality modelers new and powerful tools with which to build improved EDSSs for the water quality manager.

# REFERENCES

1. **Hansen, J., Fung, I., Lacis, A., Rind, D., Lebedeff, S., Ruedy, R., Russell, G., and Stone, P.,** Global climate changes as forecast by Goddard Institute for Space Studies three-dimensional model, *J. Geophys. Res.,* 93, 9341–9364, 1988.
2. **Barnwell, T., Vandergrift, S., and Ambrose, R. B.,** *Water Quality Modelling Software Available from US EPA,* Center for Water Quality Modelling EPA, Athens, GA, 1987.
3. **Guariso, G. and Werthner, H.,** *Environmental Decision Support Systems,* Ellis Horwood, Chichester, England, 1989.
4. **Ford, L. N.,** Decision support systems and expert systems: a comparison, *Info. Manag.,* 8, 21–26, 1985.
5. **Keen, P. G. W.,** Decision support systems: the next decade, *Dec. Supp. Systems,* 3, 253–265, 1987.
6. **Shields, F. D., Jr.,** ENDOW: an application of an expert system in technology transfer, in *WOTS Information Exchange Bulletin,* Vol E-88-3, U.S. Army Corps of Engineers, Waterways Experiment Station, Vicksburg, MS, 1988, 1–5.
7. **Lam, D. C. L., Fraser, A. S., Storey, J., and Wong, I.,** Regional analysis of watershed acidification using the expert systems approach, in *Computer Techniques in Environmental Studies,* Zannetti, P., Ed., Computational Mechanics Publications, Southampton, England, 1988, 67–81.
8. **Henderson-Sellers, B., Young, P. C., and Ribeiro da Costa, J.,** Water quality models: rivers and reservoirs, in *Procs. Int. Symp. on Water Quality Modeling of Agricultural Non-Point Sources, Utah, 1988,* U.S. Department of Agriculture, ARS-81, 381–409, 1990.
9. **Jamieson, D. G. and Fedra, K.,** The potential impact of software engineering on water-quality modeling, in *Procs. Int. Symp. on Water Quality Modeling of Agricultural Non-Point Sources, Utah, 1988,* U.S. Department of Agriculture, ARS-81, 707–713, 1990.
10. **Howard, G. S.,** Object oriented programming explained, *J. Systems Manag.,* 39, 13–19, 1988.
11. **Bird, R. and Wadler, P.,** *Introduction to Functional Programming,* Prentice Hall, New York, 1988.
12. **Ghezzi, C. and Jazayeri, M.,** *Programming Language Concepts,* 2nd ed., John Wiley, New York, 1987.
13. **Cox, B. J.,** *Object Oriented Programming. An Evolutionary Approach,* Addison-Wesley, Reading, MA, 1986.
14. **Meyer, B.,** *Object-Oriented Software Construction,* Prentice Hall, New York, 1988.
15. **Keffer, T.,** Data analysis and interactive modeling systems, *EOS,* 69(9), 133, 1988.
16. **Renguet, E. and Dubois, D. M.,** Approche stochastique de la theorie des catastrophes, in *Progress in Ecological Engineering and Management by Mathematical Modeling,* Dubois, D. M., Ed., Editions Cebedoc Sprl, Liège, Belgium, 49–86, 1981.

17. **Holden, A.,** Ed., *Chaos,* Princeton University Press, Princeton, NJ, 1986.

18. **ASCE,** Global warming poses threat to California water resources, *ASCE News,* 14(7), 16, 1989.

19. **World Meteorological Organization,** *A Report of the International Conference on the Assessment of the Role of Carbon Dioxide and Other Greenhouse Gases in Climate Variations and Associated Impacts,* Villach, Austria, 9–15 October 1985, WMO No. 661, Geneva (WMO/ICSU/UNEP), 1986.

20. **Hansen, J., Russell, G., Lacis, A., Fung, I., Rind, D., and Stone, P.,** Climatic response times: dependence on climate sensitivity and ocean mixing, *Science,* 229, 857–859, 1985.

21. **Wigley, T. M. L. and Schlesinger, M. E.,** Analytical solution for the effect of increasing $CO_2$ on global mean temperature, *Nature,* 315, 649–652, 1985.

22. **Henderson-Sellers, B.,** Modeling sea surface temperature rise resulting from increasing atmospheric carbon dioxide concentrations, *Climatic Change,* 11, 349–359, 1987.

23. **Hostetler, S.,** personal communications regarding the Pyramid Lake project, 1986, 1987.

24. **Henderson-Sellers, B.,** The impact of increasing atmospheric carbon dioxide concentrations upon reservoir water quality, in *The Influence of Climate Change and Climatic Variability on the Hydrologic Regime and Water Resources,* IAHS publ. no. 168, 571–576, 1987.

25. **Wigley, T. M. L. and Jones, P. D.,** Influences of precipitation changes and direct $CO_2$ effects on streamflow, *Nature,* 314, 149–152, 1985.

# INDEX